# The Environment and Marxism-Leninism

## Westview Special Studies

The concept of Westview Special Studies is a response to the continuing crisis in academic and informational publishing. Library budgets for books have been severely curtailed. Ever larger portions of general library budgets are being diverted from the purchase of books and used for data banks, computers, micromedia, and other methods of information retrieval. Interlibrary loan structures further reduce the edition sizes required to satisfy the needs of the scholarly community. Economic pressures on university presses and the few private scholarly publishing companies have greatly limited the capacity of the industry to properly serve the academic and research communities. As a result, many manuscripts dealing with important subjects, often representing the highest level of scholarship, are no longer economically viable publishing projects—or, if accepted for publication, are typically subject to lead times ranging from one to three years.

Westview Special Studies are our practical solution to the problem. As always, the selection criteria include the importance of the subject, the work's contribution to scholarship, and its insight, originality of thought, and excellence of exposition. We accept manuscripts in camera-ready form, typed, set, or word processed according to specifications laid out in our comprehensive manual, which contains straightforward instructions and sample pages. The responsibility for editing and proofreading lies with the author or sponsoring institution, but our editorial staff is always available to answer questions and provide guidance.

The result is a book printed on acid-free paper and bound in sturdy, library-quality soft covers. We manufacture these books ourselves using equipment that does not require a lengthy make-ready process and that allows us to publish first editions of 300 to 1000 copies and to reprint even smaller quantities as needed. Thus, we can produce Special Studies quickly and can keep even very specialized books in print as long as there is a demand for them.

# About the Book and Author

In the past two decades, environmental pollution and natural resource shortages have evoked increasing concern in the Soviet Union and Eastern Europe. The emerging ecological crisis has challenged many common assumptions in the Soviet bloc, as in the West. This book provides, for the first time, a detailed and comprehensive analysis of the ecology debate in the USSR and its highly industrialized ally, the German Democratic Republic (GDR). Based on a thorough examination of the Soviet and GDR sources, Dr. DeBardeleben explores the authorities' attempts to explain the problem to their populations. She also examines the viewpoints of scientists, writers, and scholars, with special attention to economic dimensions of the ecology debate. The study reveals the increasing sophistication of specialists in influencing public policy by adapting official values to support their positions. Through comparison of the Soviet and East German cases, the study clarifies the impact of natural resource endowment and legitimacy dilemmas on treatment of the ecology issue. The book demonstrates that Marxist-Leninist values subtly affect Soviet and GDR responses, but at the same time the environmental crisis is forcing a reevaluation of some aspects of Marxist-Leninist theory and ideology itself.

Dr. Joan DeBardeleben is an assistant professor in the Department of Political Science at McGill University. Currently she is coauthoring a book on nuclear power safety policy in the USSR and GDR.

# The Environment and Marxism-Leninism

## The Soviet and East German Experience

Joan DeBardeleben

Routledge
Taylor & Francis Group
LONDON AND NEW YORK

First published 1985 by Westview Press, Inc.

Published 2019 by Routledge
52 Vanderbilt Avenue, New York, NY 10017
2 Park Square, Milton Park, Abingdon, Oxon OX14 4RN

*Routledge is an imprint of the Taylor & Francis Group, an informa business*



Library of Congress Cataloging in Publication Data
DeBardeleben, Joan.
The environment and Marxism-Leninism.
(Westview special studies on the Soviet Union and Eastern Europe)
Bibliography: p.
Includes index.
1. Environmental policy—Communist countries.
I. Title.
HC710.E5D43 1985 363.7′056′091717 84-15219

ISBN 13: 978-0-367-29180-8 (hbk)

# Contents

# Preface

When I began this study, I was motivated primarily by the desire to unravel some mysteries about the role of Marxist-Leninist thought in the Soviet bloc. I chose to focus on environmental problems largely for theoretical reasons which I discuss in the first chapter of this book. Throughout the course of my research, however, I came to realize the great importance, in its own right, of the environmental awakening which has occurred in the Soviet Union and East Germany over the last fifteen to twenty years. Ever growing numbers of Soviet and East German citizens are devoting their energies to help ensure a livable and healthy environment for subsequent generations. I have taken the liberty of calling these individuals, most of them scholars, "environmentalists." My immersion in the internal debates over environmental problems in the USSR and German Democratic Republic (GDR) has deepened my own understanding of the nature and seriousness of the emerging ecological crisis and my commitment to the environmental movement in the West. So as I conclude this study I have come to see it as addressing two sets of issues and two, hopefully to some extent overlapping, audiences — one concerned with the nature of the evolving Soviet political system, and the other concerned with the nature of the environmental dilemmas confronting us all. In terms of both of these concerns I have become even more convinced of the profound importance of using every means possible to assure continued dialogue between Soviet-bloc and Western scholars and citizens. Even superficial exposure to the Soviet Union reveals the strong yearning of Soviet citizens for such contact. I hope that this book will whet Western appetites for such interchange as well.

While detente may seem to be dead and, as of November 1984, any talk of a "new detente" is premature, prospects for a continued thaw in relations between the GDR and the Federal Republic of Germany (FRG) remain more hopeful. There is also evidence of tension between the GDR and USSR over this "thaw," indicated by the successful Soviet pressure on East German party leader Erich Honecker to cancel his planned visit to Bonn in September of this year. The present difference in viewpoint between the GDR and USSR regarding German-German relations is reminiscent, in a backward sort of way, of the GDR's resistance to Soviet-backed detente in the late 1960s. Now, of course, the roles are ostensibly reversed, as the Soviet leaders seem more wary of any further opening between the two Germanies, and the East German leaders have experienced the potential economic benefits of continued detente. These recent developments raise intriguing questions: Are Soviet-GDR relations undergoing a fundamental change, as many West Germans would hope? Is the loyalty of the USSR's German ally in question? The present study does not deal directly with these foreign policy issues. But it does provide a perspective on the very deep link between the Soviet and East German worldviews. Hopefully this book will bring occasion for pause as some engage the wishful thought of a major shift in official GDR loyalties and aspirations. While the GDR is not subject to Soviet directives on every point, her leaders continue to define "socialist" Germany in opposition to her West German neighbor; and on most important points, East German leaders control deviations from the official Marxist-Leninist worldview with considerably more intransigence than do their Soviet counterparts. The one major exception, the existence of an "unofficial" ecology and peace movement in the GDR (linked to the Evangelical Church), became a victim of increased repression just when thousands of East German citizens were allowed to emigrate in the early months of this year. I would therefore expect this book both to inspire hope and to instill caution in interpreting the significance of recent developments in Soviet-GDR-FRG relations. For there is surely hope for international dialogue to be found in the writings of GDR and Soviet environmentalists; on the other hand, one should hardly expect a radical shift in fundamental GDR loyalties, despite the recent warming in GDR-FRG relations. The Soviet connection remains the foundation of present claims for GDR legitimacy.

A few technical notes are necessary to aid the reader. All translations are mine, except where otherwise indicated. Because most German sources consulted for the study emanated from the GDR, "Berlin" refers to East Berlin, unless I specifically designate Berlin (West) as the place of publication. I have tried consistently to use the Library of Congress transliteration system for Russian titles and names; thus, in some cases, for example in citing sources from the **Current Digest of the Soviet Press** or in citing German publications by Soviet authors, I have altered the spelling of names used in these sources to conform to the Library of

Congress system. I have also attempted to use sex-neutral language throughout; in some cases stylistic considerations, particularly in translations, led me, however, to retain the generic use of the word "man."

Of course, as always, final responsibility for any errors and misjudgments in the book remains mine. Many of the strengths of the book, however, have depended on the assistance and contributions of others. I am deeply indebted to John A. Armstrong, who provided stimulating dialogue and helpful suggestions as I completed the first version of this study, which served as my PhD dissertation. Both in those early stages and as the study went through major transformations since, others have provided both encouragement and ideas. I owe a great deal of my understanding of the East German political system to Melvin Croan. Murray Edelman, Leon Lindberg, John Straayer, Robert Hoffert, Frank Kunz, and Barbara Haskel all offered me generous encouragement. Wolfgang Müller provided extremely helpful tips on East German sources. I am particularly grateful to David Granick who helped me to puzzle out some of the economic debates treated in the book, and to Ian Matley and L. A. Kosinski for their insights regarding debates among Soviet geographers. Werner Gruhn and the Institut für Gesellschaft und Wissenschaft in Erlangen, West Germany, were of great help to me with East German sources, as was Hannsjörg F. Buck of the Gesamtdeutsches Institut in Bonn. I am also grateful to the anonymous reviewers of the manuscript who provided needed criticism and suggestions for improvement. Research grants from McGill University and the German Academic Exchange Service, as well as the opportunity to participate in the University of Illinois' Summer Research Lab on Russia and Eastern Europe, enabled me to gain access to necessary materials. The great care taken by Jeremy Grant in preparation of the final copy of the book was of inestimable value. Finally, my mother, Helen DeBardeleben, offered generous support and encouragement. My husband, Randall Droll, spent many hours commenting on and editing various versions of each chapter. Only his unremitting moral support and technical assistance made completion of this study possible.

Above all, however, this study depended on the work of many Soviet and East German citizens who have committed themselves to the cause of environmental protection. It is to them, and to their counterparts elsewhere, that this book is dedicated.

*Joan DeBardeleben*
Montreal
November 1984

# Part 1

# Ideology, Theory, and the Environment

The first part of this book addresses the broader challenges posed by the emerging ecological crisis to the prevailing world-view in the USSR and GDR. Chapter 1 might have been called an Introduction, for here the goals of the study are explained, relating them to salient issues in the broader field of Soviet studies. Chapter 2 focuses on environmental protection as a question of political legitimation. Here we attempt to discern how leaders present the issue to the public in the USSR and GDR, and how this public portrayal is related to Marxist-Leninist values. Official explanations and assessments in the mass media are contrasted with scholarly treatments. In chapters 3 and 4 the theoretical insights of Soviet scholars are analyzed in depth, with a focus on substantive viewpoints in chapter 3 and on the organization of scientific research in chapter 4.

The second half of the book (part 2) explores more concrete issues of economic policy, providing the reader both with a basic overview of the political-economic obstacles to improved environmental policy, as well as an in-depth study of economic debates. These chapters can be read as a unit for those readers who are primarily interested in economic affairs.

# 1
# Soviet Studies and Ecology

Karl Marx and Friedrich Engels envisaged the proletarian revolution and the development of communism as phenomena encompassing all nations on earth. Vladimir Il'ich Lenin, some years later, described the process preceding that revolution in which capitalism's expanding grasp would take the form of worldwide imperialism. The writings of neither Marx, Engels, nor Lenin prepared the present leaders of the Soviet bloc for the possibility that environmental pollution and depletion of natural resources might themselves become global processes, threatening both socialist and capitalist society alike and placing in question the prospect of universal material abundance in a non-exploitative world economic system.

There is no doubt, now, that the related problems of resource depletion and environmental pollution affect all major industrial nations on earth. But it is by no means clear whether any nation or international system will provide a foundation adequate for their timely resolution. Not only do environmental problems pose urgent challenges to public policy, but they have opened a new arena for both international cooperation and international competition. Furthermore, the seriousness of environmental problems has led many to question traditional value systems and established societal goals. Many intellectuals now doubt the conventional wisdom of the last two centuries, that human society is on a path of unilinear progress. No longer is increased material production an obvious component of that progress. The intensification of environmental problems has stirred controversy over the proper direction of future social and economic development, over the value of scientific and technical progress, and over the meaning of the "good" life.

The challenge presented to conventional value systems by environmental deterioration has evidenced itself in many ways in the capitalist world. If the original back-to-the-earth movements of the 1960s were a counter-cultural phenomenon, now movement from city to country has become a general demographic trend. Intense controversy over nuclear power reflects doubts about the worth of a high-energy life style when compared to unknown long-term risks to health and safety. Opposition to construction in wild life areas poses the question of the intrinsic worth of undisturbed nature and the wisdom of making human desires the measure of all things. The brown clouds covering major urban centers suggest to some that the slowed pace of collective transport and the hazards of city biking may be preferable to the privacy and speed of the automobile. Bodies of water polluted by industrial effluents challenge assumptions about the efficacy of the "invisible hand" and suggest the necessity of increased state regulation and intervention. The list could be easily expanded. But behind all these dilemmas lurks uncertainty over the possibility of combining traditional values with a healthy life: liberal values of individualism may have to be accommodated to the need for collective responsibility for nature, a public good; and continued industrial expansion may contradict maintenance of healthy and pleasant surroundings.

Just as it challenges values widely embraced in the capitalist world, environmental deterioration threatens to undermine established assumptions in the Soviet socialist world. Soviet authorities allege the superior capability of their socio-economic system to address social problems and to improve the quality of human life. But Soviet experience offers dubious support for this claim in the environmental area, even though it is precisely such social goods which collective ownership might seem best suited to protect. On the face of it similar causes (industrialization, urbanization) appear to have similar effects (resource shortages, pollution) in both socio-economic systems. The growth fetishism of the West is matched, if not exceeded, in the East. If continuation of existing growth patterns is inadvisable, if not impossible, as many environmentalists claim, the Soviet bloc is equally implicated. Science and technology have always been key elements of the Soviet definition of progress. If scientific-technical intervention in nature leads to dangerous natural imbalances, is this not equally so for Soviet science and technology? Furthermore, some Western critics believe that a common historical-cultural tradition is responsible for ecological crisis in both Western capitalist and Soviet socialist countries. Calls for a new ecological ethic, challenging anthropocentric, exploitative attitudes toward nature apply, at least implicitly, to the Soviet-bloc countries as well. Finally, the global nature of ecological problems may necessitate international cooperation on a scale comparable to that required to control the arms race. Perhaps even international class struggle should take a backseat to solving such global problems upon which the survival of humanity itself rests.

The issues addressed in this study derive from the theoretical and ideological dilemmas which the ecological crisis presents to the Marxist-Leninist states of Eastern Europe, in particular the Soviet Union (USSR) and the German Democratic Republic (GDR). It is the thesis of this study that while official Marxism-Leninism has (sometimes subtly) constrained Soviet and East German responses to the environmental challenge, at the same time the real dilemmas posed by environmental deterioration have also stimulated reevaluation of some basic assumptions in these countries, at times even reflected in official party doctrine. Marxism-Leninism may, in some ways, help shape policy-thinking, but at the same time it may be molded to legitimize policies adopted for other reasons. Furthermore, ideology may affect the **manner** in which political and economic factors influence environmental debate and policy, acting as a sort of mediating variable.[1] While we should not expect the Soviet-bloc leadership radically to alter its worldview or political priorities in response to the ecological crisis, the Soviet and East German experience does demonstrate that as communist states confront new dilemmas, the official interpretation of reality may also change to reflect the new challenges. Whether these changes in perspective will be adequate to produce a timely and appropriate response to the ecological crisis remains, however, open to serious doubt.

Since 1965, an extensive public debate has blossomed in the Soviet bloc over the implications of environmental deterioration in East and West. This debate has been accompanied by rising concern among the Soviet and East European leaders. Speaking to the 26th Party Congress of the Communist Party of the Soviet Union (CPSU) in February 1981 party leader Brezhnev acknowledged the "inevitable increase in spending on environmental protection," citing this as one factor which will "tend to make economic development more complicated" in the eighties.[2] Later in the same speech Brezhnev expressed concern about utilization of non-renewable natural resources, noting that "we are responsible for their proper and thrifty utilization not only to the present, but also to future generations. And no one has the right to forget this."[3] As these statements indicate, the Soviet leadership recognizes that policy choices confront them in dealing with environmental deterioration. From the viewpoint of Marxist-Leninist theory as well, the concerns raised by environmental deterioration and resource depletion pose genuinely unresolved questions. The theoretical mentors of the Soviet bloc provide only minimal guidance to present-day leaders as they attempt to come to terms with the broad-ranging implications of these problems. We must not therefore assume **a priori** that theoretical discussion of these issues is strictly rationalization by hack ideologues. Rather this theoretical irresolution allows us to raise questions about the limits to debate in an area where official doctrine is only emerging, and about the role of Marxism-Leninism in that debate.

Examination of environmental discussion in Eastern Europe and the Soviet Union may enable us to address three types of questions which are important in understanding the dynamics of ideological and theoretical change in the Soviet bloc:

1. What role does Marxism-Leninism play in the Soviet bloc, both as a serious theory of social analysis and as a source of regime legitimation?
2. What limits or constraints are placed on public discussion of a controversial issue like environmental deterioration? Which types of questions are legitimately raised in the public sphere and which are not?
3. In substantive terms, what answers do public officials and theorists offer to basic questions raised by environmental problems? Do they consider continued economic growth viable? How do they understand the causes of environmental problems? What solutions do they offer?

In this study these questions are addressed in the context of two countries, the USSR and the GDR. As the dominant ideological, political, and economic force in Eastern Europe, study of the Soviet approach to the environmental situation is a necessary starting point in understanding the character of Soviet-bloc attitudes. The GDR, as the USSR's most economically advanced ally, also has confronted the ecological dilemma in stark form in recent years. For reasons elaborated on below (see pages 17-23), a Soviet-East German comparison can prove particularly useful in trying to unravel the complex dynamic of ideological and theoretical evolution in Marxist-Leninist states.

First, however, let us elaborate on each of the three issues noted above and describe the manner in which our study may grant us further insight.

## MARXISM-LENINISM AS IDEOLOGY AND AS THEORY

Marxism-Leninism is the official ideology of the Soviet-bloc countries. Its actual character and function in these political systems is, however, far from obvious. In terms of content, Marxism-Leninism is not, as one might expect, the sum of Marx's and Lenin's ideas, nor even Lenin's version of Marxism. Rather the term "Marxism-Leninism" (as used here and by the Soviets themselves) actually refers to the Soviet interpretation of Marx and Lenin at any given point in time.[4] This interpretation has not remained static, but continues to evolve. In terms of function, the Soviets see Marxism-Leninism as a guide to state and party action and as the embodiment of science serving the interests of the proletariat.[5] Western observers not only disagree with this view but argue among themselves about the actual functions of Marxism-Leninism in the Soviet world. Some see it as a historical residue which the Soviet leaders rely on less and less for legitimation and which may even come to serve as a basis for dissent.[6] Others believe it is an "anti-language" which serves to confuse the population and prevent them from understanding and responding to their own oppression. In this way, the leadership attempts to legitimize increasingly conservative and oppressive policies by connecting them to revolutionary ideals.[7] Still others propose that ideological statements, having lost their rational validity outside of the system in which they operate, continue to retain pragmatic value in commanding practice directed toward their realization.[8] In sum, there is almost complete disagreement over the role of Marxism-Leninism both as a legitimizing system and as a guide to policy. To clarify how our study of environmental discussion may help to elucidate these controversies about the nature of Soviet Marxism-Leninism it is necessary to explore more fully the evolving character of this ideology and the nature of ideology in general.

Marxism-Leninism differs from ideology as we know it in the West because it is a **mandated** world-view, rather than simply a **favored** one (as liberalism or social democracy might be in the West). Although numerous definitions of ideology have been advanced in the Western social science literature, most contain the core notion that ideology exists "when a particular definition of reality becomes attached to a concrete power interest."[9] Ideology provides a perceptual screen through which social reality is apprehended and it justifies actions of government or other groups by linking them to generally accepted values. In other words, ideology has to do with political legitimation, with the process which leads people to believe that political authority (or political opposition) is just. Whereas in the Soviet Union one ideology is given official sanction, in the West a certain pluralism of ideologies is accepted, reflected in diverse party programs and periodic political crises in some Western nations. For ideologies can justify opposition as well as support for existing authority. The United States is probably the most ideologically homogeneous Western country, and in fact some basic ideological values are reflected clearly in the US Constitution (the commitment to procedural norms like free elections or judicial process and to substantive principles like the right to private property and freedom of expression).[10] Yet no Western government commits itself to a systematic and broad-ranging world-view or claims to be capable of definitive interpretation of truth as do the Marxist-Leninist parties of Eastern Europe. For this reason Darrell Hammer has wisely suggested that official Soviet ideology should be called "doctrine" to distinguish it qualitatively from Western ideologies.[11]

Yet when seeking to locate the actual content of this "doctrine" one is struck by its fluidity over time, by its flexibility in application, and by certain vague terrains, in which there appears, in fact, to be no clear doctrine. Official Marxism-Leninism has not been immutable. Lenin initiated the process of change when he adapted the insights of Marxism to what he considered to be an early and incomplete stage of monopoly capitalism in Russia. Lenin's skill lay not, however, primarily in political theory (his theoretical works on epistemology, the state, and revolutionary organization notwithstanding), but in political practice. His political theory formed, most importantly, the practical basis of operation of the Bolshevik party; thus already at this early stage Lenin's Marxism was an amalgam of commitment to basic principles and a struggle for political power.[12] Just as short-term political dilemmas molded Lenin's interpretation of Marx, so later Soviet leaders have had to adapt, augment, and interpret the original positions of Marx, Lenin, and Engels to meet new conditions. As Barrington Moore has argued, Lenin's version of Marxism not only has affected the social practice of the Soviet leadership, but official Marxism-Leninism has itself at each turn been modified both in content and in function by the conditions in which the Soviet leadership found itself.[13]

Just as official Marxism-Leninism has changed over time, so at certain periods the official position is left dangling. New guiding principles may finally be laid down in party documents or articulated by a recognized authoritative spokesperson. But even such documents may become outdated, implicitly contradicted, but never really disavowed. The 1961 **New Party Program of the CPSU** introduced some sharp breaks with Stalinist interpretations, but in the nearly two decades since Khrushchev's "hare-brained," "voluntarist" ideas were disavowed, the Soviet leadership has not yet revised this major ideological document.[14] Likewise, the Stalin constitution hung on for decades after many elements of Stalinist Marxism-Leninism had been disbanded.[15] How do we know when official theory (or "doctrine") has changed or new doctrine has emerged? When official ideological positions are definitively articulated by authoritative sources they usually reveal no ambivalence and are free of intricate elaborations or qualifications. They can be easily grasped, are simply formulated, and may in extreme cases take on the appearance of a slogan.

Such, for example, is the description of the USSR and GDR as "developed socialist societies." This phrase communicates the simplified rationalization of the position of present-day Soviet society in Marxist-Leninist eschatology.[16] The repeated assertion of the necessity "to organically fuse the achievements of the scientific-technical revolution with the advantages of the socialist system" expresses the simple rejection of any dissociation of scientific-technical progress from social progress and explains that their combination is best achieved in socialism.[17] Lacking such clear formulas and lacking authoritative statements from ideological spokespersons or party documents, one must examine the public treatment of the problem. If an explanation or interpretation is repeatedly expressed by a variety of sources with virtually no public contradiction, this provides evidence that some official sanction is granted to that position.

What makes analysis of environmental discussion so fruitful is that ideological proclamations comparable to the ones cited above have only begun to emerge in the last twenty years. This is an area in which Marxism-Leninism is developing, where doctrine is in embryo. Because this is the case, we can observe how doctrine emerges. Furthermore, we can observe Marxism-Leninism not only as an ideology used to legitimize existing conditions, but also in its role as social theory. While it would be a mistake to accept Soviet claims about the scientific character of official Marxism-Leninism, it would be equally dangerous to assume that Soviet Marxism is merely ideology. Arguments based on Marxism-Leninism in the Soviet bloc may be of various types. To be sure, Marxism-Leninism has something to do with legitimation and thus serves ideological functions. It also, however, has important cognitive content and derives from a powerful social science theory, Marxism. Natural and social scientists in the East bloc can, within limits, apply Marxist theory to the study of complex social problems. Alvin Gouldner has argued that at times Marxism acts as "a researchable model, rather than as self-evident, self-sufficient metaphysics" for Soviet social scientists.[18] This is particularly so in areas, like environmental studies, where official doctrine is evolving. Indeed scientists can hope to help shape that doctrine, and it is not unreasonable to assume that the political leadership looks to them for that help. Even if social science is influenced by its social context, by the values and assumptions of its practitioners, this need not imply that it is merely ideology. John Armstrong suggested that non-ideological arguments, theories, and doctrines can be distinguished from ideological arguments by the degree to which the latter act as a cover for group interests.[19] Further, "obviously an argument may be advanced on a non-ideological basis and then become adopted as an ideology; more rarely the reverse process occurs."[20] This distinction between ideological and non-ideological arguments may not be easy to make in practice. This is particularly the case in the Soviet bloc because scientists attempting to apply Marxist analysis to contemporary problems are shackled by political controls and must at times shape their arguments into an acceptable form. Some of the theoretical content may be lost or altered in the process. Their arguments need not however become simply instruments of legitimation. Rather they may become mutilated and masked forms of scientific inquiry. Ironically, just as Soviet officials try to make their ideology appear to be science, sometimes scientists must give their science the appearance of ideology. Nonetheless, behind the mask one often discovers serious social theory, addressing unresolved issues in a creative and courageous manner.

The foregoing should make clear that our purpose here is not to ascertain what Marx, Lenin, or Engels thought about environmental problems, nor what they would think "were they alive today."[21] Neither will we generally attempt to determine whether contemporary Soviet-bloc views are consistent with "authentic" Marxism or "authentic" Leninism.[22] If Marx and Lenin spoke to our concerns about the environment, they did so indirectly, through their general theories and through the assumptions and concepts underlying them – through assumptions about scarcity, about the potential of science and technology, and through analysis of capitalist overproduction or of the international division of labor. **Possible** interpretations of Marx's, Engels', or Lenin's ecological views are therefore just about as diverse and divergent as are possible interpretations of their theories in general. Contemporary Soviet, Chinese, Yugoslav, and neo-Marxist writings indicate just how wide that span is. Like official Soviet ideology, Soviet and East German participants in environmental debates "pick and choose," extrapolate, and adapt what they need from the original writings to substantiate their own theoretical positions. And this is precisely the process that interests us. How do concepts deriving from the classics, altered as they may be from their original purposes, enter the debate, if at all? Do these concepts constrain debate, limit alternatives, or affect the persuasiveness of arguments? Are Soviet and East German theorists able to do creative thinking and theorizing within these constraints? What other legitimizing principles are of importance to the debate? In short, how do people in the Soviet bloc try to persuade? In answering these questions we not only can better understand how Soviet Marxism helps to shape the views of scholars, intellectuals and journalists, but also what Soviet and East German censors and leaders consider to be the core values of their political systems.

To be sure, in trying to identify legitimizing principles and core political values through examination of public debate we must exercise some caution. For we do not have access to the minds and thoughts of Soviet-bloc policy-makers, intellectuals or citizens; even "traditional" methods such as elite interviews and survey research are not available to us. But we can use indirect methods to explore these questions. Comparison of different types of media, for example, can be a useful tool in trying to distinguish the differing functions of Marxism-Leninism in Soviet-bloc countries. This study thus includes comparison of debate in various media types in each country. The two major categories are: (1) relatively wide-circulation publications directed at mass audiences (newspapers, popular magazines); and (2) scholarly publications directed at much more limited and specialized audiences, presumably including potentially influential figures. (A few sources do not fit neatly into either category, such as the party press and newspapers geared to technical or managerial strata.)

While the mass media primarily reflect the attempt of the leadership to shape public opinion (and thus serve largely legitimizing functions), the same cannot be said for the more limited-circulation scholarly media. Scholarly publications do not primarily serve mass propaganda purposes, although references to official ideological precepts in such publications may reflect adherence to a "loyalty test," that is, serve as evidence of the author's commitment to the prevailing system of legitimation and power. One suspects this is the case when odes to Marxist-Leninist orthodoxy are tacked onto the opening and closing paragraphs of an otherwise scholarly article (not an uncommon or particularly subtle pattern). When, however, arguments employing Marxist-Leninist terminology and concepts are interwoven with and form the substance of scholarly analysis, this explanation is less adequate. In this case, more likely Marxist-Leninist language and concepts form part of the deeper conceptual structure defining the problem for the author **or**, alternatively, scholars **believe** that their arguments must be presented in this manner to be persuasive to other scholars and/or to policy-makers. Either explanation would indicate that the categories and values which permeate scholarly debate bear some significant relationship to those values which are perceived to motivate leadership responses and policy-making.

Thus, by comparing discussion of environmental problems in different media types, we can begin to discern different roles which Marxist-Leninist precepts may play in the Soviet bloc. Some uses of Marxism-Leninism may appear to be primarily ideological and perform legitimizing functions; others may suggest that Marxist-Leninist values, in some regards, **do** form a common foundation for basic political assumptions; in yet other cases Marxism-Leninism may represent a common "public" (and possibly persuasive) language for advocating policy options. In subsequent chapters we will observe Marxism-Leninism in each of these diverse roles (as well as some cases where it seems to have virtually no role) and seek to understand the complex functions which Marxist-Leninist political values have taken on in the USSR and GDR.

## LIMITS ON PUBLIC DISCUSSION

A state-controlled media places constraints on what can be printed in the USSR and GDR. Apart from underground publications such as **samizdat**, all journals, newspapers, and books, as well as the electronic media are responsible to state organs and ultimately to the Communist Party of the Soviet Union (CPSU) or its equivalent in the GDR, the Socialist Unity Party (SED). This monopoly control of mass communications was one of six components of the classic version of the "totalitarian syndrome" formulated in 1956 by Carl Friedrich and Zbigniew Brzezinski to characterize the Soviet and Nazi political systems.[23] In recent years, analysts of Soviet politics have come to question the usefulness of the concept of totalitarianism in describing the post-Stalinist polities of Eastern Europe and, specifically, to suggest that despite centrally controlled media, a certain pluralism of viewpoints is visible in the public press in these countries. These critics contend that monopolistic state and party control of the press need not produce a monolithic analysis nor exclude criticism of the regime. Rather, on many policy issues one finds expression of diverse viewpoints and profound disagreement in the press. This has led some Western analysts to suggest that a modified form of interest group theory is applicable to the East European polities. These interest groups are loosely defined and are made up of visible public figures like politicians (state officials and party apparatchiki at various levels), scholars, bureaucrats, and artists. These groups often have institutional bases, but in other cases may be identifiable only as a tendency of opinion observable in the media.[24] The ability of these "groups" to affect public policy is subject to dispute, but Gordon Skilling has suggested that

> sometimes the decisions of the rulers are taken as a result of the influence of groups and in a milieu of competing and conflicting tendencies. Leaders may "respond" on occasion to the threats and the arguments of rival groups, favoring one over another, or seeking a compromise between their interests.[25]

Furthermore, Skilling has concluded that apart from immediate policy impacts, public discussion by representatives of various interest groups may influence public opinion and the long-run climate of thinking.[26]

Other Soviet specialists have supported the thrust of the Skilling thesis. The economic and social complexity and interdependence resulting from a high level of industrialization have compelled the political leadership of the Soviet bloc to rely on expert opinion to facilitate sound decision-making. A freer flow of scholarly and technical communication becomes a prerequisite for the type of complex research and technological development necessary to support an interdependent socio-economic structure. According to these scholars, Stalinist repression resulted in information loss, inhibition of innovation, and excessive caution on the part of practical workers, all too costly to perpetuate.[27] On a continuum of interest group articulation, the USSR and GDR are on the more conservative end of the scale, as compared to other East European countries such as Yugoslavia, Czechoslovakia from the early 1960s to 1968, and Poland (until martial law). Nonetheless, the dependence of the Soviet regime on expert discussion by specialists was evidenced already in the late 1950s in the debate over educational reform and in the 1960s in discussion of economic reforms.[28] The role of institutional interest groups in Soviet politics has led Jerry Hough to advance the thesis that the USSR should be measured against a model of institutional pluralism. One of the components of this model involves the willingness of the political leadership "to listen to policy advice from 'society' and to permit far-ranging public discussion of policy questions."[29]

While public discussion of policy issues is freer than it was thirty years ago, this does not mean that everything is permissible. Significant constraints on free expression still remain. Overt censorship of environmental discussion in the USSR late in 1970 (lifted early in 1971) provides a striking example of the vagaries of a media monopoly.[30] Other limits on free expression are extremely predictable, however. For example, it is clearly impermissible to suggest publicly that capitalism is superior to socialism, or that the USSR is not actually a socialist society. Samuel Huntington proposes a general framework for understanding the acceptable range of public disagreement in the USSR: "The need for the party to maintain its position as a source of political legitimacy sets the limits of permissible dissent. . . . The limits of dissent are set by the fundamentals of the system, of which the party's monopoly of ultimate authority is the most sacrosanct."[31] While Huntington undoubtedly correctly relates the limits on public discussion to regime legitimacy (or perhaps more accurately to leadership **perceptions** of regime legitimacy), an important task remaining for Soviet specialists is to examine these limits in particular contexts. Such an examination will in turn provide a basis for refining our understanding of the dynamics of regime legitimation. In studying these limits one must, however, be sensitive both to the esoteric nature of communication in the Soviet bloc, as well as to differences between the form and content of that communication. The esoteric nature of communication requires the careful interpretation of subtle cues which would not appear significant to the uninformed observer. A rather mild statement may have implications beyond that contained in the statement itself, and changes in established expressions may signal significant differences of interpretation. As Leonid Il'ichev, a party theoretician, himself explained, "After all, silence also means something, also expresses some point of view."[32]

A related issue concerns sensitivity to differences between the form and content of communication. The same position may be expressed in numerous ways. Just as any argument is more persuasive if formulated in terms of the values held by the audience, so in the Soviet bloc a position is more likely to be permissible if formulated in a manner not threatening the basic foundations of regime legitimacy. The richness and demonstrated flexibility of Marxism-Leninism permits a relatively wide range of viewpoints to be expressed so as to appear consistent with fundamental Marxist-Leninist precepts. While a change in form most likely alters the content of the argument to some degree, much of the essential thrust of the position may be retained. Huntington again provides guidance here, making a similar point:

> In an established one-party system, however, the most important consideration is not what is opposed but where it is opposed. Opposition articulated through the one-party system can be much more sweeping in its criticism than opposition expressed outside the system. . . . In an established single-party system, as in a democratic competitive party system, political stability is measured by the degree to which the system possesses the institutional channels for transforming dissenters into participants.[33]

Our concern here, however, is not primarily with the availability of institutional channels, which Huntington refers to, but with the range of competing viewpoints which can be accommodated to the larger framework of values upon which the Soviet and East German systems rest. Following Huntington, a wider range of such viewpoints suggests a broader base of regime legitimacy and stability.

As is evident, the issues we are addressing here relate very closely to the problems we considered in the previous section.

The extent and basis of regime legitimacy act as a mediator both for the roles which Marxism-Leninism plays in the Soviet bloc, as well as for the range of permissible public disagreement in those systems. If Marxism-Leninism is to be more than a legitimizing doctrine, but also to serve as a serious basis of socio-economic analysis, this requires confidence of the leadership that the regime can sustain public disagreement and criticism.[34]

## SOVIET AND EAST GERMAN PERSPECTIVES ON ENVIRONMENTAL PROBLEMS

Environmental problems have been selected as the substantive focus of this study in part because these problems have both legitimacy implications and serious policy implications in Soviet-bloc countries. Attitudes toward the environment are also of intrinsic interest. Ecological effects of human activity are increasingly global in nature and will help determine the character of future life on earth. The way in which they are perceived in all major industrialized nations of the world is of critical importance in assessing solutions.

Several American and West German scholars have provided extremely useful analyses of environmental problems and policies in the USSR and GDR. The two major studies, both published in 1972 by Americans, the geographer Philip Pryde and economist Marshall Goldman, have been supplemented by more recent examinations of specific resource and pollution problems.[35] West German scholars Werner Gruhn, Ilka Norhara-Schnabel, and Gertraud Seidenstecher have provided essential background on East German responses to environmental deterioration.[36] The few available comparative studies of Soviet-bloc and Western environmental policy are also useful in assessing the similarities and differences deriving from diverse cultural, political, and economic systems.[37] Finally, some attempts have been made to analyze the inherent advantages and disadvantages of centrally-planned economies as opposed to mixed economic systems in this policy area.[38]

When many of the earlier studies of Soviet environmental policy were conducted, the debate in the USSR and Eastern Europe was just beginning to emerge. Since the mid-1970s, Soviet and East German scholars have become increasingly attentive to Western environmental writings. This, along with increased awareness of environmental deterioration, has enriched Soviet-bloc writings on the subject. As Soviet scholars have devoted greater attention to the environmental question, Western scholars have begun to identify important patterns in Soviet and East European approaches. John Kramer, for example, has determined that the most active environmentalists in the USSR come from the academic community, particularly the natural scientific fields, with social scientists less prominent, and enterprise officials and apparatchiki strongly underrepresented.[39] Robert Clawson and William Kolarik have briefly traced the stages of environmental advocacy in the USSR and have argued that increased interest articulation by environmentalists has corresponded to increased attention by the regime to these problems.[40] Keith Bush and others have noted the existence of Soviet publications particularly prone to environmental advocacy.[41] Donald Kelley has discussed use of the public media as a resource for environmental advocates and has analyzed initial reactions to Western growth debates.[42] These valuable insights underline the importance of engaging in a systematic and thorough analysis of Soviet and East European treatments of the emerging environmental crisis. For, as we shall see, over the years, Soviet and East European scholars themselves have begun to reflect more seriously on the roots of the problem in the Soviet bloc.

As Soviet and East European debate has expanded, it has become possible to explore more fully the impact of the Marxist-Leninist framework on East European approaches. As early as 1972 Philip Pryde addressed the problem, suggesting that both Marxism-Leninism and other basic value commitments of the Soviet leadership have had some impact on approaches taken to nature conservation. Pryde concluded that political ideologies generally are of lesser significance for natural resource conservation than are "implicit development priorities."[43] Nonetheless, Pryde did note that in certain cases Marxism may have an inhibitory effect on an active policy of nature conservation. Specifically he referred to "the assumption that improper natural resource exploitation is the result of the capitalist mode of production, and that a socialist economy necessarily pursues the wisest possible use of natural resources."[44] Pryde suggested that key elements of the Marxist approach, when combined with a commitment to rapid industrialization, are not well suited to guide the difficult choices required to prevent environmental deterioration. Furthermore, "Marxist ideology very definitely suggests that 'technological' solutions exist to all environmental problems. . . ."[45] Goldman and others also have identified other ideological hindrances to environmental responsibility in the USSR, in particular the resistance to placing charges on natural resource use because of their alleged incompatibility with certain ideological precepts.[46] Josef Füllenbach reached a similar conclusion, noting that "ideological hurdles" are too high to allow consistent solution of the problem of natural resource evaluation.[47] My study was stimulated by the suggestions of these earlier investigations. As I shall seek to demonstrate in subsequent chapters, the interaction of Marxism-Leninism with other influences appears to be considerably more complex than we had previously realized. Responses to new policy challenges, such as intensified environmental deterioration, are not only shaped by Marxist-Leninist assumptions, but they also generate new interpretations of Marxist-Leninist concepts. A thorough examination of Soviet and GDR writings can aid us in unravelling this complex interaction of Marxism-Leninism with other policy considerations.

As we seek to determine the ideological-theoretical influences on environmental thinking in Eastern Europe we must carefully distinguish Soviet interpretations and applications of Marxism-Leninism from Marxist theory as such. The history of the USSR and Eastern Europe makes clear that Marxism-Leninism has been repeatedly reinterpreted to serve legitimizing functions as the situation has changed in these countries. We have here a case of a non-ideological theory transformed into an ideology in the course of historical development. As in other such transformations, however, the ideological role which has been assigned to Marxism does not necessarily vitiate its original utility as a social theory. Moreover, as suggested earlier, ideologization of Marxist theory in the Soviet-bloc countries has not deprived the theory of all flexibility, especially in areas which are not immediately related to salient political concerns. Hence the caveat concerning the ideological role of Marxism in Eastern Europe does not minimize the importance of studying Soviet Marxist interpretations of environmental problems, but merely places that study in proper perspective as regards general questions of Marxist theory. Thus this study does not claim to assess the value of Marxist analysis itself in analyzing the environmental crisis, but rather seeks to clarify the evolution of Soviet interpretations of Marxism and their impact on policy-thinking.

The well-documented descriptions of East European environmental problems and policies provided in previous Western works form the concrete groundwork for this study. While much of the discussion of the environment in the USSR and GDR is surely not visible, it behooves us to scrutinize carefully voices which have reached the surface.

## THE COMPARATIVE DIMENSION

The GDR and USSR share many important characteristics. Yet in some ways they exhibit important differences. It is this combination of similarities and differences which makes a GDR-Soviet comparison so fruitful.

The GDR and USSR are both highly industrialized countries. Both face growing ecological problems and both have experienced a public awakening to environmental concerns since the mid-1960s. In addition they share a common ideology in Marxism-Leninism. This last characteristic is particularly important for our study. The GDR has been among the most, if not the most, ideologically orthodox of the East European satellite countries. Soviet ideological leadership, important in all of the Soviet-bloc countries, has been strongly enforced in the GDR for a variety of reasons: the unique status the GDR held after World War II; her economic importance to the USSR due to her high level of industrial development; her pivotal position in the development of detente and trade relations with the West; and her strategic location for defense purposes. Despite some strains in the late Ulbricht period (the late 1960s), the East German leadership has generally accepted Soviet leadership. This loyalty very likely derives from the unique challenge which the Federal Republic of Germany (FRG) presents to the GDR's legitimacy. Lacking a national base of legitimacy (due to the split of Germany), the GDR leadership depends on practical and ideological support from the USSR for continued stability.[48]

Soviet Marxism-Leninism posits the same general path of development for the GDR and USSR. Both are now described as developed socialist societies in official Marxist-Leninist parlance, with the promised transition to communism lying at some undefined point in the future.[49] Because of the close economic, political, and ideological ties between the two countries, one might expect that the theoretical and ideological discussion of environmental problems would be virtually identical in the two countries. The leaders of the GDR and USSR are, however, influenced by different political and economic pressures. Two of these deserve particular attention: (1) the differing challenges to regime legitimacy, closely linked to differences in political culture; and (2) the tighter natural resource situation of the GDR.

## Legitimation and Political Culture

The East German regime faces more intense challenges to its legitimacy vis-a-vis the population as a whole than does the Soviet regime. In the Soviet Union, there is intense dissatisfaction among selected groups, in particular among some intellectuals of minority nationalities, religious activists, and some Jews. While the discontent of these minorities may be high relative to generalized dissatisfaction in the GDR, the potential for widespread disruption in East Germany is probably greater. Several factors create this potential for broad popular discontent in the GDR. First, East Germany has stronger historical and cultural bonds to Western Europe and of course especially to West

Germany. A common cultural history makes Western European political ideals and institutions potentially more appealing to East German citizens. The cultural and historical bond to West Germany eliminates national identity as a source of GDR legitimacy, thus the hefty ideological campaign against West German policies which reinforce a **German** national identity.[50] The split between the two Germanies encourages citizens in both countries to compare the strengths and weaknesses of East and West. While the GDR may appear superior to the FRG in some regards (e.g., more equal educational opportunity, health care, day care facilities, and some aspects of social welfare), the material prosperity of West Germany is an ideological magnet for East German citizens.[51]

Western views and perspectives can also infiltrate the GDR much more easily than they can the USSR, through family, friends, and, above all, through penetration by the West German electronic media. The recent rise of the West German environmental Green Party as an active political force in the FRG exemplifies the dilemma that this potential "infiltration" poses for GDR authorities in the ecological area. Many positions of the West German Greens, e.g., challenges to economic growth as a social priority, critiques of nuclear power safety, and calls for economic and political decentralization, are anathema to GDR leadership goals. Concern that such "Green" values might infect East German youth is reflected in GDR attempts to restrict contacts between GDR citizens and West German Greens, as well as in efforts to ridicule the Greens as "parsley parties."[52] Whatever such defensive measures GDR authorities may take, however, their actions and claims are subject to constant public refutation from the West.

The relative newness of the East German regime is another factor affecting the dynamics of legitimation. The GDR has been ruled by a Marxist-Leninist party for a much shorter time period than the USSR. This reduces the life-long impact of socialization processes and the legitimacy deriving from the "givenness" of existing social and political institutions. Although many Soviet citizens or their parents suffered brutalities in the Stalinist period, they also shared the successful patriotic struggle of World War II and rapid economic strides under Soviet institutions and leadership. Historic Russia has become a superpower under Soviet guidance and that is bound to instill pride, at least among ethnic Russians. In contrast, the strong Soviet presence in the GDR may be repulsive to many Germans. Many Germans suffered brutality and violence at the hands of the Soviets after World War II. The continued and obvious Soviet presence provides a potential basis for popular national resentment against the present regime.

The factors discussed in the preceding paragraphs affect not only the population as a whole, but also scholars in both countries. Gouldner has suggested that Soviet sociologists need to provide less evidence of their loyalty to the regime as the political leadership becomes more confident of that loyalty. According to Gouldner, sociology takes on a more non-partisan appearance precisely "when the elites of a society are confident that its social scientists are, in fact, not neutral."[53] East German authorities may well be more sceptical of the loyalty of their citizens and scholars than are Soviet authorities. One must remember that only some twenty-five years ago the East German middle classes were streaming from their homeland by the thousands, seeking refuge in the Federal Republic of Germany, the GDR's clearest adversary.

One should not overdramatize the dissatisfaction of East German citizens with their political system nor over-estimate the satisfaction of the Soviets with theirs. The potential disruptive effects of nationality discontent cannot be discounted in the USSR. There is ample evidence of Soviet dissidence. Conversely, certain limited evidence indicates considerable citizen pride in certain aspects of the East German political system.[54] Furthermore, East Germans enjoy a high standard of living relative to most East European countries (with the possible exception of Hungary, and, of course, Yugoslavia). Data is simply not available to permit a fool-proof assessment of popular East German and Soviet attitudes. While it seems likely that the dynamics of regime legitimacy are considerably more sensitive in the GDR than in the USSR, it is beyond doubt that the nature of the legitimacy challenges facing the countries are very different.

**Natural Resource Endowment**

As discussed above, the dynamics of regime legitimacy affect both the role of Marxism-Leninism and the breadth of public debate. The intensity of policy dilemmas may also have an impact on these issues. The GDR's problems of resource depletion, and most likely pollution as well, are, on the whole, more intense than are the same problems in the USSR. The cause lies in the higher population density and scarcity of essential resources like land, water, and mineral deposits. Of course in some geographic areas the USSR also faces severe resource shortages (e.g., of water) and intense pollution,[55] but the USSR has more flexibility in dealing with these regional problems because of her richer resource endowment and greater geographical expanses. This difference is particularly marked regarding energy sources and their environmental effects, which we will examine more closely below.

Regarding agricultural land, the GDR has only .28 hectares of arable land per capita as compared to .85 for the USSR. A relatively high proportion of the GDR's better agricultural land in the districts of Magdeburg, Halle, and Leipzig has been removed from agricultural production for such activities as brown coal mining. The USSR faces similar but less pressing problems. Soviet sources indicate that plowland per capita in the USSR declined from 1.0 to .85 hectares from 1960 to 1980.[56] Despite differences in the quality of land and climate, it is clear that (with irrigation, reclamation, and melioration possibilities) the USSR has options in land use which do not exist for the GDR.

Likewise the GDR faces an immense strain on its water resources. Maier and Roos have estimated that after 1990 the usable supply of water will be lower than the demand in dry years if the use of water resources is not decisively intensified. Indeed, since 1982, following passage of a new water law, further measures have been taken to increase the effectiveness of water use in the production process.[57] Nonetheless water shortages place restrictions on the placement and development of industry in the GDR (as well as on the development of future nuclear power installations). Water shortages also, of course, exist in several regions in the USSR. For example, in the Southern Ukraine, Southern Caucasus, the central Chernozem regions, and many parts of Central Asia, water shortages limit the development of many types of production. Expenditures for irrigation projects, reclamation projects, and water supply systems are high. These measures, as well as the building of major canals to solve local water shortages and the proposed river diversion projects, offer alternatives for the USSR not available to the GDR.[58] In the GDR there are no significant extra sources of water to be exploited, other than through expensive desalinization. Pollution of available water sources thus becomes an immediate policy concern. As the GDR economist Streibel points out, the only real options available are improvement of purification processes to allow reuse of water and the development of water-free technology.[59]

In terms of other resources as well, the USSR is relatively well supplied. The GDR, on the other hand, does not have extensive mineral deposits or forest resources and must import fuel and raw materials. In fact the GDR imports large amounts of crude oil, natural gas, iron ore, zinc, aluminum, lead, and copper from the USSR.[60] Since the mid-1970s hard currency debt, rising raw material costs, and trade deficits have led the GDR to attempt to reduce import dependence through increased reliance on domestic raw materials, more efficient use of material inputs, and recycling of waste.[61] Two natural resources which the GDR does have in abundant measure are brown coal and potash. Both are heavily exploited; brown coal is the key energy resource in the GDR and potash is important for export to hard-currency markets. But use of both of these domestic resources has serious environmental effects. Potash processing has resulted in heavy pollution of the Werra River.[62] The GDR's brown coal has very high sulfur content, making it a powerful pollutant, contributing to alarming damage to forests in the southern part of the GDR, to respiratory disease, and to shortened life span in some industrial regions.[63] Unlike the USSR, the GDR cannot turn to clean domestic sources of energy such as hydroelectric power or natural gas to any large extent. To be sure, development of Western Siberian gas and oil will produce localized ecological disturbances in the USSR, especially important because of the sensitivity of ecosystems in such extreme climatic zones. But neither this nor plans to burn low-grade coal near the point of extraction and transmit the power by high-voltage electrical lines to the western part of the country will mean an added load of air pollution in the heaviest populated areas in the USSR. Hydroelectric generation produces its own ecological disturbances, such as flooding, erosion, disruption of spawning patterns, and loss of valuable agricultural and forest land. These problems again, however, generally do not affect the most heavily-polluted industrial areas. In short, the USSR has more options in resource substitution and in location of industry (difficult as these options may be to realize), thus affording greater flexibility in dealing with the environmental problem.

The tight resource situation facing the GDR makes it imperative that East German authorities receive good policy advice from experts, as do the complexities of effective resource management in the USSR. Effective use of natural resources requires sophisticated application of scientific knowledge spanning many disciplines, including biology, geography, chemistry, engineering, demography, law, and economics. The scientific cooperation required means that relevant information must be disseminated among a broad range of participants in different disciplines and institutions. This is most easily accomplished by letting the discussion go public. While these pressures may encourage the GDR leadership to allow airing of diverse technical viewpoints, considerations of regime legitimacy may pressure the leadership to constrain public debate of a controversial or disturbing nature. Both of these countervailing pressures affect the Soviet leadership as well, but in different degrees and forms. Soviet leaders may feel less compelled to restrict scientific and popular debate, due to the factors discussed in the previous section, unless this debate is likely, for example, to incite minority ethnic sentiments. In addition the size of the Soviet Union makes informal inter-disciplinary dialogue less feasible than in the GDR; this increases the potential utility of scholarly publications.

## Conclusion

It is the combination of similarities and differences between the GDR and USSR treated in this section which makes comparison of environmental debate in the two countries so promising. In essence, they allow us to apply a simple "most-similar-systems" approach, attempting to assess the importance of a few key differences in systems with otherwise similar ideological, political, and economic traits.[64] I try to demonstrate in the following chapters that two salient areas of contrast between the USSR and GDR (differing legitimacy dilemmas and differing natural resources endowments) are related to differences in the interpretation of Marxism-Leninism, the nature of public debate, and, in some cases, policy outcomes in the environmental area. While it is not possible to establish a definite causal link between these variables, this study does provide arguments and evidence for such linkages and offers support for the findings of other scholars by relating the nature of policy conflict to the dynamics of regime legitimation. In the future, perhaps a study encompassing all countries in Eastern Europe will further reinforce our understanding of these relationships.

## A NOTE ON SOURCES

The volume of material relevant to this study, particularly in the USSR, exceeded initial expectations. The literature on environmental problems appears to be expanding nearly exponentially each year. Therefore, selection has been necessary. Choice of adequately comparable sources for the USSR and GDR involves judgment. In some cases a thorough and systematic perusal of pertinent academic journals seemed appropriate, particularly regarding philosophy and economics; for other journals, especially in the legal and geographic fields, a more selective approach was taken, focusing on particular issues. Mass media sources included a thorough reading of relevant articles in **Current Digest of the Soviet Press** (**CDSP**), supplemented by **Priroda** (Nature), a Soviet popular science magazine which has gained a reputation for environmental advocacy, and by selected articles from Soviet newspapers. For the GDR, there is no summary comparable to the **CDSP**, so I was forced to sample various popular publications like **Neues Deutschland** (the daily newspaper of the SED), **Urania** (a popular science magazine), **Die Wirtschaft** (The Economy), and

**Forum** (the organ of the Free German Youth). While this study does not include examination of the electronic media, major discrepancies in content between that and the printed media seem unlikely. There is always the danger in this type of study, where statistical sampling methods are not appropriate, that source selection may blind one to whole schools of thought. I feel confident, however, that my sources reflect no systematic bias. The careful scrutinizing of secondary references in all the Western, Soviet, and East German sources consulted would make such a systematic bias highly improbable.

Although the primary concern of this study is with the political-ideological dynamics of the Soviet and East German systems, the substantive focus of the study necessitates an interdisciplinary approach. Environmental problems involve complex issues, impinging on the natural, social, and philosophical sciences. This very complexity belies the modern divisions of human knowledge. Among the natural sciences, biology, medicine, ecology, physics, genetics, geography, and chemistry all have obvious importance. The technical sciences must also be participants in the consideration of practical solutions to environmental problems. The fields of economics, sociology, and law face challenges in determination of the optimal patterns of material production, legal responsibility, and social organization to deal effectively with the threat of environmental deterioration. Questions of human values involve philosophers, journalists, and artists in the dialogue. A part of the task here has been to scrutinize the varying perspectives of this diversity of participants. While this study does not examine technical aspects of the contributions of natural scientists, the ideological-theoretical import of their positions is analyzed. The contributions of economists have received the closest scrutiny, for even technical economic issues reveal the impact of Marxist-Leninist theory. Furthermore, an important basis of present regime legitimacy rests on the economic accomplishments achieved under a centrally-planned system. Finally, many of Marx's most important writings treat complex economic matters. The diversity of materials involved in undertaking an interdisciplinary study of this type may produce somewhat simplified portrayals of theoretical disagreements in particular scientific disciplines, for it is necessary to extract relevant segments from a vast body of material in each case. A concerted effort has been made to be sensitive to the diversity of backgrounds and contexts from which various participants in the environmental exchange speak.

## NOTES

1. Thus, although I focus on the importance of Marxism-Leninism for environmental attitudes in the USSR and GDR, I do not necessarily conceive of it only as the independent variable in the study. At times the official ideological position may be shaped by other factors (economic or social forces) and should be conceived of as the dependent variable. More commonly it may act as a mediating variable. This same problem arises in consideration of the role of political culture in communist systems. For a view similar to my own, but this time dealing with political culture, see Stephen White, **Political Culture and Soviet Politics** (London: Macmillan Press, 1979), pp. 19-21.

2. **Report of the Central Committee of the CPSU to the XXVI Congress of the Communist Party of the Soviet Union and the Immediate Tasks of the Party in Home and Foreign Policy,** delivered by L. I. Brezhnev on February 23, 1981 (Moscow: Novosti Press Agency Publishing House, 1981), p. 50.

3. Ibid., p. 55.

4. Of course the official Soviet position would see this as synonymous with Lenin's adaptation of Marx.

5. Soviet theorist Kelle explains that Marxism-Leninism is a "scientific ideology" fusing cognitive and ideological functions. Because Marxism-Leninism expresses the interests of the working class, it is ideology; because the working class is the universal class of future human society, Marxism-Leninism takes on a scientific character. See V. Zh. Kelle, "The Leninist Conception of Scientific Ideology and Its Critics," **Voprosy filosofii**, no. 4 (1970), trans. in **Soviet Studies in Philosophy** 9 (Fall 1970): 114.

6. This seems to be the implication of Samuel Huntington's argument in "Social and Institutional Dynamics of One-Party Systems," in Samuel Huntington and Clement Moore, eds., **Authoritarian Politics in Modern Society** (New York: Basic Books, 1970), pp. 28-29, 40.

7. Pyotr Eguides, a prominent Soviet dissident associated with the emigré publication of **Poiski**, described official ideology as anti-language in a speech at McGill University on Nov. 18, 1981. A similar argument is made by the Polish writer Michael Szkolny (pseud.), in "Revolution in Poland," **Monthly Review** 33 (June 1981): 1-4. Szkolny calls this process "ideological embezzlement," and points out that propaganda does not have to be believed to be effective; it can just as easily protect the existing order by making a whole range of ideas inexpressible because the terms (like revolution and socialism) have been coopted by the regime. Robert Tucker's argument is more straightforward, that intensified verbal support for Marxist-Leninist goals is an attempt to legitimize conservative policies.

"The Deradicalization of Marxist Movements," **American Political Science Review** 61 (June 1967): 343-58, especially p. 350.

8. See Herbert Marcuse, **Soviet Marxism: A Critical Analysis** (New York: Columbia University Press, 1958), pp. 87-88; and Charles Ziegler, "Soviet Environmental Policy Parameters: The Macro-Value Framework," **Studies in Soviet Philosophy** 23 (April 1982): 187-88.

9. Peter L. Berger and Thomas Luckman, **The Social Construction of Reality** (Garden City, NY: Doubleday & Co., 1966), p. 113. One should point out that the concept of ideology itself derived in large part from Marx's own writings. Marx analyzed the manner in which concepts and explanations predominant in capitalist society, while parading as universal truths, are actually veiled justifications for the continued rule of the bourgeoisie. See Karl Marx and Frederick Engels, **The German Ideology**, Part I, C. J. Arthur, ed. (New York: International Publishers, 1970), pp. 39-96; and **The Communist Manifesto** (New York: Monthly Review Press, 1964), pp. 36-38. A frequent criticism directed at this Marxist understanding of ideology is that it must logically be applied to Marxism itself. If the theory is to be consistent, Marxism itself becomes an ideological expression of a certain group's interest. See Karl Mannheim, **Ideology and Utopia: An Introduction to the Sociology of Knowledge** (New York: Harcourt, Brace, and World, 1968), for a discussion of this problem. The Soviet response to this charge was discussed in note 5 above.

10. For a discussion of liberalism in America see Louis Hartz, **The Liberal Tradition in America** (New York: Harcourt, Brace, and World, 1955). In contrast to the US, the effects of ideological diversity are illustrated by the French case.

11. Darrell P. Hammer, **USSR: The Politics of Oligarchy** (Hinsdale, IL: Dryden Press, 1974), pp. 95-96.

12. Excellent discussion of the evolution of Leninist thought is found in Marcel Liebman's **Leninism under Lenin** (London: Merlin Press, 1975). Alfred Meyer's **Leninism** (Cambridge, MA: Harvard University Press, 1957) remains an outstanding treatment of Lenin's thought, although Meyer is less sympathetic to Lenin's position than is Liebman.

13. Barrington Moore, **Soviet Politics: The Dilemma of Power: The Role of Ideas in Social Change** (Cambridge, MA: Harvard University Press, 1950). See especially p. 412 and remaining parts of the concluding chapter which summarize his findings in this regard.

14. On Khrushchev's innovations see Carl A. Linden, **Khrushchev and the Soviet Leadership** (Baltimore: Johns Hopkins Press, 1966), pp. 17-18, 215-17. At the 26th Party Congress in 1981 Brezhnev indicated that based on changes which had occurred since Khrushchev's **New Party Program** was adopted in

1961, we "should introduce the necessary amendments and additions into the present Programme. If this proposal is approved by the delegates to the Congress, the Central Committee could be instructed to redraft the Programme of the CPSU" (**Report to the XXVI Congress**, p. 104). As Brezhnev noted in his report, important new concepts such as "developed socialist society" have been introduced since 1961, and these are not reflected in the present party program (p. 103).

15. For analysis of changes introduced in the 1977 Soviet Constitution see Robert Sharlet, **The New Soviet Constitution of 1977: Analysis and Text** (Brunswick, OH: King's Court Communications, 1978).

16. On the concept of developed socialist society see Alfred B. Evans, Jr., "Developed Socialism in Soviet Ideology," **Soviet Studies** 29 (July 1977): 409-28.

17. There is also an extensive theoretical literature on the scientific-technical revolution (STR). Simple ideological formulas often derive from serious theoretical analyses, selected ideas from which are appropriated by the regime and simplified as a legitimizing explanation for popular consumption. The literature on the scientific-technical revolution, as I note in the next chapter, is particularly interesting to us because consideration of ecological problems forms one component of that larger literature. Western analyses of the vast theoretical literature on the scientific-technical revolution are not numerous. See, for example, Julian Cooper, "The Scientific Technical Revolution in Soviet Theory," in Frederic J. Fleron, ed., **Technology and Communist Culture** (New York: Praeger Publishers, 1977), pp. 146-75; Clemens Burrichter, Eckart Förtsch, and Hans-Joachim Müller, "Die wissenschaftlich-technische Revolution – Kriterien und Konsequenzen," **Deutschland Archiv** 9 (1976): 516 ff; and Frederic J. Fleron, Jr., "The Western Connection: Technical Rationality and Soviet Politics," **Soviet Union/Union Soviétique** 4, Part 1 (1977): 58-84. Robert F. Miller tries to relate the STR to Soviet agricultural policy in "Communist Agricultural Organization: The Future of the Kolkhoz," **Problems of Communism** 25 (March-April 1976): 34-50. See also Erik P. Hoffmann and Robbin F. Laird, **"The Scientific-technological revolution" and Soviet Foreign Policy** (Elmsford, NY: Pergamon Press, 1982). The Soviet and East German literature on the STR is so vast that I must refer the reader to the secondary sources cited above for bibliographic guidance. For a good introduction see Academy of Sciences of the USSR and Academy of Sciences of Czechoslovakia, **Man, Science, and Technology: A Marxist Analysis of the Scientific-Technological Revolution** (Prague: Academia Prague, 1973); Kurt Hager, **Wissenschaft und Technik im Sozialismus** (Berlin: Dietz Verlag, 1974); bibliography provided in "Sotsial'no-

filosofskie problemy nauchno-tekhnicheskoi revoliutsii (issledovaniia 1971-75)," **Voprosy filosofii**, no. 2 (1976), pp. 37-53; and numerous articles in the journals **Voprosy filosofii** and **Deutsche Zeitschrift für Philosophie.**

18. Alvin Gouldner, **The Coming Crisis in Western Sociology** (New York: Basic Books, 1970), pp. 461-62.

19. John A. Armstrong, **The European Administrative Elite** (Princeton, NJ: Princeton University Press, 1973), p. 9.

20. Ibid.

21. Howard L. Parsons tries to deal with the first question in **Marx and Engels on Ecology** (Westport, CT: Greenwood Press, 1977). Parsons has also published articles in the Soviet Union. See his "Mal'tuzianstvo i sotsializm," **Voprosy filosofii**, no. 1 (1977), pp. 60-70; and his review of **Man, Society, and the Environment**, edited by I. G. Gerasimov, **Filosofskie nauki**, no. 1 (1978), pp. 172-75. Charles E. Ziegler attempts to identify "macro-value parameters" for Soviet environmental policy based on his reading of Marx, Engels, Lenin, Stalin, and contemporary Soviet writers. Ziegler doesn't discuss, however, some of the bolder and more imaginative approaches taken by contemporary Soviet writers. See Ziegler, "Soviet Environmental Policy Parameters," **Studies in Soviet Thought** 23 (April 1982): 187-204. See also Zigurds L. Zile, "Lenin's Contribution to Law: The Case of Protection and Preservation of the Environment," in Bernard W. Eissenstat, ed., **Lenin and Leninism: State, Law and Society** (Lexington, MA: Lexington Books, 1971).

22. More broadly, the relationship between social theory and practical political considerations in Marx's writings has been a topic granted considerable attention by Western scholars. This debate also involves the degree to which Marx's general analysis can legitimately be amended to accommodate differing concrete historical settings, without compromising the theory itself. See, for example, Alan Gilbert, "Social Theory and Revolutionary Activity in Marx," **American Political Science Review** 73 (June 1979): 521-53.

23. Carl J. Friedrich and Zbigniew K. Brzezinski, **Totalitarian Dictatorship and Autocracy** (Cambridge, MA: Harvard University Press, 1956), p. 10. Friedrich and Brzezinski also included in their six features of the syndrome "an official ideology, consisting of an official body of doctrine covering all vital aspects of man's existence to which every one living in that society is supposed to adhere, at least passively" (p. 9).

24. H. Gordon Skilling, "Groups in Soviet Politics," and Franklyn Griffiths, "A Tendency Analysis of Soviet Policy-Making," provide two somewhat different conceptualizations of the nature of interest groups. Both are published in H. Gordon Skilling and Franklyn Griffiths, eds., **Interest Groups in Soviet**

**Politics** (Princeton: Princeton University Press, 1971), pp. 19-46 and 335-77. For application of the interest group approach to Soviet environmental policy see Donald R. Kelley, "Environmental Policy-Making in the USSR: The Role of Industrial and Environmental Interest Groups," **Soviet Studies** 28 (October 1976): 570-89. William Mandel provides a rather uncritical discussion of mass participation in conservation activities in "The Soviet Ecology Movement," **Science and Society** 36 (Winter 1972): 385-416.

25. H. Gordon Skilling, "Group Conflict in Soviet Politics: Some Conclusions," in Skilling and Griffiths, p. 391.

26. Ibid.

27. See, for example, Chalmers Johnson, "Comparing Communist Nations," Chalmers Johnson, ed., **Change in Communist Systems** (Stanford, CA: Stanford University Press, 1970), pp. 20-26; and Alexander Dallin and George W. Breslauer, "Political Terror in the Post-Mobilization Stage," in the same volume, pp. 196-214. On the impact of these processes in Eastern Europe see A. Korbonski, "The Prospects for Change in Eastern Europe," and M. Croan, "Some Constraints on Change in Eastern Europe," **Slavic Review** 33 (June 1974): 219-39, 240-45.

28. Joel J. Schwartz and William R. Keech, "Group Influence and the Policy Process in the Soviet Union," in Frederic J. Fleron, ed., **Communist Studies and the Social Sciences** (Chicago: Rand McNally & Co., 1969), pp. 298-317; and Richard W. Judy, "The Economists," in Skilling and Griffiths, pp. 213-45.

29. Jerry Hough, **The Soviet Union and Social Science Theory** (Cambridge, MA: Harvard University Press, 1977), p. 25. Hough's interpretation of Soviet politics has been sharply debated. See, e.g., Zbigniew Brzezinski, "Soviet Politics: From the Future to the Past," Paul Cocks, Robert W. Daniels, and Nancy Whittier Heer, eds., **The Dynamics of Soviet Politics** (Cambridge, MA: Harvard University Press, 1976), p. 343.

30. Donald R. Kelley, Kenneth R. Stunkel, and Richard R. Wescott, **The Economic Superpowers and the Environment: The United States, the Soviet Union and Japan** (San Francisco: W. H. Freeman & Co., 1976), p. 134. In an important samizdat publication, Boris Komarov (pseud.) alleges that in the USSR "since 1975 you will not find a single reference to pollution of the air, the water, or the soil even in special articles." Komarov's book was published in West Germany in 1978 (as **Inichtozhenie prirody, obostrenie ekologicheskogo krizisa v SSSR** by Possev Verlag, Frankfurt am Main), thus presumably he wrote it in 1976 or 1977. Based on my research, however, there is no obvious lacuna in writings on pollution during 1976 or following. Either the prohibition was very short-lived or narrow in scope. See Komarov, **The Destruction of Nature in the Soviet Union** (White Plains, NY: M. E. Sharpe, 1980), p. 16.

31. Huntington, in Huntington and Moore, p. 43.

32. Quoted by Skilling, "Group Conflict," in Skilling and Griffiths, p. 390.

33. Huntington, in Huntington and Moore, p. 44.

34. The political leadership may indeed not accurately assess the degree and basis of its own legitimacy in the eyes of the population. McAuley suggests that "one of the consequences of being a self-appointed leadership is that it is difficult to assess the degree and nature of support from below." This may result in "the ambivalent attitude of the leadership, confident one moment, insecure the next." Mary McAuley, **Politics and the Soviet Union** (Harmondsworth, Middlesex, ENG: Penguin Books, 1977), p. 262.

35. Philip R. Pryde, **Conservation in the Soviet Union** (Cambridge, ENG: Cambridge University Press, 1972); Marshall Goldman, **The Spoils of Progress: Environmental Pollution in the Soviet Union** (Cambridge, MA: MIT Press, 1972). Also published in the early 1970s were: Keith Bush, "Environmental Problems in the USSR," **Problems of Communism** 21 (July-August 1972): 21-31; and David Powell, "The Social Costs of Modernization: Ecological Problems in the USSR," **World Politics** 23 (July 1971): 618-34. The following books contain articles on particular aspects of environmental problems in the USSR: Ivan Volgyes, ed., **Environmental Deterioration in the Soviet Union and Eastern Europe** (New York: Praeger, 1974); Fred Singleton, ed., **Environmental Misuse in the Soviet Union** (New York: Praeger, 1976); and W. A. Douglas Jackson, ed., **Soviet Resource Management and the Environment** (Columbus, OH: American Association for the Advancement of Slavic Studies, 1978). See also Thane Gustafson, **Reform in Soviet Politics: Lessons of Recent Policies on Land and Water** (Cambridge, ENG: Cambridge University Press, 1981); and Charles Ziegler, "Policy Alternatives in Soviet Environmental Protection," The Carl Beck Papers in Russian and East European Studies (Pittsburgh: University of Pittsburgh, n.d.).

36. Hans-Hermann Höhmann, Gertraud Seidenstecher, and Thomas Vajna, **Umweltschutz und ökonomisches System in Osteuropa: Drei Beispiele: Sowjetunion, DDR, Ungarn** (Stuttgart: Verlag W. Kohlhammer, 1973); Ilka Norhara-Schnabel, "Zur Entwicklung der Umweltpolitik in der DDR," **Deutschland Archiv** 9 (1976): 809-28; Werner Gruhn, "Umweltschutz in der DDR," **Deutschland Archiv** 5 (1972): 1038-50; idem, "Aktuelle Aspekte der DDR-Umweltpolitik," **Deutsche Studien** 19, no. 76 (1981): 424-40; idem, **Umweltpolitische Aspekte der DDR-Energiepolitik** (Erlangen, FRG: Institut für Gesellschaft und Wissenschaft, 1982); Christiane Busch-Lüty, "Zur Umweltproblematik in sozialistischen Systeme: Ideologie und Realität," **Aus Politik und Zeitgeschichte**, no. 27 (1981), pp. 8-46; Hannsjörg F. Buck and Bernd Spindler, "Luftbelastung in der DDR durch Schadstoffemis-

sionen," **Deutschland Archiv** 15 (1982): 943-58; and Bernd Spindler, **Zum Stand des Umweltschutzes in der DDR** (Bonn: Gesamtdeutsches Institut, 1979).

37. See Kelley et al., **The Economic Superpowers**; Cynthia H. Enloe, **The Politics of Pollution in a Comparative Perspective** (New York: David McKay Company, 1975); and Josef Füllenbach, **Umweltschutz zwischen Ost und West** (Bonn: Europa Union Verlag GmbH, 1977). Füllenbach concerns himself primarily with questions of international cooperation, but also touches on a comparison of causes of environmental problems in East and West, esp. 90-132.

38. See Judith Thornton, "Resources and Property Rights in the Soviet Union," in Jackson, pp. 1-12; Leonore Shever Taga, "Externalities in a Command System," in Singleton, pp. 75-100; Arthur Wright, "Environmental Disruption: An Attempt at an Analytical Framework," **ASTE Bulletin** 13, no. 1 (Spring 1971): 1-16; Robert J. McIntyre and James R. Thornton, "On the Environmental Efficiency of Economic Systems," **Soviet Studies** 30 (April 1978): 173-92; Charles Ziegler, "Soviet Environmental Policy and Soviet Central Planning: A Reply to McIntyre and Thornton," **Soviet Studies** 32 (January 1980): 124-34; and the subsequent exchanges in **Soviet Studies** 33 (January 1981): 146-49 and 34 (April 1982): 296-99.

39. John Kramer, "The Politics of Conservation and Pollution in the USSR," (Ph.D. Dissertation, Charlottesville, VA, University of Virginia, 1973), pp. 24-28. A GDR environmental advocate, now an emigre, whose work on the environment has not been published in the GDR, supports Kramer's conclusion regarding the role of natural scientists. See Wolfgang Harich, **Kommunismus ohne Wachstum? Babeuf und der 'Club of Rome'** (Reinbek bei Hamburg: Rowohlt, 1975), p. 77.

40. Robert W. Clawson and William Kolarik, "Soviet Resource Management: Political Aspects of Water Pollution Control," in Jackson, pp. 105-22.

41. Keith Bush, "The Soviet Response to Environmental Disruption," in Volgyes, p. 28.

42. Kelley et al., **The Economic Superpowers**, pp. 171-72; and Donald R. Kelley, "Economic Growth and Environmental Quality in the USSR: Soviet Reactions to 'The Limits to Growth,'" **Canadian Slavonic Papers** 18 (1976): 266-83.

43. Pryde, **Conservation**, p. 8. Goldman indicates agreement with this in **Spoils of Progress**, p. 5.

44. Pryde, **Conservation**, p. 136. Pryde also cites commitment to economic growth as an inhibiting ideological factor, p. 164. Füllenbach cites the primacy of material production as a cause, pp. 105-11.

45. Pryde, **Conservation**, p. 165.

46. Goldman, **Spoils of Progress**, pp. 46-48; Pryde, **Conservation**, pp. 42-43; Busch-Lüty, pp. 21-22; Höhmann, Seidenstecher, and Vajna, pp. 29-30; and Füllenbach, pp. 97-105.

47. Füllenbach comes to this conclusion after a brief discussion of the Soviet debate over the issue in ibid., pp. 101-4. Goldman, while acknowledging that "there was considerable debate as to whether to sanction a meaningful charge for water" before 1970 (p. 112), at another point states that "to this day Russians do not seem able to understand what the difficulty is" regarding lack of charges for mineral resource use (**Spoils of Progress**, p. 48).

48. See Melvin Croan, **East Germany: The Soviet Connection**, The Washington Papers, Vol. 9 (Beverly Hills: Sage Publications, 1976), pp. 38-42.

49. For a thorough analysis of changes in the policy of the Soviet Union to the East European satellites, with a focus on ideological questions, see Zbigniew K. Brzezinski, **The Soviet Bloc: Unity and Conflict** (Cambridge, MA: Harvard University Press, 1967). See "Programm der Sozialistischen Einheitspartei Deutschlands" (May 22, 1976), in **Programm und Statut der SED vom 22. Mai 1976** (Cologne: Verlag Wissenschaft und Politik, 1976), pp. 54-57 for a description of developed socialist society. See the introductory commentary on the program by Karl Wilhelm Fricke for discussion of the concept, pp. 20-23 in the same volume. Peter Christian Ludz also discusses the East German acceptance of the concept of developed socialist society in **Deutschlands doppelte Zukunft: Bundesrepublik und DDR in der Welt von morgen** (Munich: Carl Hanser Verlag, 1974), p. 84. For a more general discussion of the idea of developed socialist society see Alfred B. Evans, Jr., "Developed Socialism in Soviet Ideology," **Soviet Studies** 29 (July 1977): 409-28.

50. For example the East German authorities strongly object to the West German policy of granting immediate citizenship to emigrés from East Germany. East Germany also receives concrete benefits from the West German approach, for example the extension of EEC tariff benefits due to East Germany's implied membership in the European Economic Community. The East Germans have not, out of principle, felt compelled to reject such advantages.

51. See Ludz, **Deutschlands doppelte Zukunft**, pp. 80-86, for discussion of the GDR's "legitimacy deficit" vis-a-vis the FRG.

52. Of course the primary leadership concern about contact between the FRG Greens and GDR young people may revolve around the peace issue rather than environmental problems.

53. Gouldner, **The Coming Crisis in Western Sociology**, p. 470.

54. See, for example, results of interviews conducted by

Hans Apel between 1962 and 1966 in the GDR which indicate increased loyalty to the regime. Reported in Arnold Heidenheimer, **The Governments of Germany** (New York: Thomas Y. Crowell Co., 1971), p. 268. Kurt Sontheimer and Wilhelm Bleek also assess more positively the legitimacy of the GDR regime in the eyes of the citizenry. See their **The Government and Politics of East Germany** (London: Hutchinson University Library, 1975), pp. 44-49.

55. For comparisons of CMEA countries, demonstrating the tight resource situation of the GDR, see V. Gzovskii, "Ispol'zovanie vozobnovliaemykh prirodnykh resursov v stranakh SEV," **Voprosy ekonomiki**, no. 5 (1983), pp. 108-18. For a disturbing portrayal of environmental deterioration in the USSR see the **samizdat** publication by Boris Komarov (pseud.), **Destruction of Nature in the Soviet Union** (White Plains, NY: M. E. Sharpe, Inc., 1980).

56. Gzovskii, p. 113; S. Bobylev, "Puti uluchsheniia ispol'zovaniia zemel'nykh resursov," **Voprosy ekonomiki**, no. 8 (1980), p. 49; and V. Sidorova, "Ratsional'no ispol'zovat' prirodnye resursy," **Planovoe khoziaistvo**, no. 1 (1980), pp. 119-22.

57. Harry Maier and Hans Roos, "Die Mensch-Umwelt Beziehung als politökonomisches Problem," in Werner Gringmuth, ed., **Reproduktion der natürlichen Umwelt** (Berlin: Akademie Verlag, 1974), p. 28. See also Horst Paucke and Adolf Bauer, **Umweltprobleme – Herausforderung der Menschheit** (Frankfurt am Main: Verlag Marxistische Blätter, 1980), p. 164. These authors claim that the GDR is poorer in water resources than any other industrial nation (measured per capita/per year). Gzovskii, however, indicates that Hungary is even worse off, pp. 110, 115. For the new water law see **Gesetzblatt der DDR**, Teil 1, no. 26 (July 21, 1982), pp. 467-94. Recently on the water problem see "Süsswasser," **Horizont**, no. 4 (1983), pp. 13-20.

58. For the politics of river diversion projects see Gustafson, **Reform**, chap. 5. For a time it appeared that the major Siberian river diversion project had been dropped. See Theodore Shabad, "Soviet, After Studies, Shelves Plan to Turn Siberian Rivers," **New York Times**, Dec. 16, 1983, p. A7. Now, however, it is again being discussed in the Soviet press. See V. Zakharko, "Northern Water for the South," **Izvestiia**, June 22, 1984, trans. in **CDSP**, vol. 36, no. 25, pp. 1-3.

59. Günter Streibel, "Die ökonomische Stimulierung der rationellen Nutzung natürlicher Gebietsressourcen," **Wirtschaftswissenschaft** 14 (December 1966): 1971.

60. See Hans Sandig and Siegfried Wolf, "Einige Probleme der Wirkung verstärkter Produktion einheimischer mineralischen Rohstoffe auf die Effektivität der gesellschaftlichen Produktion in der DDR," **Wirtschaftswissenschaft** 27 (January 1979): 61-70.

61. See, e.g., Gerhard Schürer, "Die Energie- und Rohstoffbasis unseres Landes stärken," **Einheit** 33 (1978): 1245-51.

62. On potash see Gesamtdeutsches Institut, **Die Rohstoffwirtschaft der DDR** (Bonn, Feb. 1980), p. 44; and Marlies Menge, "Für Filter fehlen die Devisen," **Die Zeit** (Hamburg), March 25, 1983, p. 21 (Canada Edition).

63. Buck and Spindler, pp. 949-50; Martin Jänicke, "Umweltpolitik in Osteuropa," in Hannelore Horn, Alexander Schwann, and Thomas Weingartner, eds., **Sozialismus in Theorie und Praxis**, Festschrift für Richard Lowenthal (Berlin [West]: Walter deGruyter, 1978), pp. 180-83; Gruhn, **Umweltpolitische Aspekte**, pp. 73-81, 88-96; Conrad Schwartau, "Zunehmende Luftverschmutzung in der DDR durch Renaissance der Braunkohle," **DIW-Wochenbericht** 50, no. 4 (January 1983): 43-49; and "Baumsterben in der DDR am weitesten vorangeschritten," **Frankfurter Allgemeine Zeitung**, March 9, 1984, p. 10. Other central European countries, including Czechoslovakia, Poland, and the FRG, also are both sources and recipients of polluted air which is causing widespread damage to forests. Damage is particularly intense in the Erzgebirge in Czechoslovakia, near the GDR and FRG borders. See K. Thiele, "Sonderheft: Waldsterben," **Information 1/83** (Bayerishce Staatsforstverwaltung, 1983); and Hannelore Rudisch, "Bald so gespentisch wie im Erzgebirge," **Frankfurter Allgemeine Zeitung**, Dec. 15, 1983, p. 10.

64. On the strengths and weaknesses of the "most-similar-systems" approach see Adam Przeworski and Henry Teune, **The Logic of Comparative Social Inquiry** (New York: John Wiley & Sons, 1970), pp. 32-34.

# 2
# Political Legitimation and Environmental Problems*

We know very little about the process of political legitimation in the USSR and Eastern Europe. We do not know why citizens of the USSR and the GDR generally obey their governments. If we have a somewhat better idea of why Polish citizens often do not, this is simply because their disobedience has periodically allowed public articulation of grievances. More commonly, however, limited information and restricted access in applying sophisticated research techniques relegate assessments of regime legitimacy in the Soviet bloc to the realm of speculation and journalism. Political scientists have suggested diverse explanations for political compliance in the Soviet bloc, only some of which imply effective regime legitimation: the totalitarian school focused on ideological indoctrination, personal charisma, and terror;[1] the political culture school emphasizes nationalism and congruity between traditional and modern value systems;[2] Samuel Huntington cites institutional stability as a source of system legitimacy;[3] and others note pervasive depoliticization of the population induced by political repression and ideological confusion.[4]

If we lack the tools to assess the efficacy of the legitimation process, we can, however, analyze the way in which regimes try to legitimate their rule. The official rationalization system in Eastern Europe involves more than the tedious repetition of Marxist-Leninist phraseology which one can observe so readily in Soviet-bloc propaganda. Preoccupation with the role of Marxism-

---

*The original version of this chapter appeared under the title "Political Legitimation in the USSR and GDR: Environmental Deterioration as a Case Study," by Joan DeBardeleben published in **Soviet Union/Union Soviétique**, vol. 11, part 2 (1984): 212-42, and is reprinted by permission of Charles Schlacks, publisher.

Leninism can lead one to overlook the multi-leveled and complex picture of reality implicitly presented in the media of these relatively controlled political systems. Furthermore, possible differences in patterns of legitimation between the USSR and even her most orthodox allies, such as the German Democratic Republic (GDR), require further examination. Alfred Meyer has suggested that Soviet leaders have successfully indoctrinated their citizens, that, in effect, Soviet citizens do talk like **Pravda** editorials.[5] The first step in assessing this claim (and counter-claims which assert an incongruence between official and popular conceptions) is to clarify how the Soviet and East German media **do** portray the world. To this end, in this chapter, we examine the "public image" of environmental problems and policies, that is, how they are presented to the citizen. In this way, we can also begin to unravel the system of official rationalization in the USSR and GDR and bring at least a modest advance in our understanding of the dynamics of political legitimation in Soviet-type systems.

It is difficult to examine a system of rationalization in a general way. Such a task would be unmanageable. Furthermore, a general examination would exclude, a priori, the identification of micro-level explanations of social problems which form a part of the legitimation system. Therefore, a focus on a specific issue area like environmental problems is useful.[6] At first glance environmental problems may seem peripheral to the question of regime legitimacy. Environmental deterioration most likely does, however, pose disturbing problems of system legitimacy to Soviet and East German authorities (although to a degree this must remain a matter of speculation for reasons noted above). We will address this issue further in the section which follows.

## ENVIRONMENTAL DETERIORATION AS A LEGITIMACY CHALLENGE

Some forms of pollution and environmental deterioration are obvious; others require sophisticated instruments to detect. Even a casual visitor to the East German industrial city of Leipzig will immediately detect the effects of the burning of high-sulfur brown coal, the GDR's major energy resource. Indeed, one West German study concluded that levels of sulfur dioxide and dust in the air in Leipzig, Halle, and Zwickau are double those usually found in Chicago, Dusseldorf, or Tokyo.[7] An estimated twelve per cent of the forests in the GDR are already damaged, in large part a result of air pollution. Furthermore, in the industrial region of Halle, life expectancy is reportedly five years less and the incidence of heart and respiratory disease ten to fifteen per cent greater than elsewhere in the GDR.[8] Some Soviet industrial cities have equally obvious air pollution problems. **Meditsinskaia gazeta** reports that even in a medium-size city like Semipalatinsk (Kazakhstan) three hundred small boiler houses pollute the air enough to make breathing difficult on windless winter days.[9] A recent dissident account indicates alarming rates of lung cancer and respiratory disease in the Novokuznetsk and Kemerov industrial regions, while numerous other cities all over the USSR register concentrations of noxious gases far exceeding the official maximum permissible concentrations. Oil contamination off the Black Sea coast is another form of pollution obvious to Soviet vacationers.[10] Several bodies of water in the USSR have suffered significantly declining water levels, with damaging effects on recreational activities, fish habitats, and the surrounding ecosystems. For example, diversion of water for irrigation purposes has had this effect on Lake Issyk-Kul and the Aral Sea, while extraction of river-bed gravel has lowered the level of the Tom River.[11] The GDR also has water problems. Reportedly, most surface waters in the districts of Halle and Dresden show pollution levels exceeding the legally established maximum values. Some 30,000 tons of potassium salt are dumped into the Werra River daily.[12] These are only some of the more obvious types of environmental damage which occur in the USSR and GDR. Other kinds of damage are more subtle, such as DDT and PCB contamination, extinction of some types of plants and wildlife,[13] slow salinization of soil, the effects of lead pollution on public health, and potential hazards of nuclear waste transport and disposal.

Because environmental damage is both ubiquitous and sometimes invisible, even public awareness is hard to estimate. We simply do not know how Soviet and East German citizens view environmental deterioration. We do know that Soviet and East German authorities feel compelled to restrict information about the extent of the problem and to provide popular reassurance about the ability of the socialist system to deal with it.[14] Official Soviet commentators occasionally refer obliquely to disturbing indications of back-to-the-earth sentiments.[15] We can also observe significant expressions of concern over environmental destruction in the official and dissident media. Particularly in the GDR, Marxist dissidents such as Rudolf Bahro and Wolfgang Harich have made ecological considerations a pivotal point in their critiques of the regime.[16] In addition, since the late 1970s, the ecology question has gained considerable attention within the Evangelical (Lutheran) Church organizations and a form of grassroots environmental activism has become visible among young people in the GDR.[17]

These indications notwithstanding, familiarity with Western environmental movements and arguments might lead one to doubt whether environmental deterioration really poses a legitimacy problem for the Soviet and East German authorities. Western environmental debates (both regarding concrete policies as well as the "limits to growth" debate) suggest that increased attention to environmental protection will necessitate a slow-down in economic growth and thus a certain sacrifice of material well-being in the short run. (The validity of this assertion is not at stake here, rather its widespread dissemination.)[18] By extension, one might assume that the Soviet and East German populations would have little concern with environmental protection, given more pressing problems of consumer goods supply. In fact, however, this tendency to extrapolate from Western assumptions to Soviet-bloc assumptions is probably incorrect. Unlike in the West, prices in the USSR and GDR do not directly reflect increased costs of environmental protection measures in the production process. The Soviet and East German citizen is therefore less likely to believe that increased expenditures for environmental protection will mean a decrease in the supply and quality of consumer goods, or indeed that there is any kind of real trade-off here. Rather, both environmental deterioration and poor consumer goods supply are probably viewed as symptomatic of the pervasive inefficiency and skewed priorities in the economic sphere. Environmental deterioration is simply an additional burden, a reflection of the unresponsiveness of the political-economic system to popular demands and needs. Indeed, the low quality and poor distribution of consumer goods in the USSR and GDR may in many regards result from the same deficiencies which contribute to misuse of natural resources – poor incentive structures for enterprises, uncoordinated plans, inadequate information flow, and faulty evaluation and accounting techniques. Even the Soviet press emphasizes that money is often available, but it "is being used poorly and unwisely" and is at times left unspent.[19]

Oddly enough, the Soviet and East German media generally do not propagate the view that there is a trade-off between environmental protection and an improved standard of living. Probably the best way Soviet authorities could deflate environmental concern would be to adopt such a strategy. Yet such an argument would grant even greater legitimacy to consumer complaints and imply a promise of improvement. Furthermore, it might imply a built-in limitation on socialism's ability endlessly to improve people's lives. In the USSR more environmental protection would probably have little effect on the consumer's situation. Soviet citizens probably correctly understand this reality, as do Soviet planners and ideologists. Soviet authorities instead focus on a different trade-off; above all the capitalist-inspired arms race forces environmental expenditures down.[20] Consumer goods are rarely mentioned in this context. Simultaneously, Soviet and East German commentators reject Club of Rome no-growth arguments on the grounds that no-growth would condemn the Third World to perpetual poverty. These commentators imply that an improved standard of living and adequate environmental protection would be compatible goals in a socialist world economic order, as they are in socialist society.[21] Soviet-bloc authorities claim that the socialist system is at its root superior in all regards, even if that superiority has not yet been realized; the obverse of this logic is the citizen's perception that problems derive from the system, not from an economic trade-off.

On another level environmental deterioration may challenge basic values of the Soviet and East German systems. Deficiencies in pollution control undermine claims of socialism's superiority in social policy. Creditable gains in education, health care, vacation time, job security, and price stability are marred by visibly contaminated rivers and air, causing both aesthetic and health deterioration. Furthermore, allusions in the Soviet press and in literature indicate that environmental deterioration may evoke disillusionment with urban, industrial, anthropocentric values. Indicative of this sentiment is a surprising transposition of Marx's famous statement about "the idiocy of rural life," published in the literary review, **Literaturnoe obozrenie:** "Already long ago the phrase 'the idiocy of rural life' was popular, times have changed sharply, and now it is much more appropriate to speak of the 'idiocy of urban life,' with its polluted air and smog, with the expenditure of several hours a day only for transportation to and from work, and with constant noise."[22] Such criticism of the polluted urban environment has been linked in the USSR with rural romanticism, with all the potential ethnic, religious, and anti-industrial undertones this can invoke.[23]

Although we do not know the extent of these anti-industrial sentiments, they pose a direct challenge to the very rationale of Soviet economic and political modernization. The historical role of the Communist Party of the Soviet Union (CPSU) as a modernizing force (not to mention the party's Marxist-Leninist foundations) rests on an urban, industrial, scientific, and atheistic world-view. In addition, any subtle appeal to rural ethnic traditionalism poses a challenge to the fragile multi-ethnic Soviet "nation." A healthy wave of ecological dissent would be neither trivial nor easy to deflate in the USSR.

The dilemma for the East German leadership is somewhat different. The GDR stands at the interface of the East-West conflict. With the flick of a switch, the East German citizen can receive West German television and radio broadcasts, enabling him/her to view former compatriots enjoying material prosperity, suffering environmental deterioration, and engaging in environmental activism. Furthermore, many East German citizens no doubt hear the views of their own "dissident ecologists" (such as

Rudolf Bahro) through the publicity given them in the West German electronic media.[24] Not only are East German citizens exposed more often to arguments of ecological activists, but they are also more likely than Soviet citizens to be receptive to the idea that increased environmental expenditures would slow down consumer goods production. Privileged in the Soviet bloc, the East German consumer is simultaneously the poor cousin to his/her West German counterpart. Both the relative success of the East German economy (in the East bloc) and West German images of affluence suggest a frustrated potential for increased material prosperity. The West German media also exposes the East German to the "trade-off" argument. Although the absence of survey data and the strictness of GDR censorship make it difficult to demonstrate, the consumer goods-environmental protection trade-off may well be more credible to the East German citizen than to the Soviet citizen.[25] Also obvious, however, is the failure of the GDR to surpass the FRG either in the consumer or the environmental sphere. If there is a trade-off, a good bargain has not been struck. Thus GDR authorities, like their Soviet comrades, must justify this deficiency.

Industrial, urban values are deeply rooted in the GDR and rural romanticism does not threaten ethnic revivalism, as it does in the USSR. Furthermore, the FRG presents an example of a more palatable urban life-style, pollution and all. But on the margins of GDR society, in church-related groups and in literature, one finds dissent from the predominant vision of progress which prevails in both East and West, a vision of abundant material wealth and of nature subjected to human desires. Particularly since the late 1970s there is increasing evidence of disillusionment among some GDR youth with the impersonal life of a consumerist society alienated from nature. These critiques of modern "civilization" challenge the emphasis placed by GDR authorities on economic growth and rationalization; they suggest that the spiritual and human values of socialism have been sacrificed in the name of a material-technical notion of progress. As one West German analyst has suggested, such sentiments do not "in any way involve a resort to 'pre-industrial ideals,' but rather much more the insight that **this** path of industrial development can no longer be followed if one wants to preserve the earth from its own gradual destruction."[26] If Soviet authorities seek to cement a multi-ethnic nation to industrial values, in the GDR the problem may be more complex. East German authorities must distinguish the GDR from the more prosperous FRG, but without calling into question the priorities required to emulate the FRG's material success and without spurring consumer appetites which cannot be satisfied.

The persistence of environmental deterioration implies a

failure of public policy, particularly in centrally-planned economies, where the party-state expressly assumes responsibility for economic and social decision-making. How then do the Soviet and GDR authorities respond? They could simply initiate a comprehensive and effective program of environmental protection, thus demonstrating socialism's superiority and their own competence. Indeed, both regimes have instituted programs of modest proportions since the mid-1960s. In recent years, as costs for raw materials have risen, particular efforts have been devoted to recycling of waste products and their use as secondary raw materials. The scope of these efforts and obstacles to their expansion are examined further in chapter 5.[27] Suffice it to say that barring fundamental shifts in economic priorities and significant reform of enterprise incentives, such measures are unlikely to bring marked improvement, except in target areas. This leaves us with the process of political legitimation. How are problems publicly defined, justified, and explained? With all forms of public communication at its disposal, the political leadership can hope to shape public acceptance of the status quo.

## The Environmental Problem As a Case Study

Justification of polluted air may seem a minor concern compared to the arms race, the Polish situation, and poor agricultural performance. But this particular issue can provide a workshop for examining more general questions of regime legitimation. First, pollution is not simply a problem of legitimation; pollution is also an important policy problem. In this regard it resembles most other sensitive issues in Eastern Europe, e.g., consumer goods shortages, ethnic representation, population growth, and agricultural production. Specifically, environmental deterioration has disturbing economic implications. For example, salinization and erosion destroy agricultural land; water pollution reduces caviar output, and thus hard currency exchange; chemical pollutants increase health costs and job absences; and valuable resources spew from smokestacks as industrial waste.[28] If environmental problems are an arena of ideological struggle, they are also an arena for disinterested scientific and technical investigation. Restricting analysis of such problems to ideological justifications introduces the risk of restraining their resolution. A political balancing act is required in this area, as in many others.

Second, within the realm of legitimation, environmental problems represent a microcosm for study of the extensive Soviet-bloc literature on the scientific-technical revolution (STR). This literature has both an ideological and a theoretical side: it seeks to justify and to understand. In this sense it embodies the dilemma discussed above. The STR literature tries to come to terms with the radical changes introduced in the productive forces since World War II, including automated production systems, increasingly sophisticated computer technology, molecular and atomic processes, and intensified nature-society interaction.[29] Intense and irreversible ecological disturbances are one of the most pervasive and admittedly ambiguous consequences of the STR. Ecological problems provide a wedge to pry open this major body of Soviet theorizing-ideologizing. As Frederic Fleron has pointed out, systematic analysis of this type of literature provides a means of "seriously scrutinizing the Soviet system of rationalization."[30]

In this chapter we look at two components of the legitimation structure regarding environmental deterioration. First, what is the nature of the problem? Second, what are its causes and solutions?

## POLITICS AS LANGUAGE: THE NATURE OF ENVIRONMENTAL PROTECTION

Language is the spontaneous expression of thought patterns. Conversely, reshaping language may subtly shape perception of reality. Not surprisingly, East German linguists announce the emergence of a new "socialist" German terminology.[31] Less dramatic and ambitious is the attempt to impose a definite meaning on familiar terms. Both the East German and Soviet authorities have sought to do this in shaping public understanding of environmental deterioration and protection.

By defining and employing the major terms of discourse in a particular way, Soviet and East German authorities seek to establish a particular understanding of what constitutes the environmental problem. Is, for example, the extinction of animal and plant species in itself evidence of environmental deterioration, or only if loss of such genetic diversity imposes costs on the human species? Is the environment improved or damaged by irrigation and land reclamation projects, if the result is higher agricultural productivity **and** disturbance of the previous balance in the local ecosystem? Answers to such questions depend on one's fundamental understanding of concepts like "environmental deterioration" and "environmental protection." These terms may be what William Connolly calls "essentially contested concepts."[32] The ability to impose a particular definition implies the establishment of an entire normative viewpoint.

Because of their unique legitimation problems, East German authorities have introduced a linguistic distinction in the area of environmental protection which the Soviets have not. While

Soviet terms for "protection of nature" (**okhrana prirody**) or "protection of the environment" (**okhrana okruzhaiushchei sredy**), on the face of it, seem similar to their English equivalents, in the GDR two qualitatively different concepts are used to describe the task of environmental protection – **Umweltschutz** and **Landeskultur**. The term **Umweltschutz** means, quite straightforwardly, protection of the environment, and is the word generally used in the Federal Republic of Germany. As the West German analyst Gertraud Seidenstecher has pointed out, however, the word **Umweltschutz** did not come into common usage in the GDR until after the 1972 United Nations Conference on environmental problems held in Stockholm.[33] Until 1972 and since, the somewhat puzzling term, **Landeskultur**, has also been employed. The comprehensive law dealing with environmental protection passed in 1970 is called the **Landeskulturgesetz.**[34]

Although in general usage the two terms – **Umweltschutz** and **Landeskultur** – seem to be almost interchangeable, the distinction between them is theoretically and ideologically significant. An East German handbook on socialist **Landeskultur** explains that **Umweltschutz** is only a part of **Landeskultur**, referring primarily to short-term efforts to address technological sources of environmental deterioration (e.g., replacement of old technology, application of emission control equipment), while **Landeskultur** involves a more comprehensive and long-term program to shape the environment in accordance with the demands and lifestyle of socialism. The term **Umweltschutz** was avoided until 1972 but following the GDR's admission into the United Nations and her increased involvement in international programs, "the internationally conventional expression **Umweltschutz** [environmental protection] had to be used in the GDR as well."[35] Placing environmental protection (**Umweltschutz**) within the context of **Landeskultur** allows East German authorities to distinguish socialist politics and practice from that of capitalist nations: "Socialist **Landeskultur** includes the rejection of every implication of convergence theory in the area of environmental protection."[36] Seidenstecher suggests that the delineation of a unique socialist principle in the GDR is a response to the direct challenge posed by the FRG.[37] Establishment of a linguistic distinction between East and West German environmental policy presumably provides de facto refutation of convergence theory, and thus discourages East German citizens from making unflattering comparisons with West German policies.

Examination of the **Landeskulturgesetz** reveals the broad-ranging scope of socialist **Landeskultur.** In addition to pollution control, the law provides a framework for treatment of a wide range of problems: development of recreational areas; beautification of cities; establishment of protected objects; rational use of forest resources; and melioration of agricultural land. Even this list is not inclusive. References to **Landeskultur** in the popular press encompass healthier working and living conditions generally.[38] **Landeskultur** is portrayed as a positive, constructive process of environmental protection, forming an integral component of the building of socialism. It contrasts with the allegedly defensive posture characteristic of capitalist environmental protection. As the West German analyst Norhara-Schnabel points out, defining **Landeskultur** to encompass such diverse activities as rational resource use, public health, and city planning may allow the East German authorities to mask inadequacies in ecological policy proper.[39] It also begs the question of deeper inconsistencies between an ecological life-style and the East European brand of socialism. Conversely, such a close association of environmental protection with the goals of socialism unwittingly provides East German environmentalists with the opportunity to argue their case in terms of officially affirmed values.[40]

There is no Soviet equivalent for the East German concept of **Landeskultur** nor have the Soviets passed a comprehensive national framework law on environmental protection.[41] The Soviet party/state does not face a legitimacy challenge from a linguistically identical country, as does the GDR. Hence there is no need for a specifically "socialist" Russian language and there is a lesser urgency in propagating an integrated socialist approach to environmental protection. In other regards, however, common attitudes underlie East German and Soviet definitions of the problem. The media in both countries assert the inextricable link between protection and use of nature. Probably for this reason the Soviet word most closely akin to the English word "conservation" (**sokhranenie**) is used infrequently when discussing general problems of nature protection. The definition of nature protection offered by Blagosklonov et al. in 1967 still summarizes the official understanding of the term in the Soviet Union and the GDR very well: "a complex of governmental, international, and social measures directed towards rational use of nature, renewal, increase, and protection of natural resources for the good of human society."[42] Such a definition signifies that the maintenance of ecological balance is not of value in itself, but only as a means to assure that natural resources will be available for future human use. This definition is not simply a tautological designation of the referent of a phrase, but rather the explication of the correct understanding of a social and economic task. This task extends beyond protection of the environment for future human use to actual transformation or development of the environment (**Umweltgestaltung, preobrazovanie prirody**). This longer-term process of environmental transformation involves human intervention to bring actual improvement in the organization of natural processes to realize a greater effectiveness of production and better working and living conditions. As one Soviet analyst put it, "protection, use, and transformation of nature cannot be viewed in isolation from one another – they are an indivisible pursuit."[43]

Soviet state structures embody this link between the protection and use of nature. In the USSR users of resources often have primary responsibility for their production and reproduction. Soviet critics see this scheme as ineffective for environmental protection and argue for separate provision for use and protection.[44] Assignment in 1979 of supervisory tasks to the newly-created State Committee on Hydrometeorology and the Environment (Gosudarstvennyi Komitet po Gidrometeorologii i Kontroliu Prirodnoi Sredy) in the USSR was an effort to provide a source of environmental oversight autonomous of particular ministries. The committee is, however, responsible primarily for air quality, a resource whose users are ubiquitous.[45] The status quo operates to benefit productive users of natural resources and these powerful bureaucratic interests form the greatest obstacle to further administrative separation of use and protection of nature. But Soviet administrative patterns also reflect and reinforce the anthropocentric, functionalist rationale for environmental policy.

## CAUSES AND SOLUTIONS: CAPITALISM AND SOCIALISM

An early Western study identified a problem underlying environmental planning in the USSR to be the "political-philosophical assumption that improper natural resource exploitation is the result of the capitalist mode of production, and that a socialist economy necessarily pursues the wisest possible use of natural resources."[46] Although Soviet and GDR writers continue to hold capitalism responsible for much environmental deterioration in the world, by now most have abandoned the view that environmental problems are automatically solved in socialist society. Rather, explanations of causes and solutions to environmental deterioration have become increasingly complex, involving several levels of analysis, at times seemingly contradictory. Soviet and GDR commentators do not, of course, explicitly distinguish between these levels of explanation, and indeed the explanations overlap and are linked together by key concepts.

One level of explanation (the "relations-of-production" level) does stress the responsibility of capitalism for environmental deterioration and the potential of socialism for lasting solutions. Even here, however, Soviet and GDR analysts emphasize that the solution in socialism is not automatic, but depends on scientifically-based policy. Another level of explanation (the "forces-of-production" level) sees environmental deterioration as growing out of contradictions produced by rapid scientific-technical progress itself. A third level of explanation (the "micro-level factors" level) deals with micro-level causes of environment deterioration within socialist society such as specific policy deficiencies, incorrect attitudes, or inadequate knowledge. The interaction of these various levels of explanation accounts for the existence of environmental deterioration in state socialist societies, asserts the superiority of socialism over capitalism, and allows leeway for specialists to debate policy responses. The "relations-of-production" explanation alone would not be adequate to account for the continued existence of environmental problems in the Soviet socialist world. The "forces-of-production" explanation alone might suggest technological determinism or offer support for convergence theory, both ideas explicitly rejected in the GDR and USSR. The micro-level explanations allow debate on particular policy issues, and communicate that present inadequacies in the USSR and GDR are not systemic, but rather aberrations in an otherwise sound system.

Careful study of the East German and Soviet media indicates that attention given to each type of explanation varies between the USSR and GDR, and between media type, as illustrated in Table 2.1. As Jerry Hough has observed, the larger the circulation of a publication, the more restrictive the range of discussion of policy issues.[47] Thus the more ideological explanation (stressing the responsibility of capitalism for environmental deterioration) is found more often in the mass circulation press than in the limited circulation scholarly media. Furthermore, this type of explanation is stressed more in the East German press than in the USSR. This "relations-of-production" explanation is never openly contradicted, but is sometimes omitted or treated in a perfunctory manner in scholarly writings, particularly in the USSR. Conversely, scientific-technical progress itself as a source of environmental deterioration is dealt with most explicitly in the Soviet academic press and in a leading Soviet popular science magazine, **Priroda** (Nature). In the mass circulation media there are occasional references to common environmental problems in the West and Soviet bloc, and attention to common technical solutions also implicitly rests on this type of analysis. This type of explanation is found less frequently in the East German press than in the Soviet (the mirror image of the "relations-of-production" explanation). Micro-level explanations, involving practical and subjective sources of problems in present socialist policy, reveal disagreement among specialists. The academic press in both countries contains significant debate about particular policy options, although in general this public debate is more restricted in the GDR than in the USSR. In the USSR even the mass media devote considerable attention to problems with the organization of planning and production; newspapers like **Pravda** and **Izvestiia** contain sharp criticisms of particular enterprises or ministries in carrying out their environmental obligations.[48] Here we find another sharp contrast with the GDR mass media. Despite some greater public admission of problems in the last two years, in general the GDR mass media emphasize policy successes, only obliquely referring to policy failures; greater importance is attributed to popular attitudes and technological change than to policy reform.

Debate carried on in academic journals, particularly that which rests on forces-of-production or micro-level explanations, does not primarily serve legitimizing functions. Its audience in the academic community is likely to have an already established stake in the continued stability of the system. More important than the legitimizing role of such academic discourse is its advocacy function, as scholars address controversial issues of regime priorities and budgetary allocation. As in the West, such scholarly debate also furthers communciation between experts engaged in research into problems important to the regime. While this academic debate may indirectly influence policy-makers, Soviet and East German academics also express frustration that their viewpoints are too rarely heeded (a phenomenon not unique to the Soviet bloc). Nonetheless, examination of scholarly journals aids us in our study of legitimation. By comparing positions taken in the mass media with those in scholarly publications, one can more clearly distinguish legitimizing arguments from advocacy arguments. Finally, the very existence of such scholarly debate, which often spills over into the popular media in the Soviet Union, may well reinforce popular belief in the scientific and rational nature of policy deliberation.

It remains now to examine more closely each type of explanation and the role each performs in the legitimation process in the two countries.

## The "Relations-of-Production" Premise

Soviet and East German authorities, scholars, and journalists present capitalism and imperialism as the fundamental sources and socialism as the fundamental solution of continued environmental deterioration in the world today.[49] According to this argument, the primacy of the profit motive in capitalist industry results in crude indifference to resource needs and to maintenance of a healthy environment for present and future generations. Capitalist production produces consequences which are neither planned nor foreseen. Only when pollution and resource

**TABLE 2.1 Relative Frequency of Explanations of Environmental Deterioration by Media Types**[a]

| Explanation | Nature of Explanation | Media Types | | | | | |
|---|---|---|---|---|---|---|---|
| | | GDR popular[b] | GDR scholarly[c] | Soviet popular[d] | Soviet scholarly[e] | GDR, both types[f] | Soviet both types[g] |
| **Macro-Level**<br>"Relations-of-production" | Cause: Capitalism<br>Solution: Socialism | more frequent | ← | more frequent | ← | more frequent | ← |
| "Forces-of-production" | Cause: Scientific-technical progress<br>Solution: Scientific management with an adequate scientific-technical base | → | more frequent | h<br>→ | more frequent | → | more frequent |
| **Micro-Level Factors in Socialism**<br>Policy factors | Causes: Poor incentives, laws, organization<br>Solutions: New laws; policy changes | k<br>→ | more frequent | → | more frequent | → | more frequent |
| Inadequate Scientific-Technical knowledge | Causes: Inadequate research and application of knowledge<br>Solutions: More research; better coordination | → | more frequent | → | more frequent | → | more frequent |
| Incorrect Attitudes | Causes: Elite attitudes (ministries, managers) | | i | → | more frequent | → | more frequent |
| | Mass attitudes<br>Solution: Education | more frequent | ← | more frequent | ← | more frequent | ← |

Comparisons are made for three distinct pairs of media types, but no comparisons are drawn between pairs. The arrow indicates the direction of greater frequency. For example, the relations-of-production explanation appears more frequently in the GDR popular media than in the GDR scholarly media; more frequently in the Soviet popular media than in the Soviet scholarly media; and, considering both scholarly and popular sources, more often in the GDR media than in the Soviet media.

a. This chart is based on careful study of articles on environmental deterioration from a broad spectrum of East German and Soviet media sources in the period from 1965 to 1983. Publications consulted include those indicated in the following notes, as well as numerous other sources listed in the bibliography.

b. Sources consulted include: **Leipziger Volkszeitung, Neues Deutschland, Urania, Die Wirtschaft** and **Forum**. For time periods consulted see the bibliography, pages 316-17.

In general, time periods were selected at random to include articles from a broad spectrum of sources over a reasonably long period of time. On the other hand, the issues of **Leipziger Volkszeitung** were included because of the special series carried on environmental issues in 1973.

c. Sources consulted include: **Deutsche Zeitschrift für Philosophie**, 1965-1983; **Staat und Recht**, 1965-83; **Wirtschaftswissenschaft**, 1965-1983; and selected articles from **Petermanns Geographische Mitteilungen.**

d. Sources consulted include: **Current Digest of the Soviet Press (CDSP)**, 1965-1983; and **Priroda**, 1965-March 1982. Although CDSP does not represent a random sample of the Soviet popular press, this should not bias our conclusions since presumably unorthodox and critical articles appearing in the Soviet press would very likely be translated in **CDSP**. Nonetheless, these articles were less critical and more ideological than Soviet scholarly articles, on the whole. On the other hand, in the East German popular media I found virtually no articles as critical as numerous ones translated from **Pravda** and **Izvestiia** in CDSP.

e. Sources consulted include: **Voprosy filosofii**, 1965-1983; **Sovetskoe gosudarstvo i pravo**, 1965-1981; **Voprosy ekonomiki**, 1965-1983; and selected articles translated in **Soviet Geography: Review and Translation.**

f. This includes sources noted in (b) and (c).

g. This includes sources noted in (d) and (e).

h. As noted in the text, the popular magazine, **Priroda**, gives an exceptional degree of attention to this theme compared to other mass circulation sources.

i. This argument is found too infrequently in the East German media to discern a pattern.

k. In these cases, East German sources emphasize achievements in socialism (i.e., solutions) rather than causes of environmental deterioration in socialism. E.g., the East German popular media carry frequent references to the importance of the **Landeskulturgesetz** as evidence of wise socialist policy.

shortages threaten profits and endanger the quality of life of the bourgeoisie itself, or when militant popular demands (generally described as led by the working class and communist party) force concessions do capitalist states begin to institute limited, but finally ineffective, measures to control environmental deterioration. Environmental decline is part of the general crisis of capitalism.[50] In capitalist society, ecological degradation is rooted in the social class structure; in socialist society such social class obstacles have been eliminated. Socialism, in this view, does not automatically eliminate pollution, but it provides a potential to do so. Public ownership of the means of production allows decision-making in the interests of the whole population and central planning permits scientific decision-making which can take environmental factors into account.[51] This explanation for environmental problems takes a more sophisticated form in scholarly contributions than in mass publications.[52]

This "relations-of-production" premise serves a dubious legitimizing function in Soviet socialist countries. However adequate its critique of capitalism, it obviously poses a dilemma for the leadership of state socialist countries which admittedly experience ecological disturbances. The obvious inference is either that the explanation is wrong (or at least incomplete) **or** that the USSR and GDR themselves are capitalist nations (or at least share their essential traits). Despite these ambiguous implications the explanation is asserted even more persistently in the GDR than in the USSR and more frequently in the popular media than in academic sources. To deal with the ambiguous implications, capitalism is claimed to be indirectly responsible for environmental problems in socialist countries as well. In the GDR, old conceptions persist and old productive facilities, built before World War II, pollute the environment and utilize raw materials wastefully.[53] Furthermore, in both the GDR and USSR destruction from World War II (rooted in imperialism) necessitated diverting funds to repair extensive damage.[54] Other burdens imposed by capitalism have limited the feasibility of extensive environmental expenditures: the backwardness of Russia before the revolution, necessitating heavy investments for rapid economic growth; Western intervention in the Civil War; and the arms race.[55] On this basis, one Soviet analyst asserts "the historical responsibility of capitalism for the arising of ecological conflict."[56]

On a global level as well, capitalism exports problems, according to Soviet and East German analysts. Rapacious exploitation of Third World resources is one example.[57] Those ecological problems considered most serious by Soviet scientists are global. The proportionally high level of oxygen consumed by the United States is seen as ecological imperialism. Capitalist society also threatens potential abuse of humanly induced climatic changes.[58] Although the capitalist system is incapable of solving environmental problems at home, on an international scale, peaceful coexistence can provide a context for the amelioration (although not the solution) of such global problems. The most important precondition for effective action is, however, in the Soviet view, a reduction in the arms race. The importance attributed to the arms race suggests that one function of the "relations-of-production" explanation in the USSR is to appeal implicitly to patriotic sentiments, although the explicit language is Marxist-Leninist. In the Soviet Union, of course, Marxist-Leninist concepts have been used to justify patriotism ever since the articulation of the doctrine of socialism in one country by Stalin in 1924.

If capitalism bears heavy responsibility for environmental deterioration, from the Soviet and East German viewpoint, socialism creates the possibility for an effective remedy, but that remedy **will not occur automatically.**[59] Schmidt-Renner, an East German economist, argues that without directed action, socialist society faces the same dangers as capitalist society.[60] Other analysts emphasize the need for consciously applied efforts if the advantages of socialism are to be realized. Thus while environmental problems are described as inherent to capitalism, and therefore insoluble, they represent "aberrations" or "mistakes" when they appear in socialism.[61] These "mistakes" can only be rectified over a fairly extended period of time, with the help of an adequate technical base and implementation of scientific management.[62] As one East German analyst explains, "finding socialist solutions to environmental development is a cognitive process like every other, with all its struggles and mistakes."[63]

The contingency of progress and the role of technology lead directly to consideration of the second more interesting level of explanation.

### The "Forces-of-Production" Premise

In an insightful and provocative essay, Frederic Fleron has noted that Soviet theories of the scientific-technical revolution (STR) generally attribute a subjective and an objective character to technology.[64] Material advances achieved under capitalism (e.g., computer technology) are necessary for large scale industrial production, and these advances can (and indeed must) be taken over by socialist society. This is the objective side of technology. At the same time, within capitalist society, technology is applied to reinforce exploitation of the working class and promote the interests of the bourgeoisie (e.g., computer technology enables bankers and capitalist managers to manipulate funds and people more effectively so as to make a higher profit). This is the subjective side of technology. Only when applied in the context of socialist productive relations does technology reveal its full liberating potential. This same dichotomy between subjective and objective factors pervades Soviet explanations of environmental deterioration. The objective factor behind environmental deterioration (the "forces-of-production" premise) is the ever intensifying nature-society interaction. This "objective" effect of the STR is unavoidably felt to some extent in socialist society (but in this case, unlike computer technology, for example, the objective effect is partially negative, and thus more problematic). The subjective factor (the "relations-of-production" premise) is the tendency of capital to sacrifice the environment for the sake of profit. This "subjective" factor does not affect socialist society, as noted above. On the contrary, socialism provides conditions (relations of production) enabling resolution of environmental problems. Soviet analyses of the causes of environmental problems thus parallel Soviet analyses of the STR and of technology in general.

Soviet theories (most recently theories of the STR) weave together these objective and subjective strands of causality. As noted earlier, both strands are necessary for political legitimation. One explains why socialism is superior to capitalism (by implication why the USSR is superior to the USA, and why East Germany is superior to West Germany); the other explains why, nonetheless, socialist societies seem to be just as mired in malodorous murk. The editors of the leading Soviet philosophical journal articulate the theoretical synthesis of the two strands in explaining how the relations of production (the subjective factor) mediate the effects of the forces of production (the objective factor):

> In itself the scientific-technical revolution appears, on the one hand, as a factor increasing the intensity of the use of natural resources, and on the other as a means of their economical expenditure and utilization, but the development of technology and the growth of production do not only contribute to the adaptation of nature to the needs of society, but serve as a source of pollution of the environment and the disturbance of ecological equilibrium. The character of the interaction of the noted factors is essentially mediated by the nature of the social structure, although it also depends on the level of scientific knowledge attained.[65]

This argument clarifies why ecologically sound technology (e.g., waste-free production systems) or ecologically significant advances can be produced in capitalist countries[66] and, if appropriate, adapted to socialist conditions, or vice versa. It also explains why socialism must **struggle** to solve environmental problems.

Once we see how Soviet theories of the STR synthesize the two strains of argument (the "forces-of-production" and "relations-of-production") and why both strains are important for legitimation, we must recall that our concern here is not with Soviet theory as such, but with legitimation. The theory as a whole may or may not be intellectually persuasive. Its legitimizing function does not depend on that, since legitimation is not such a rational process and since the theory is not usually presented to the Soviet citizen in its full sophistication. Indeed we must disaggregate the two distinct causal patterns implicit in the theory in order to understand the role each plays in political legitimation. The theory is not always presented as a whole. Strands appear and emphases vary. The Soviet reader need not understand the whole to absorb the part.

The ambivalence of the theory has several consequences. The first consequence is potentially subversive. The ambivalence of STR theory allows particular writers to emphasize one strand or the other, either the capitalist origin of environmental deterioration or its origin in industrial progress itself. Writers who emphasize the "objective" sources of environmental deterioration intimate, unwittingly, or perhaps intentionally, a type of technological convergence between East and West. Technological convergence is a more subversive idea in East Germany than in the USSR for reasons already discussed; this probably explains the scantiness of references to the "objective" sources of environmental deterioration in the East German popular media. In an unusual exchange of letters published in a GDR literary journal the critic Günter Kunert suggests a "symmetry" of negative environmental effects resulting from a "system-neutral" physical process. But he carefully warns: "Please do not confuse this with convergence, here there is no talk of the assimilation of opposing systems!" Often the East German academic media either mimic Soviet theories or publish translated versions of Soviet articles on this point.[67] Surprising, however, is the attention given to the "objective" sources of environmental deterioration in widely circulated Soviet publications, notably **Priroda**, a leading popular science magazine. This magazine has rightfully gained a pro-environmental reputation (along with **Komsomol'skaia pravda** and **Literaturnaia gazeta**) in the USSR. There are not comparably "**umweltfreundliche**" magazines **or** journals in the GDR.[68] Most articles published in **Priroda** are written by scholars, but in popular style.

Their scientific content is relatively sophisticated, compared to newspapers, and also compared to the leading East German popular science magazine, **Urania.** The implications of the STR receive significant attention. Two Soviet scientists, for example, write in **Priroda** that the STR increases the possibilities "for the rational and more effective use of natural wealth in the interests of mankind," but "at the same time with the development of scientific-technical progress the danger of harmful effects of contemporary technology on nature grew and continues to grow."[69] This statement is a typical "forces-of-production" explanation. These authors do soften the effect of this statement by citing numerous examples of environmental destruction in capitalist countries, but the implicit message is fairly clear. Many other scientists writing in **Priroda** hardly mention differences between socialism and capitalism, but cite urbanisation and changes in technology and production as sources of pollution.[70]

Such explicit attention to "objective" sources of environmental deterioration occurs rarely in newspaper articles. Occasionally examples of similar types of deterioration in socialism and capitalism are juxtaposed, or measures for environmental protection in capitalist countries are favorably reported.[71] But, mass circulation publications generally either discuss the responsibility of capitalism for environmental problems ("relations-of-production" explanation) or discuss practical problems and successes in dealing with environmental problems in socialist society ("micro-level factors"). Furthermore, the relatively open discussion in **Priroda** suggests that the emphasis on natural science in the magazine and the presumed interest of its readers in science (as opposed to politics) mitigates the danger of any intimated technological convergence theory.

The ambivalence of STR theory has another consequence. Scholars, in their media, may explore somewhat dispassionately the roots of environmental crises. Soviet acceptance of basic research in the physical sciences as a basis for technological innovation is well-established in the post-Khrushchev period.[72] Environmental issues stand at the crossroads between the natural and social sciences, and natural scientists have been the most active participants in environmental debates. Natural and social scientists, as well as philosophers, are permitted to reflect relatively openly on the implications of the STR, so long as this discussion remains within the scholarly media.[73] The contradictory effects of the STR are explored in more depth in Soviet scholarly journals than in the popular media or than in East German journals. Arakelian, an economist, explicitly connects the development of environmental problems to the STR: "In other words, there is a contradiction between human activity as an agent of the STR and its consequences."[74] Gudozhnik, a systems analyst and expert on the STR, explains the interconnection more explicitly: The growth of human intervention in natural processes and the application on a large scale of processes which were previously applied only on a small scale lie behind the ecological crisis. "The reasons mentioned for the arising of the ecological crisis situation flow out of the essence of the scientific-technical revolution itself; they must be taken into account in the prognosis of the changes in the area of the relationship of society and nature."[75] Gudozhnik does not totally ignore the "relations-of-production" argument; another cause of ecological crisis lies in the profit-oriented direction of scientific-technical progress in capitalism, and indeed capitalism is the "ecological-crisis society."[76] But the ambivalence of STR theory allows Gudozhnik to explore macro-level sources of global and Soviet environmental problems. Other scholars follow suit. Marakhov poses the possibility of a new scientific-technical revolution, precisely because of ecological dislocations produced by the present STR. An optimal relationship between society and nature on a global scale would result from this new revolution, based on technological processes such as closed cycle production.[77] Scholars may thus direct their attention to the "objective" sources of environmental deterioration, and in this way perhaps make a contribution to their resolution. This type of debate also, however, allows an opportunity for ecological lobbying justified by scientific expertise.

Frederic Fleron directs our attention to another potential consequence of the ambivalent understanding of causality rooted in Soviet STR theories. In his study of technology and communist culture, he concludes that Soviet theories of the STR have implications similar to technological determinism:

> Inasfar as the STR perpetuates the Cartesian and Kantian subject-object distinction, it perfectly sets up the problematic for a resolution by an "objective" technology that is the embodiment of reason itself . . . human choice no longer exists, and indeed, is no longer necessary, since subjective calculations of individual human rationality can only be brought into alignment with the objective rationality of large-scale social production.[78]

Indeed, within the Soviet socialist world itself, solutions offered to environmental problems do consist largely of technological and managerial innovations. Closed-cycle technology and nuclear energy may well become the Soviet "embodiment of reason itself." Tendencies in this direction are already evident.[79] This conditional technical determinism explains why Soviet and East German analysts attack both technical definitions of environmental problems and technological pessimism when they emanate from the West.[80] There is, of course, no real inconsistency, assuming that all social roots of ecological disturbances have been truly eliminated in the Soviet world. Fleron may overstate his case, in that the Soviets do affirm some choice, for socialist society must take affirmative action to achieve a cleaner environment. But this is not a choice about fundamental social relations or values (along the lines of a Gorz or a Lovins);[81] at this level "subjective calculations of individual human rationality" must indeed be brought into line with the "objective rationality" of Soviet productive relations, as Fleron suggests. Because Soviet and East German social relations are avowedly free of fundamental fault, the solution to environmental deterioration becomes primarily conditional on adequate technological and managerial fixes. This is the important meaning for political legitimation of the "forces-of-production" explanation: the gains of technology have brought us far, and we need more of the same.

**Micro-Level Explanations: "Mistakes" in Socialism**

Were environmental discussion in the USSR and GDR limited solely to the types of macro-level explanations considered above, law-makers and planners would be left adrift in muddied waters with little orientation for practical response. Such is, however, not the case. Rather, scholars provide a plethora of potentially contradictory proposals through which the policy-maker must navigate. Specific proposals will not be dealt with in depth here; the reader is referred to detailed treatments elsewhere.[82] But brief examination of types of micro-level arguments provides revealing insights into political legitimation.

As indicated in Table 2.1, pages 48 and 49, micro-level explanations for the persistence of environmental problems in socialism fall into three conceptually-distinct categories: (1) **policy deficiencies**, for example inappropriate enterprise and branch incentives, the absence of charges for natural resource use, and lack of legal coordination;[83] (2) presently **inadequate** levels of **scientific-technical knowledge**, including deficient research coordination and delayed translation of knowledge into productive equipment;[84] and (3) **improper attitudes** among managers, ministry officials, and other responsible figures, as well as among the population at large.[85] Debates center, in turn, on appropriate responses to each problem.

As with macro-level explanations, the various types of micro-level arguments are differentially present in the various media types. Both the Soviet and East German scholarly media focus at length on policy deficiencies and on the level of scientific-technical knowledge. (The specific problems identified and solutions offered do differ between the two countries, however.) Policy factors also make their way into the Soviet mass circulation media, including major newspapers. A frequent target of attack in these media is "departmentalism," which refers to the tendency of particular enterprises or economic branches to ignore externalities of production in order to fulfil plan quotas, thus sloughing off environmental costs to other users of natural resources. "Departmentalism" can be taken either to indicate a skewed incentive structure (a policy problem) or incorrect attitudes of responsible officials. Both interpretations are presented, depending on the particular article.[86] Other criticisms more clearly refer to bad attitudes, for example references to poaching, engaged in even by high officials.[87]

In the GDR the popular media rarely contain policy debate, but frequently praise collective achievements.[88] A slight opening in discussion of environmental deterioration seems to have developed in the GDR since 1982, but allusions to policy deficiencies remain indirect and guarded compared to the Soviet pattern.[89] Public treatment of forest damage may provide an illustration. As noted above, an estimated twelve per cent of the GDR's forested areas are damaged, especially near the Czechoslovak border and near industrial centers like Leipzig, Halle, Bitterfeld, Cottbus, and Karl-Marx-Stadt, where air pollution from the chemical industry and brown coal combustion is severe.[90] Since 1982 several articles in **Neues Deutschland** (the major daily newspaper) and **Urania** (a popular science magazine) have treated this problem. In an April 1982 article **Urania** made no mention of air pollution as a source of damage to city trees, although it was conceded that "unfortunately often trees have been treated in a very careless and irresponsible manner."[91] In February 1983, amidst much discussion of the forest problem in West Germany, **Neues Deutschland** reported that the Politburo and the Council of Ministers had devoted careful attention to the protection of trees. Although the article referred to snow-(**Schneebruch**) and smoke-damaged trees, the source of the smoke was left unnamed. Among measures outlined to deal with the damage – extra fertilization, control of pests, decontamination of fallen wood, planting of smoke-resistant species, and international cooperation – control of air pollution was absent. On a positive note, the article concluded: "Outstanding achievements for the protection and use of forests will be recognized with additional state honors in the future."[92]

Later in 1983 the link between brown coal combustion and smoke-damaged trees was made more explicit, but not until a technological solution to the problem could also be presented to the public. **Urania** and **Neues Deutschland** both published articles describing new technological processes for de-sulfurization of combustion gases, finally linking these emissions to the forest damage. Even here, however, **Neues Deutschland** tried to divert public attention from the dangers of brown coal by noting that oil and hard coal also can have damaging effects and that "the relatively low combustion temperatures used with brown coal compared to other fuels produce quite favorable consequences regarding nitric oxide production."[93] In both articles the emphasis was placed on technological solutions and policy measures aimed at minimizing damaging pollutants. Finally, in March 1984 a brief article in **Neues Deutschland** described as "pure inventions" reports which had appeared in Western newspapers calling the GDR's forest damage "the most advanced." (This assessment was reportedly derived from a study from the United Nations Economic Commission on Europe.)[94] GDR authorities were obviously concerned that their citizens had heard these reports via the West German media. While GDR authorities seem prepared to admit the existence of a problem, there is still no debate in the GDR mass media regarding the real causes, significance, or proper solution. The articles cited above portray citizen commitment and efforts of scientific-technical cadres as the key in limiting forest damage. As with other environmental problems, the GDR mass media focus on policy initiatives (not deficiencies), the promise of scientific-technical advances, and the need for proper citizen involvement.

In general the East German popular media are distinguished by their emphasis on instilling an ecological consciousness and encouraging active participation of the population in socialist **Landeskultur**. "Thinking ecology" involves heightened awareness that natural resources are exhaustible and must be used judiciously, with a particular emphasis on recycling of wastes. The extreme consumerism and throw-away mentality of the West are discouraged,[95] but so is an ascetic ethos popular among some Western environmentalists. The desired psychological reorientation does not place in question anthropocentric, industrial values, but rather draws attention to the dependence of material production on a healthy environment and on frugal husbanding of resources. The mass media remains an instrument to mobilize and publicize popular commitment to socialist **Landeskultur**. Examples abound in East German newspapers: the Freie Deutsche Jugend (the communist youth organization) organizes waste product collection for recycling; "Mach-mit" (Participate!) campaigns and "The Week of Socialist Landeskultur" involve citizens in planting greenery, caring for public parks, and creating new recreational areas;[96] and the 50,000 member Society for Nature and the Environment, formed in 1980, assists local councils, enterprises, and scientific organizations with activities such as recultivation of waste-heaps, planting of trees and shrubs, caring for recreational areas, and educating the public regarding the importance of the rational use of natural resources.[97] Likewise, the professional organization of East German engineers and technicians (Kammer der Technik) uses its publications to communicate to its members the "deep moral and ethical content" of socialist **Landeskultur** and urges them "to apply more spirit in order to gain new ideas."[98]

This assertive effort to mobilize collective action through the media contrasts with more passive forms of citizen involvement in the USSR. Some voluntary projects to protect natural areas are reported. But most calls for popular involvement in the USSR resemble those which one would expect to find in Western liberal states: reprimands against those who litter or damage recreational areas; criticism of illegal hunting; and encouragement of sensitivity to nature in one's work. Soviet recycling programs appeal almost exclusively to individual self-interest. Initially, a citizen received a coupon for a hard-to-get novel in exchange for wastepaper. **Pravda** reported repeated frustrations with this program: promised books were not available; collection depots maintained unpredictable hours; and coupons ran short. Thus the program was altered, awarding books by lottery to coupon holders.[99] Reports of such bungling are likely to discourage rather than to mobilize participation. Incentives are material, not ideological. In the GDR, on the other hand, success in gathering waste is hailed as a contribution to socialism, as an important component of the class struggle; while payment is offered for certain types of recycled waste, public emphasis is placed on the collective long-run benefit as recycling aids the GDR's socialist development.

These examples indicate that in the environmental area the East German system exhibits more characteristics of a mobilization regime than does the Soviet.[100] While scholars do debate policy issues, the East German citizen hears less of this and more of citizen responsibility. From the regime's viewpoint, mobilization is an instrument of political legitimation, although Western scholars, influenced by the totalitarian school, have traditionally emphasized political mobilization as an instrument of political control and coordination (**Gleichschaltung**). Aryeh Unger, a contemporary representative of this school of thought, offers a typical definition of mobilization: "Any process by which totalitarian regimes seek to activate their peoples in support of official norms and goals. It involves changes in opinion, attitude, and behavior; it involves control. It is an organized, directed practice."[101] This definition reveals that, like political socialization, mobilization is aimed at gaining popular compliance with regime directives and goals. Analysts like Unger see political mobilization as more coercive and more active than is political socialization in Western countries because the political authority has a virtual monopoly over education, the media, and the coercive apparatus; the goal is not merely compliance but commitment. But socialization also involves involuntary molding of preferences, and socialization is a key part of regime legitimation. Since political legitimation refers to an empirical process (i.e., the process which leads a population to **believe** a regime legitimate) rather than a normative determination (i.e., whether the regime **deserves** to be considered legitimate, from the viewpoint of political philosophy), political mobilization is relevant to our concern.

The emphasis on mobilization in the area of environmental protection in the GDR, as contrasted with the USSR, most likely represents a general pattern which is related to particularities of the East German situation. First, compared to the USSR, the GDR is a new state, so legitimacy has not been as firmly established on the basis of longer term socialization processes. Thus the media is used more assertively to mobilize support for regime goals. Second, in response to the West German challenge, the emphasis on popular participation represents an attempt to develop a collective identity and responsibility, diverting attention from inadequacies in policies. Furthermore, it provides a democratic facade for centrally directed public action.[102] Finally, the intensity of nature use and resource shortages (particularly since GDR citizens are likely to have higher consumer expectations) may leave few alternatives to the East German leadership. As the East German analyst Grundmann says, without much time or money, much can be done with the participation of every individual.[103]

In the GDR mobilization is based on an ideological appeal, typical of "mobilization" systems. This is consistent with the pattern observed at the macro-level, namely the emphasis on the "relations-of-production" explanation, which focuses on the differences between capitalism and socialism. In the USSR micro-level explanations rely less on ideological values and more on substantive analysis of policy defects; this is consistent with the greater attention given to the "forces-of-production" explanation at the macro-level.

## CONCLUSION

In setting the terms and limits of public discussion of sensitive issues like environmental deterioration, East European authorities must consider that political legitimacy rests on effective policy, as well as on effective political socialization. This is particularly so for a political system which bases its right to rule on substantive claims, rather than on decision-making procedures. The multi-level structure of rationalization of environmental deterioration reflects this need to provide opportunities for substantive inquiry as well as for political rationalization. Scholars, particularly in the Soviet Union, are granted leeway to seek explanations, rather than simply to explicate justifications.

In the environmental area, differences between the Soviet and East German system of rationalization are of degree, not kind. The explanations are the same, but the emphasis varies. This common framework of legitimation in the two countries is, as the Soviets would say, "not accidental." East German legitimation does, however, rely more explicitly on Marxist-Leninist values of international class conflict and socialist commitment. East German citizens are treated to praise of their polity and censure of capitalism; Soviet citizens receive a more varied diet, including self-criticism and policy debate. Both regimes must mediate multiple dilemmas: on the one hand, the ideological challenge posed by more affluent Western societies; on the other hand, the need for efficacious responses to declining economic growth and resource shortages. The GDR confronts both dilemmas in starker form, due to the proximity and lure of the FRG and the dependence on the USSR for natural resources. The challenge of the capitalist FRG probably produces a greater tendency toward a defensive posture by the East German leadership. For looser control might allow West German ideas to serve as a catalyst for organization of political opposition in East Germany.

Any legitimizing principle contains some ambiguities or costs. The ideological emphasis in the GDR and the atheistic, scientific, anthropocentric emphasis in both countries are not inevitable responses to objective challenges. The psychology and socialization of the leadership, of which Marxism-Leninism is one component, play a role. The cliched, ideological response in the East German media is not the only option available to the East German leadership. Given that East German censorship is largely ineffectual, due to Western media penetration, more open dialogue with Western viewpoints might reinforce leadership credibility rather than undermine political control. The ideological emphasis of political legitimation in East Germany poses its own risks, after all. An emphasis on Marxist-Leninist principles (as expressed in the GDR **Landeskultur** concept) raises the specter of a principled opposition appealing to the humanistic values of socialism: witness Bahro.[104] It also makes political cynicism more likely: witness Biermann.[105] Likewise, the emphasis in both the USSR and GDR on scientific expertise and values grants a certain popular credibility and political influence to the experts. It is thus hardly accidental that the most vocal environmental advocates are scientists.

This study does not finally tell us what the Soviet or East German citizen believes. Answers to that question await better times. But it does provide a view of what these citizens are reacting to.

## NOTES

1. See Hannah Arendt's treatment of the psychology of the movement, **The Origins of Totalitarianism** (New York: Harcourt, Brace, Jovanovich, 1951), Part III, chapter 11. Also Z. Brzezinski and C. Friedrich, **Totalitarian Dictatorship and Autocracy** (Cambridge: Harvard University Press, 1956); and, more recently, Aryeh Unger, **The Totalitarian Party: Party and People in Nazi Germany and Soviet Russia** (London: Cambridge University Press, 1974).

2. For example, as applied to Stalinism, Robert Tucker, "Stalinism as Revolution from Above," in Robert Tucker, ed., **Stalinism: Essays in Historical Interpretation** (New York: W. W. Norton & Co., 1977), pp. 77-108.

3. Samuel Huntington, "Social and Institutional Dynamics of One-Party Systems," in S. Huntington and Clement Moore, eds., **Authoritarian Politics in Modern Society** (New York: Basic Books, 1970), p. 28.

4. See the insightful discussion of "ideological embezzlement" by Michael Szkolny (pseud.) in "Revolution in Poland," **Monthly Review** 33 (June 1981): 1-4.

5. Alfred Meyer, "The Functions of Ideology in the Soviet Political System," **Soviet Studies** 17 (January 1966): 276.

6. In this chapter, unlike in other chapters, I focus on pollution and ecological disturbances (e.g., humanly induced climatic changes, disturbance of spawning patterns, destruction of natural habitats for animals) and exclude problems of resource depletion from explicit consideration. Issues related to resource depletion are closely tied to broader problems of the intensification of economic production, as I discuss in chapter 6 (pp. 181-82). Therefore their public treatment is often difficult or impossible to extricate from general economic discussion. Furthermore, prospects of resource depletion are not evident to the average Soviet citizen in the way pollution often is.

7. Martin Jänicke, "Umweltpolitik in Osteuropa," in Hannelore Horn, Alexander Schwan, and Thomas Weingartner, eds., **Sozialismus in Theorie und Praxis**, Festschrift für Richard Löwenthal (Berlin [West]: Walter de Gruyter, 1978), p. 182.

8. Marlies Menge, "Für Filter fehlen die Devisen," **Die Zeit** (Hamburg), March 25, 1983, p. 21 (Canada Edition); and "Baumsterben in der DDR am weitesten vorangeschritten," **Frankfurter Allgemeine Zeitung**, March 9, 1984, p. 10. For the GDR reaction to such reports see note 94 below.

9. **Meditsinskaia gazeta**, Oct. 23, 1981, p. 2, trans. (abstract) in **Current Digest of the Soviet Press** (hereafter **CDSP**), vol. 33, no. 52, p. 14.

10. Boris Komarov (pseud.), **The Destruction of Nature in the Soviet Union** (White Plains, NY: M. E. Sharpe, 1980), pp. 20-28, 37-39. On the authenticity of this publication see Marshall Goldman's "Foreword" to this edition.

11. On the Tom River see "Will the Tom in Tomsk Continue to Exist" (letter), **Izvestiia**, May 6, 1982, p. 3, trans. (condensed) in **CDSP**, vol. 34, no. 18, p. 21. On Lake Issyk-Kul see Chingiz Aitmatov, "The Fragile Pearl of Issyk-Kul," **Pravda**, Oct. 12, 1981, p. 7, trans. (condensed) in **CDSP**, vol. 33, no. 41, p. 13. On the Aral Sea see B. Samoilenko, "The Aral Takes an Exam," **Kazakhstanskaia Pravda**, Apr. 1, 1983, p. 4, trans. (abstract) in **CDSP**, vol. 35, no. 13, pp. 13-14.

12. The hectographic reproduction of the report of the Kirchliche Forschungsheim Wittenberg, "Die Erde ist zu retten – Umweltkrise, christliche Glaube, Handlungsmöglichkeiten," 1982, contains some information on pollution levels, esp. pp. 3-4. For more information on this source see chapter 3, pp. 87-88; Menge; and Reiner Raestrup and Thomas Weymar, " 'Schuld ist allein der Kapitalismus': Umweltprobleme und ihre Bewältigung in der DDR," **Deutschland Archiv** 15 (1982): 832. On efforts to control pollution of the Werra and its effect on rivers in the Federal Republic of Germany see Konstantin Pritzel, "Die Umweltpolitik in den intereuropäischen und innerdeutschen Beziehungen," **Deutschland Archiv** 13 (1980): 842-43.

13. On PCB and DDT contamination see Komarov, pp. 32-36. On endangered species see, for example, Siegfried Hamsch, "Niederlausitzer Floristen – engagierte Freizeit-Forscher," **Urania**, no. 6 (1982), p. 15; and V. E. Sokolev, A. M. Borodin, and K. D. Zykov, "A New USSR Red Book," **Priroda**, no. 2 (February 1983), pp. 36-39, trans. (abstract) in **CDSP**, vol. 35, no. 14, p. 17.

14. See Komarov, pp. 16-17, 21 ff. Also Donald Kelley, Kenneth R. Stunkel, and Richard R. Wescott, **The Economic Superpowers and the Environment: The United States, the Soviet Union, and Japan** (San Francisco: W. H. Freeman & Co., 1976), p. 134.

15. See chapter 3, pp. 84-86.

16. An East German scholar, Gerhard Schmidt-Renner, alludes to political problems associated with environmental deterioration. He explains that because environmental damage

affects the population directly, it may consequently affect popular consciousness. "There are therefore also negative political-ideological factors with the associated control problems." See "Über Regelungen zum Schutz von Naturressourcen und -potenzen der gesellschaftlichen Reproduktion und Rekreation vor Störfunktionen in der DDR," **Petermanns Geographische Mitteilungen** 116, no. 2 (1972): 110. Likewise Soviet philosopher Frolov notes that "today the problem of the environment in many cases even takes on dominant meaning in comparison with many others, at least if one has in view how global problems are reflected in mass consciousness." I. T. Frolov, "Filosofiia global'nykh problem," **Voprosy filosofii**, no. 3 (1980), p. 36. **Neues Deutschland** refers to "numerous letters of our readers" concerning protection of nature in "Umwelt zum Nutzen der Menschen erschlossen," July 15, 1974, p. 3. Hans Jürgen Box, Jorgen Haalck, and Siegfried Joachim also indicate that the problem of environmental protection has "quickly forced its way into the public consciousness," in "Maritimer Umweltschutz (Küstenbereich DDR) in rechtlicher Sicht," **Wissenschaftliche Zeitschrift der Universität Rostock: Gesellschafts- und sprachwissenschaftliche Reihe** 22, Heft 4 (1973): 371. M. Mullagaliamov, chairperson of the Sterlitamst City Executive Committee, notes that air and water pollution arouse anxiety in the city in "The City's Air," **Trud**, Dec. 2, 1977, trans. (condensed) in **CDSP**, vol. 29, no. 49, p. 7.

Dissident sources dealing with environmental problems include Komarov; Andrei Sakharov, **Progress, Coexistence, and Intellectual Freedom** (New York: W. W. Norton & Co., 1968), pp. 48-49; Zhores Medvedev, **Nuclear Disaster in the Urals** (New York: W. W. Norton & Co., 1979); Wolfgang Harich, **Kommunismus ohne Wachstum? Babeuf und der 'Club of Rome'** (Reinbek bei Hamburg: Rowohlt, 1975); Rudolf Bahro, **The Alternative in Eastern Europe** (London: Verso, 1978), pp. 262 ff; Robert Havemann, **Morgen: Die Industriegesellschaft am Scheideweg** (Munich: R. Piper, 1980); essays in P. Wensierski and W. Büscher, eds., **Beton ist Beton: Zivilisationskritik aus der DDR** (Berlin [West]: Scandica Verlag, 1981); and the novel by GDR author Monika Maron, **Flugasche** (Frankfurt am Main: S. Fischer Verlag, 1981).

17. On local environmental activism and attention to the ecology problem in Evangelical Church organizations in the GDR see Wensierski and Büscher, eds., **Beton ist Beton;** Wolfgang Büscher and Peter Wensierski, **Null Bock auf DDR: Aussteigerjugend im anderen Deutschland** (Reinbek bei Hamburg: Speigel Verlag, 1984), chapter 3; and the paper by Hubertus Knabe, "Gesellschaftlicher Dissens im Wandel. Ökologische Diskussionen und Umweltengagement in der DDR," presented at the XVII. DDR-Forscher-Tagung in the Europäische Akademie Lerbach,

FRG, June 12-15, 1984. This paper, along with others presented at the conference, will be published by Verlag Wissenschaft und Politik (Cologne).

18. Although the early Club of Rome publications, e.g., Donella H. Meadows et al., **The Limits to Growth** (New York: Universe Books, 1972) do not conclude that living standards must decline with growth rates (rather, if growth is not limited there will be a rather "sudden and uncontrollable decline," p. 29), the association in recent decades between growth slow-downs and rising unemployment can easily encourage this impression. For a refutation see André Gorz, **Ecology as Politics** (Montreal: Black Rose Press, 1980), Parts I and II.

19. Iu. Khrenov, "In Debt to Nature," **Izvestiia**, July 29, 1983, p. 3, trans. (condensed) in **CDSP**, vol. 35, no. 30, p. 7. Along similar lines to my argument here, Roy and Zhores Medvedev suggest that "Soviet defense spending, contrary to widespread belief, is not generally perceived as a major cause of their economic difficulties by the Russian people. The ordinary Soviet citizen blames the omnipresent bureaucracy for shortages, inefficiency, and poor quality products, not the army." See "The USSR and the Arms Race," **New Left Review**, no. 130 (Nov.-Dec. 1981), pp. 18-19. For a different view on the GDR see Martin Jänicke's comments in Raestrup and Weymar, p. 842.

20. E.g., Harry Nick, "Für rationelle Beziehungen zwischen Gesellschaft und Natur," **Urania**, no. 7 (1977), p. 2; **Leipziger Volkszeitung**, no. 3 (1973), p. 10; and E. K. Fedorov and I. B. Novik, "Problemy vzaimodeistviia cheloveka s prirodnoi sredoi," **Voprosy filosofii**, no. 12 (1976), p. 57.

21. E.g., Erhard Vagt, "Harmonieren statt erobern," **Forum** 33, no. 1 (1979): 5. Occasionally there is, however, a suggestion that in the Third World "there exists the danger that an overemphasis on environmental demands will come into variance with the necessity to make economic use of resources in order to assure provision of the population." See Oskar Hugler, "Rechtsfragen der internationalen Zusammenarbeit zum Schutz der Umwelt," **Neue Justiz** 37 (1983): 235.

22. Vadim Khozinov, "Avtoritet istorii," **Literaturnoe obozrenie**, no. 3 (March 1977), p. 67.

23. For references to the connection between environmental concern and rural romanticism, with religious and ethnic overtones, see also chapter 3, pp. 85-86.

24. E.g., the radio program dealing with environmental problems in the GDR transmitted by Sender Freies Berlin (West Berlin) on March 6, 1982 included among its participants Rudolf Bahro, as well as West German specialists. See Raestrup and Weymar, pp. 832-44.

25. Indirect response to the trade-off argument is found in

Vagt, p. 5. Evidence of more widespread belief that there is a trade-off between environmental protection and a rising standard of living is provided by statements in documents emanating from Evangelical Church organizations in the GDR, published only in the West. See, e.g., "Kein Sonnenschein ohn' Unterlass," from a document produced by the Church Research Facility in Wittenberg (das Kirchliche Forschungsheim Wittenberg, cited hereafter as Forschungsheim [Wittenberg]), published in Wensierski and Büscher, eds., **Beton ist Beton,** p. 58. In such documents there are also frequent calls for personal sacrifice of luxury goods to further environmental protection. See chapter 3, pp. 89-90.

26. See Hubertus Knabe, " 'Der Mensch mordet sich selbst': Ökologiekritik in der erzählenden DDR-Literatur," **Deutschland Archiv** 16 (1983): 954-72. See also chapter 3, pp. 87-91.

27. See also Kelley, Stunkel, and Wescott; Thane Gustafson, **Reform in Soviet Politics** (Cambridge, ENG: Cambridge University Press, 1981), chap. 8; and Charles E. Ziegler, "Policy Alternatives in Soviet Environmental Protection," The Carl Beck Papers in Russian and East European Studies, Paper 102 (Pittsburgh: University of Pittsburgh, n.d.). On the GDR see Ilka Norhara-Schnabel, "Zur Entwicklung der Umweltpolitik in der DDR," **Deutschland Archiv** 9 (1976): 806-29; and Werner Gruhn, "Aktuelle Aspekte der DDR-Umweltpolitik," **Deutsche Studien** 19, no. 76 (1981): 424-40. On recent changes in environmental policy and on the use of productive wastes in the GDR and USSR see chapter 5, pp. 157-65.

28. For damage to agriculture in the USSR see Gustafson, chapter 3.

29. See Julian Cooper, "The Scientific Technical Revolution in Soviet Theory," in Frederic J. Fleron, ed., **Technology and Communist Culture: The Socio-Cultural Impact of Technology under Socialism** (New York: Praeger, 1977), chapter 3. For East Germany see Clemens Burrichter et al., "Die wissenschaftlich-technische Revolution — Kriterien und Konsequenzen," **Deutschland Archiv** 9 (1976): 516-29.

30. Fleron, "Introduction," in Fleron, **Technology and Communist Culture,** p. 55.

31. Based on lectures at the "Hochschulferienkurs für Germanistik," at the Herder Institut, Karl-Marx Universität, Leipzig, Summer 1980.

32. See William Connolly, **The Terms of Political Discourse** (Lexington, MA: D.C. Heath, 1974), chapter 1.

33. Gertraud Seidenstecher in Hans-Hermann Höhmann, Gertraud Seidenstecher, and Thomas Vajna, **Umweltschutz und ökonomisches System in Ost-Europa** (Stuttgart: Verlag W. Kohlhammer, 1973), pp. 84-85.

34. Stephan Supranowitz et al., **Landeskulturgesetz: Kom-**

**mentar zum Gesetz über die planmässige Gestaltung der sozialistischen Landeskultur in der DDR vom 14. Mai 1970** (Berlin: Staatsverlag der DDR, 1973).

35. **Sozialistische Landeskultur: Umweltgestaltung – Umweltschutz mit einem ABC**, Brockhaus Handbuch (Leipzig: VEB F.A. Brockhaus Verlag, 1977), pp. 55-56. Soviet analyst Polianskaia considers **Landeskultur** to include both passive (conserving) and active (regulating use) aspects of the nature-society relationship, while **Umweltschutz** presumably is concerned more directly with avoidance and redress of damage done to nature. See G. N. Polianskaia, "Poniatie okhrany prirody i prirodookhranitel'noe zakonodatel'stvo," **Sovetskoe gosudarstvo i pravo**, no. 9 (September 1975), p. 74. For a less ideological view see Erhard Schuster, "Charakter und Stellung der Umweltprobleme und einige sich daraus ergebende Konsequenzen," **Wissenschaftliche Zeitschrift der Technischen Universität Dresden** 31 (1982): 211. Schuster even suggests that perhaps a new term is needed to provide a clear and broader depiction of the problem.

36. Supranowitz et al., pp. 18, 22.

37. Seidenstecher in Höhmann et al., pp. 84-85.

38. See Karlheinz Lohs and Sonnhild Döring, eds., **Im Mittelpunkt der Mensch: Umweltgestaltung-Umweltschutz** (Berlin: Akademie Verlag, 1975); Erhard Krack, "Bürger, Gemeinde, Umwelt," **Stadt und Gemeinde** 18 (August-September 1974), p. 14; and Gerhard Mohs, Konrad Billwitz, and Hans Richter, "Umweltschutz," **Urania**, no. 5 (1973), pp. 25-28.

39. Norhara-Schnabel, p. 811.

40. Samuel Huntington, in Huntington and Moore, eds., p. 28, writes: "At some point in the consolidation of a one-party state, ideology ceases to be a way of achieving consensus and instead becomes a threat to consensus."

41. Many Soviet legal scholars advocate such a comprehensive law on the environment, e.g., V. V. Petrov, "Ob"ekt i predmet pravovoi okhrany prirody v SSSR," **Sovetskoe gosudarstvo i pravo**, no. 4 (April 1976), p. 64; Polianskaia, who in 1975 specifically cited the East German law as an example, pp. 73-74; and A. Ia. Sukharev, then USSR First Deputy Minister of Justice (interview), "The Law and Nature," **Izvestiia**, Aug. 23, 1979, p. 3, trans. (condensed) in **CDSP**, vol. 31, no. 34, pp. 6-7.

42. K. N. Blagosklonov et al., **Okhrana prirody** (Moscow: Vyshaia Shkola, 1967), p. 7.

43. E. A. Kliushkin, "Chto takoe okhrana prirody?" Akademiia Nauk Turkmenskoi SSR, **Okhrana prirody Turkmenii i nashi zadachi** (Ashkhabad: Ylym, 1972), pp. 15-16. See also Günter Streibel, "Umweltschutz und Umweltgestaltung als volkswirtschaftliche Aufgabe," **Wirtschaftswissenschaft** 23 (August 1975): 1140.

44. See M. T. Meleshkin, A. P. Zaitsev, and Kh. Marinov, **Ekonomika i okruzhaiushchaia sreda: vzaimodeistvie i upravlenie** (Moscow: Ekonomika, 1979), p. 108. On water, see "Protect Water Purity," **Izvestiia**, July 12, 1979, p. 3, trans. in **CDSP** (condensed) vol. 31, no. 28, p. 19. In the GDR a special Ministerium für Umweltschutz und Wasserwirtschaft was created in 1971. Delegation of responsibility to relevant production ministries in terms of planning apparently remains, however. For a critique see Petrov, pp. 58, 62-63; and N. P. Fedorenko and K. G. Gofman, "Upravlenie sotsialisticheskim prirodopol'zovaniem," in N. N. Nekrasov and E. Mateev, eds., **Sotsialisticheskoe prirodopol'zovanie** (Moscow: Ekonomika, 1980), p. 20.

45. "On Additional Measures to Intensify Conservation and Improve Utilization of Natural Resources," **Pravda**, Jan. 6, 1979, p. 1, trans. (condensed) in **CDSP**, vol. 31, no. 1, p. 1. See also Ziegler, "Policy Alternatives," pp. 13-14.

46. Pryde, **Conservation**, p. 136. See also Komarov, p. 17.

47. Jerry Hough and Merle Fainsod, **How the Soviet Union is Governed** (Cambridge: Harvard University Press, 1979), pp. 293-97.

48. E.g., for recent criticism of Gossnab (State Committee on Material-Technical Supplies) and the State Committee on Prices see A. Pimenov, "Ne po-khoziaiski," **Pravda**, Feb. 2, 1982, p. 3; for criticism of the Ministry of the Chemical Industry see A. Kucherenko, "Ozero trebuet zashchity," **Pravda**, Feb. 8, 1982, p. 3; for criticism of the Ministry of Nonferrous Metallurgy see V. Nevelskii, "Pure Sky, Pure Waters," **Izvestiia**, Sept. 12, 1983, p. 12, trans. (condensed) in **CDSP**, vol. 35, no. 37, pp. 22-23 and Iu. Khrenov, "In Debt to Nature," **Izvestiia**, July 29, 1983, p. 3, trans. (condensed) in **CDSP**, vol. 35, no. 30, pp. 7-8; and for criticism of the Ministry of the Timber, Pulp-and-Paper and Lumber Industry see Valentin Krasnogorov, "We'll Have to Ask Forgiveness Yet Another Time," **Literaturnaia gazeta**, May 26, 1983, p. 15, trans. (abstract) in **CDSP**, vol. 35, no. 31, p. 12.

49. E.g., Alfred Kosing, "Mensch-Gesellschaft-Umwelt," **Einheit** 31 (1976): 797; G. I. Tsaregorodtsev, " 'Tekhnizatsiia' sredy i zdorov'e cheloveka," **Voprosy filosofii**, no. 10 (1972), p. 70; E. K. Fedorov in "Chelovek i sreda obitaniia," **Voprosy filosofii**, no. 1 (1973), p. 50; and Supranowitz, p. 14.

50. G. S. Gudozhnik, **Nauchno-tekhnicheskaia revoliutsiia i ekologicheskii krizis** (Moscow: Mezhdunarodnye otnosheniia, 1975), p. 112; Siegfried Grundmann, "Mensch und Umwelt," **Deutsche Zeitschrift für Philosophie** 21 (1973): 196-97; and Bernd Bittighöfer, Herbert Edeling and Hans Kulow, "Theoretische und politisch-ideologische Fragen der Beziehung von Mensch und Umwelt," **Deutsche Zeitschrift für Philosophie** 20 (1972): 72. Jürgen Kuczynski warns against the illusion that the working class

is leading the struggle in imperialist countries, for it actually is "in the hands of the middle class." **Gleichgewicht der Null** (Frankfurt am Main: Verlag Marxistische Blätter, 1973), p. 67. (This book was published in both the GDR and FRG, but only the West German edition is available in North America.)

51. Bittighöfer et al., p. 61.

52. E.g., "Obshchestvo i priroda" (editorial), **Voprosy filosofii**, no. 4 (1973), pp. 47-48; and Horst Paucke and Günter Streibel, "Zu Versuchen kapitalistischer Systemstabilisierung unter dem Aspekt der Umweltreproduktion," **Wirtschaftswissenschaft** 26 (August 1978): 924-38.

53. Krack, p. 16; Gerhard Schmidt-Renner, "Sozialistischer Umweltschutz unter volkswirtschaftlichen Gesichtspunkten," **Petermanns Geographische Mitteilungen** 118 (1974): 10; and "Landeskulturgesetz. ein Jahr Danach," **Urania**, no. 5 (1971), pp. 2-3.

54. E.g., **Neues Deutschland**, May 15, 1970, p. 4; Bittighöfer et al., p. 75; Günter Streibel, "Die ökonomische Stimulierung der rationellen Nutzung natürlicher Gebeitsressourcen," **Wirtschaftswissenschaft** 14 (December 1966): 1975; and M. N. Rutkevich and S. S. Shvarts, "Filosofskie problemy upravleniia biosfery," **Voprosy filosofii**, no. 10 (1971), p. 54.

55. Adolf Bauer and Horst Paucke, "Umweltprobleme in der Sicht des historischen Materialismus," **Deutsche Zeitschrift für Philosophie** 24 (1976): 787; Krack, pp. 12-13, Streibel, "Ökonomische Stimulierung," p. 1775; and B. Gorizontov, "Sotrudnichestvo stran SEV v okhrane okruzhaiushchei sredy i ispol'zovanii prirodnykh resursov," **Voprosy ekonomiki**, no. 9 (1974), p. 69.

56. Makliarskii, in "Nauka i global'nye problemy sovremennosti" (round table), **Voprosy filosofii**, no. 9 (1974), pp. 68-69.

57. Hans Reichelt, "Der Beitrag der KdT bei der Gestaltung der sozialistischen Landeskultur und zum Schutz der Umwelt," **Die Wirtschaft**, Ausgabe A, Beilage 19-72, no. 42 (Oct. 18, 1972), p. 4; Manfred Scheuffer, "Mehr Energie-Woher?" **Urania**, no. 1 (1976), p. 28; Gudozhnik, pp. 107-17; and E. K. Fedorov, **Vzaimodeistvie obshchestva i prirody** (Leningrad: Gidrometeoizdat, 1972), pp. 47-48.

58. E. K. Fedorov, "Man and Nature," **World Marxist Review**, no. 5 (May 1968), p. 31; and Peter Richter, "Veto gegen Missbrauch der Natur," **Forum** 29, no. 3 (1975): 10-11.

59. Even as early as the late 1960s this was the clear implication of much public discussion in the USSR. See, e.g., I. Eliseev and A. Kondratenko, "Nashe krovnoe delo," **Kommunist**, no. 5 (1968), pp. 112-17; I. Gerasimov, "Nuzhen general'nyi plan preobrazovaniia prirody nashei strane," **Kommunist**, no. 2 (1969), pp. 68-79; Vladimir Chivilikhin, "How is Your Breathing, City Dwellers?" **Literaturnaia gazeta**, no. 32 (Aug. 9, 1967), p. 10,

trans. in **CDSP**, vol. 19, no. 33, p. 12; and A. Inozemtsev, "Nature Awaits a Solicitous People," **Komsomol'skaia pravda**, Jan. 6, 1968, p. 2, trans. in **CDSP**, vol. 20, no. 3, p. 29. For a more recent East German source see Helmar Hegewald, "Moralische Triebkräfte bei der rationellen Nutzung der Naturressourcen im Sozialismus," **Deutsche Zeitschrift für Philosphie** 30 (1982): 1460.

60. Gerhard Schmidt-Renner, "Über Regelungen zum Schutz von Naturressourcen und -potenzen der gesellschaftlichen Reproduktion und Rekreation vor Störfunktionen in der DDR," **Petermanns Geographische Mitteilungen** 116 (1972): 101. See also Bittighöfer et al., p. 61.

61. I. V. Komar, "Resource Cycles, Their Improvement to Optimum Levels and Forecasting," in Gerasimov et al., **Man, Society, and the Environment**, p. 240. Also A. Bauer and H. Paucke, "Umweltprobleme in der Sicht des historischen Materialismus und bürgerliche Weltmodelltheorien," **Deutsche Zeitschrift für Philosophie** 24 (1976): 786-87.

62. Bittighöfer et al., pp. 75-76; and Bauer and Paucke, "Sicht des historischen Materialismus," p. 790.

63. Grundmann, "Mensch und Umwelt," p. 205.

64. Fleron, "Afterword," in Fleron, **Technology and Communist Culture**, p. 464.

65. "Sotsial'no-filosofskie problemy nauchno-tekhnicheskoi revoliutsii," **Voprosy filosofii**, no. 2 (1976), p. 46. Adolf Bauer and Horst Paucke elaborate what I call the "forces-of-production" and "relations-of-production" arguments quite distinctly in "Einheit und Kampf zwischen Natur und Gesellschaft," **Deutsche Zeitschrift für Philosophie** 27 (1979): 595-98. See also Horst Paucke, "Zum Verhältnis von Wirtschaftswachstum und Umweltbelastung," **Zeitschrift für den Erdkundeunterricht** 34, no. 2/3 (1982): 49-55.

66. The section "Novosti Nauki" in **Priroda** provides good examples of the reporting of technological advances or scientific advances in capitalist countries. See, for example, no. 5 (1976), p. 135; no. 4 (1975), p. 114; no. 5 (1977), pp. 137-38.

67. For the Kunert quote see "Anlässlich Ritsos – Ein Briefwechsel zwischen Günter Kunert and Wilhelm Girnus," **Sinn und Form** 31 (1979): 850-51. I am grateful to Hubertus Knabe for this reference. Translations of Soviet scholarly articles appear in the GDR journal **Sowjetwissenschaft: Gesellschaftswissenschaftliche Beiträge.**

68. The GDR satirical newspaper, **Eulenspiegel**, does occasionally contain pointed references to environmental deterioration in the GDR, but not the kind of serious and sustained criticism found in the Soviet newspapers cited here. See, e.g., Manfred Strahl, "Die Insel," **Eulenspiegel** 28, no. 29 (1981), pp. 8-9; and idem, "Mensch, Natur, wie haste dir verändert!" **Eulenspiegel** 27, no. 36 (1980), p. 13. The popular science magazine of the

Academy of Sciences of the GDR, **Wissenschaft und Fortschritt**, also carries frequent articles touching on environmental problems. In general, however, these articles are not as critical as are articles in **Priroda**. For example, P. L. Kapitsa's article in **Priroda** on energy and physics was reprinted in **Wissenschaft und Fortschritt**, but with the deletion of his entire critical commentary on nuclear power safety. Compare "Energiia i fiziki," **Priroda**, no. 2 (1976), pp. 75-76 with "Energie und Physik," **Wissenschaft und Fortschritt** 27 (1977): 444-49.

69. F. V. Konstantinov and G. A. Aksenenok, "Chelovek i priroda v epokhu nauchno-tekhnicheskoi revoliutsii," **Priroda**, no. 10 (1974), p. 2.

70. F. G. Krotkov, "Zagriaznenie okruzhaiushchei sredy i problemy gigieny," **Priroda**, no. 5 (1975), pp. 63-65; and M. Strong and V. N. Kunin, "Prirodnaia sreda – edinoe tseloe," **Priroda**, no. 5 (1973), pp. 27-39.

71. For praise of capitalist control of noise pollution, see P. Rogozinskii, "The Right to Quiet," **Izvestiia**, Apr. 24, 1973, p. 5, trans. (condensed) in **CDSP**, vol. 25, no. 17, p. 24. Some articles in **Forum**, the biweekly newspaper of the Freie Deutsche Jugend, suggest that environmental problems are correlated with industrial development, although a "relations-of-production" explanation is emphasized. See Vagt, p. 5. Unfortunately, early in 1983 publication of **Forum** (along with other GDR newspapers and journals) was ceased, reportedly due to a paper shortage!

72. As an example, see R and D in the energy sector in Robert W. Campbell, **Soviet Energy Technologies: Planning, Policy, Research and Development** (Bloomington, IN: Indiana University Press, 1980), chapter 2.

73. Particularly revealing was the "round table" discussion on "Man and His Environment" sponsored by and published in **Voprosy filosofii** (no. 1-4, 1973), the philosophical journal of the Soviet Academy of Sciences. See chapter 3 for discussion of viewpoints expressed in the round table.

74. A. Arakelian, "The Scientific-Technical Revolution and the Biosphere," **Voprosy ekonomiki**, no. 5 (1976), translated in **Problems of Economics** 19 (March 1977): 68.

75. Gudozhnik, p. 63.

76. Ibid., p. 32.

77. V. G. Marakhov, "Nauchno-tekhnicheskaia revoliutsiia i prirodnaia sreda," **Voprosy filosofii**, no. 8 (1974), p. 100.

78. Fleron, "Afterword," in Fleron, **Technology and Communist Culture**, p. 468.

79. For discussion of nuclear power debates in the USSR and GDR see Joan DeBardeleben, "Esoteric Policy Debate: Nuclear Safety Issues in the USSR and GDR," forthcoming in the **British Journal of Political Science**. On Soviet nuclear energy

policy see Gloria Duffy, **Soviet Nuclear Energy: Domestic and International Policies**, prepared for the U.S. Department of Energy (Santa Monica: Rand Corp., 1979); M. Messengiesser, "Die Atomenergiepolitik der UdSSR," **Osteuropa** 30 (1980): 1207-19; and Philip Pryde, "Nuclear Energy in the Soviet Union," **Current History** 77 (October 1979): 115-18. On East German nuclear policy see Werner Gruhn, "Kernkraftsicherheit aus DDR-Sicht," **Deutschland Archiv** 10 (1977): 806-10. On closed-cycle technology see chapter 6, pp. 187-90.

80. Contrast Shilin's criticism of technological optimists with Nick's criticism of technological pessimism. K. I. Shilin, " 'Ekologicheskaia revoliutsiia' v sovremennoi zapadnoi nauke," **Voprosy filosofii**, no. 11 (1972), pp. 126-27; and Harry Nick, "Mensch und Umwelt," **Einheit** 34 (1979): 709-10.

81. For Gorz see note 18; also Amory Lovins, **Soft Energy Paths** (Cambridge, MA: Ballinger Publishing Co., 1977).

82. See, for example, Ziegler, "Policy Alternatives"; and chapters 5 through 8.

83. For a more detailed examination of specific policy deficiencies see Part 2 of this book.

84. For further discussion of this type of problem see chapter 4.

85. Some articles mention all factors. See, e.g., the **Pravda** editorial on recycling wastes, "Iz vtorichnykh resursov," Jan. 13, 1982, p. 1.

86. E.g., on attitudes see M. Mullagaliamov, p. 8 (see note 16). On incentive structures and organizational deficiencies see "In the CPSU Central Committee and the USSR Council of Ministers," **Izvestiia**, Feb. 4, 1976, p. 3, trans. (condensed) in **CDSP**, vol. 28, no. 5, p. 5. Further examples of this type of critique are in V. Brovkin, "Pustyri pod oknami," **Pravda**, Jan. 27, 1982, p. 3; and P. Bulatov, "Kto khoziain vody?" **Pravda**, Feb. 17, 1982. See also chapter 7, pp. 218-19.

87. E.g., R. Esenov, "Shots Along the River Bank," **Pravda**, Oct. 12, 1970, p. 4, trans. (condensed) in **CDSP**, vol. 22, no. 41, pp. 19-20; "Called to Account for Violating Discipline," **Sovetskaia Moldaviia**, March 19, 1983, p. 1, trans. (excerpts) in **CDSP**, vol. 35, no. 12, pp. 21-22; and "Concern for Nature," **Izvestiia**, Sept. 10, 1983, p. 3, trans. in **CDSP**, vol. 35, no. 36, p. 25.

88. This includes numerous reports on effective action by enterprises, e.g., **Neues Deutschland**, Sept. 17, 1980, p. 8. The emphasis, however, is usually on improved production rather than on environmental amelioration. The Soviet press also contains reports of success, of course. See, e.g., "Sokhraniat lesa," **Pravda**, Feb. 9, 1982, p. 1; and "Sluzhba chistoi vody," **Pravda**, Apr. 22, 1982, p. 6.

89. In addition to the articles cited below see L. Kowalsky,

"Grundwasser geht uns alle an," **Urania**, no. 5 (1982), p. 26; and Elke Rath, "Was tut sich gegen Ungesundsein der Luft," **Leipziger Volkszeitung**, Aug. 14/15, 1982, from **Pressespeigel**, no. 13, Sept. 10, 1982, p. 10. See also "Auch in der DDR melden sich die Umweltschützer," **Frankfurter Allgemeine Zeitung**, Feb. 8, 1983.

90. See especially Martin Jänicke, "Der Tod aus dem Schornstein," **Frankfurter Rundschau**, (Zeit und Bild), Jan. 28, 1984; and "Das Waldsterben stellt auch die 'DDR' vor erhebliche Probleme," **Die Welt**, Aug. 30, 1983. See also note 7.

91. Hans-Joachim Dreger, "Bäume in unseren Städten," **Urania**, no. 5 (1982), pp. 2-3.

92. "Massnahmen zum Schutz der Wälder der DDR," **Neues Deutschland**, Feb. 26/27, 1983, p. 1.

93. B. Kahn and F. Grossman, "Erfolgreiche Verfahren zur Rauchgasentschwefelung erprobt und angewandt," **Neues Deutschland**, Oct. 14, 1983, p. 4; and Volker Hoffman, "Weniger Schwefel in der Luft," **Urania**, no. 5 (1983), pp. 12-16. See especially the insert on "Zum Schutz unserer Wälder," pp. 14-15.

94. "Erfindungen," **Neues Deutschland**, Mar. 10/11, 1984, p. 2. For Western reports see **Frankfurter Allgemeine Zeitung**, March 9, 1984, p. 10.

95. Vagt, p. 5; and Hermann Klare, "Briefe," **Spektrum**, June 1973, pp. 2-3.

96. E.g., **Neues Deutschland**, Jan. 28, 1981, p. 8 on collection of waste by children; and "Schlusspunkt nur im Seminar – die Umwelt braucht uns weiter," **Leipziger Volkszeitung**, Dec. 15, 1973, p. 10. See also Norhara-Schnabel, p. 815.

97. The Society for Nature and the Environment is part of the Kulturbund der DDR. See Harald Thomasius (chairperson of the Society), "Natürliche Reichtümer der DDR sinnvoll nutzen und schützen," **Neues Deutschland**, Nov. 27, 1981, p. 3; and "Landeskultur ist das Anliegen der gesamten Gesellschaft," **Neues Deutschland**, Dec. 1, 1983, p. 7. On its role in forest protection see "Appell an alle Vorstände, Fachausschusse, Interessengemeinschaften und Fachgruppen der Gesellschaft für Natur und Umwelt der Kulturbund der DDR," **Urania**, no. 8 (1983), pp. 18-19. See also Peter Wensierski, "Die Gesellschaft für Natur und Umwelt," presented at the XVII. Forscher-Tagung, Europäische Akademie Lerbach, FRG, June 12-15, 1984, to be published with other conference papers by Verlag Wissenschaft und Politik (Cologne).

98. Reichelt, "Beitrag der KdT," **Die Wirtschaft**, Ausgabe A, Oct. 18, 1972, p. 6; Peter Zimmermann, in "Mit beiden Augen," **Technische Gemeinschaft** 28, no. 1 (1980), pp. 14-15; and "Aufruf der Kommission Umweltschutz beim Präsidium der Kammer der Technik und Ministerien . . . zur Durchführung eines Ideenwettbewerbs 'Veränderung vorhandener technologischer Verfahren im Sinne der abproduktarmen/-freien Technologie," **Technische**

**Gemeinschaft** 31, no. 3 (1983), p. 10.

99. See Vil Dorofeev, "The New Adventures of 'Queen Margot,'" **Literaturnaia gazeta**, Dec. 25, 1974, p. 12, trans. (condensed) **CDSP**, vol. 27, no. 9, pp. 16-17; L. Ivchenko, "Wealth in Trash," **Izvestiia**, Apr. 15, 1976, p. 2, trans. **CDSP**, vol. 28, no. 15, p. 9; L. Velikanova, "In a Passion Over Coupons," **Literaturnaia gazeta**, Aug. 25, 1976, p. 13, trans. (abstract) in **CDSP**, vol. 28, no. 33, pp. 16-17; and idem, "The Man with Scrap Paper," **Literaturnaia gazeta**, July 18, 1979, p. 1, trans. (abstract) in **CDSP**, vol. 31, no. 35, pp. 13-14.

100. On the mobilization concept and its application to Soviet-type systems see Chalmers Johnson, "Comparing Communist Nations," in Johnson, **Change in Communist Systems** (Stanford: Stanford University Press, 1970), pp. 8-10; and David E. Apter, **The Politics of Modernization** (Chicago: University of Chicago Press, 1965), chap. 10.

101. Unger, p. 5.

102. E.g., Krack, pp. 19-20; and Hegewald, p. 1465.

103. Grundmann, "Mensch und Umwelt," p. 206.

104. See note 16.

105. Wolf Biermann is a popular East German folksinger. Until 1976 his songs of political protest were distributed underground or heard in Western radio broadcasts in the GDR. The East German authorities blocked his return to the GDR in 1976 after he presented a concert in Cologne. For texts of his early songs see Wolf Biermann, **Mit Marx- und Engelszungen**, and **Für meine Genossen** (Berlin [West]: Verlag K. Wagenbach, 1968 and 1972).

# 3
# Philosophical, Theoretical, and Theological Perspectives*

The previous chapter explored the "public image" of environmental problems, an image shaped by Soviet and GDR authorities seeking to legitimize their rule. In this chapter, we examine in depth the more serious analyses of the ecological crisis which emerge mainly in limited-circulation publications. The goal here is not, as in chapter 2, to identify principles of legitimation, but rather to uncover serious theoretical inquiry. For example, does Marxism provide theoretical guidance as experts attempt to understand changes in the natural environment? How serious do scientists consider the situation to be? Is the ecological crisis undermining the belief of Soviet and GDR intellectuals in the possibility of continued "progress"?

In both the GDR and USSR we are, of course, dealing with a censored media, so we must assume that people think thoughts which never appear in print. Some of these make their way into dissident or samizdat documents which are published in the West. Our purpose here is not to examine these dissident viewpoints in any depth, but rather to explore the debate which is allowed at least limited dissemination by the political authorities. This will provide us with a better measure of the adaptability of these societies in reconciling socialist values to the challenges of the ecological crisis. Of course, the reality of a state-controlled press places severe limits on the ability of thinkers to discuss such issues in a "scientific" or open manner. The authorities' concern with political legitimation forces itself into scholars' theorizing, or at least into their published writings. Part of the task in this

---

*Parts of this chapter originally appeared under the title "Optimists and Pessimists: The Ecology Debate in the USSR," by Joan DeBardeleben, published in **Canadian Slavonic Papers**, vol. XXVI, nos. 2-3 (1984), pp. 127-40, and are reprinted by permission of **Canadian Slavonic Papers**.

chapter, therefore, must be to read between the lines in interpreting the meaning which analysts are striving to communicate, i.e., to filter out some of the ideological static which the political environment generates.

Scientists, humanistic intellectuals, and writers have been in the vanguard of the environmental movement in both the USSR and GDR. These individuals are in a strategic position to understand the grave implications of ecological imbalance and to address those implications publicly and persuasively. Scientists have the expertise to assess the condition of the natural environment. Scholars often have access to sources of information denied the general public (including Western publications). Finally, scholars and writers are accustomed to putting their ideas on paper and are experienced in shaping their public positions to the demands of a censored press. Reading Marx, Engels, and Lenin provides few ready-made solutions to the approaching environmental crisis. Many Soviet scholars and a few in the GDR are strikingly aware of this theoretical lacuna in the ecological area and have set themselves the conscious and explicit task of addressing the myriads of unresolved ethical, scientific, and practical dilemmas that have arisen.[1] If Soviet and East German scholars have not produced any revolutionary insights into the sources and nature of the ecological crisis, consideration of the problem has been surprisingly widespread and at times frank. Alarm over the scope of ecological imbalance and disagreement over its deeper causes and implications may seem retarded in comparison with Western developments, but the ecological pessimism evident in the thinking of some GDR, but especially Soviet, intellectuals is striking in the context of official optimism about human progress.

The party/state in the USSR, less so in the GDR, appears to sanction efforts of its scholars to come to terms with the theoretical implications of environmental problems. Official Soviet doctrine does not claim to have resolved all scientific and theoretical problems but rather enounces a methodology (dialectical materialism) and a perspective (the interests of the working class) for their analysis.[2] Since the death of Stalin the leeway allowed scholars for applying these parameters has greatly expanded. In the past, Soviet attempts to harness scientific theory to Communist Party policy have sometimes had negative practical consequences. For example, Stalin's efforts to assure conformity between his policies and natural scientific theory contributed to the rejection of Mendelian genetics in favor of Lysenko's theory of genetic adaptation to environmental stimuli, resulting in serious setbacks for both science and agriculture.[3] Since the death of Stalin, however, the Soviet political leadership has gradually loosened its grip over scientific inquiry, and this has been reflected in the East European states as well. Perhaps the debacle of Lysenkoism taught the present generation of leaders a lesson. Natural scientists are no longer required to design scientific theories to confirm the Party's understanding of historical development. The philosophy discipline is not as closely watched as during the Stalinist period, when philosophical positions were expected to conform more explicitly to party policy and official Marxist-Leninist doctrine. On the contrary, the post-Stalinist leadership has supported and encouraged fundamental research.[4] The Academy of Sciences, a prestigious institution of scholars, has come to enjoy considerable autonomy from direct political intrusion.

This greater autonomy of scientific research has brought an upsurge in genuine scientific contributions to world knowledge in such areas as nuclear fusion, mathematics, and certain aspects of ecology. At the same time it has allowed philosophers and natural scientists to address practical and controversial issues in speculative form, without incurring severe restrictions from political authorities. The interdisciplinary nature of the environmental debate, which touches on areas of the social, physical, and technical sciences, allows these scholars to transgress the boundaries of their disciplines in their capacity as scientists. Social scientists can, however, do this less easily than philosophers and natural scientists. They are subject to greater political control because their endeavors have more clear and direct political implications. Probably for this reason fewer social scientists have focused boldly on the broader implications of environmental problems, preferring consideration of concrete economic and legal policies. Accordingly, the unique theoretical contributions of the Soviet social sciences have generally been less significant than Soviet contributions in the natural sciences.[5]

It is not surprising, then, that the most open and controversial discussion of environmental problems in the USSR has appeared in the philosophical journal of the Soviet Academy of Sciences, **Voprosy filosofii**. Particularly important was a "round table" on "Man and His Environment" sponsored and published in **Voprosy filosofii** in 1973 and 1974.[6] Striking in this discussion and the debate which ensued were the efforts of philosophers and scientists to escape their own narrow specialization and link theoretical inquiries to practical problems facing Soviet society. In the late 1970s and early 1980s the journal also published numerous articles under the rubrics of "Man-Society-Nature" and of "Contemporary Global Problems — Socio-Philosophical and Methodological Aspects," covering such issues as pollution, resource depletion, the population explosion and international health problems.[7] The journal has thus become an arena for reflection on the most pressing issues confronting humanity. I. Frolov, the former editor of **Voprosy filosofii**, who helped make the journal a focal point of ecological concern, has captured the sentiment which moves this interdisciplinary discussion of global problems:

> Ever more now we talk about protection of nature for example. Unfortunately, however, we recognize to a significantly lesser degree that perhaps today the problem of the protection of man himself is primary. The acuteness of this question is explained by the ever growing influence of various carcinogenic factors on man: ionizing radiation, chemical mutagens, and also experiments on man.[8]

For Frolov the environmental question is above all a humanistic question;[9] like the arms race, it is a question of human survival – for citizens of the Soviet and capitalist world alike.

As we have seen, broad-ranging analyses of the ecological situation have appeared in various Soviet scholarly journals. While such scholarly journals have a limited circulation, they are, in principle, accessible to any interested citizen, can often be bought at public newsstands, and represent a part of the "official" media, supported and approved by the party/state. In the GDR, scholars are also concerned about environmental deterioration, but the openness of general theoretical discussion in official scholarly publications is meagre and inferior by comparison with Soviet writings. Since 1979 there has been a surge of attention to the ecological problem in the leading GDR philosophical journal, but even here the debate is less bold, frequently relying on Soviet concepts and writings. For example, discussion of "global problems" seems clearly derivative from and stimulated by Soviet writings.[10] The more limited GDR debate is presumably due to the stricter political control which we already observed in chapter 2. Even in the scholarly media, GDR authorities control expression of ecological pessimism to reinforce their claim that the GDR is superior to the FRG and to prevent their own media from serving as a catalyst for esoteric communication of attitudes found in the West. GDR authorities may also be cautious in allowing their scholars to engage in broad-ranging speculation in this area since Ulbricht's removal as First Secretary of the Socialist Unity Party (SED) in 1971 is often seen, in part, as a response to Soviet dissatisfaction with his encouragement of ideological innovation in the 1960s.[11] In two areas GDR scholarly analyses do stand out: in treatment of the environmental crisis in capitalist countries (which we do not analyze in depth here) and in more practical analyses of economic planning methods and technological development (dealt with in Part 2). Such studies do not threaten Soviet ideological hegemony, nor do they question the systemic foundations of socialist society.

Despite greater control of scholarly publications, GDR authorities have permitted the emergence of thoughtful commentary on the ecological crisis in the belles-lettres[12] and inside the Evangelical Church. Although the church became involved with the environmental issue already in the early 1960s,[13] since the late 1970s burgeoning interest, especially among young people, has generated a plethora of activities (e.g., tree planting actions, environmental exhibitions, seminars, bicycle excursions)[14] and documents emanating from church-related groups. Church writings are disseminated on a limited basis. Some of these, e.g., regional church newspapers, are subject to censorship and are available only through mail subscription, making their distribution subject to state-imposed limits. Evangelical Church newspapers include five weeklies with additional editions of **Die Kirche** (The Church), published in East Berlin, to cover other areas of the GDR.[15] (These church newspapers should be distinguished from the organ of the GDR's Christian Democratic Union, a National Front party and thus part of the GDR's official network of mass mobilization organizations. **Neue Zeit**, the party's newspaper, has also devoted some attention to environmental problems but the tone is typical of SED-controlled publications, despite some attention to Christian perspectives.[16]) Other church documents, hectographically reproduced reports, are intended "only for official inner-church use," but these may also be distributed on a wider basis through the mail.[17] These latter sources have been tolerated, if not encouraged, by the party/state, and to a degree they represent the emergence of a semi-autonomous "unofficial" source of information, subject to self-censorship rather than state-censorship. Comparable documents dealing with the ecological crisis are not available in the USSR and probably do not exist. The greater leeway offered GDR, as opposed to Soviet, church activities is a question beyond the scope of this study, but in the ecological area it seems that the relatively open inner-church discussion provides a means of containing the potential for widespread youth alienation. At the same time many of the activities and values furthered within these church-based initiatives support regime goals and programs.[18] Church authorities help to deflate any anti-systemic tendencies as the price for retaining some degree of organizational autonomy. Frequent references to Marxist insights and positive regime initiatives, along with encouragement of support for state efforts dampen the effect of the implicit rejection of some regime values in these writings.

These different patterns of control in the two countries make a balanced comparison of Soviet and GDR viewpoints difficult. For we are forced, to a large extent, to compare apples (Soviet scholarly debate in official journals) with oranges (GDR church-based debate). This very asymmetry, however, confirms that the official GDR media do not accurately reflect the full range of GDR theorizing. It also tells us something about leadership perspectives. Soviet leaders are more confident in allowing open expression of ecological pessimism, but place strict controls over organizational initiatives. GDR authorities, on the other hand, maintain strict control over the visible public world, while tolerating some islands of semi-autonomous activity, so long as they do not take on an anti-systemic character.

Let us now examine in some greater depth the attitudes of Soviet and GDR scholars as expressed in official journals. We will then seek counterpoints in the unofficial church-related discussion in the GDR. In the section which follows we explore reactions to Western analyses of the cultural and religious roots of the ecological crisis. Following this, diverse judgments about the seriousness of the crisis are explored.

## WESTERN CULTURE, MARXISM, AND ENVIRONMENTAL ETHICS

Soviet leaders have consistently pursued policies intended to bring nature under human control. This "mastery-of-nature" mentality was perhaps most dramatically evident in the huge engineering feats of Gidroproekt (the hydro-power engineering agency) under Stalin.[19] But the same zeal for mastering nature is no less striking in contemporary Soviet plans to harness the rich natural resource potential of Siberia and the Far East, including proposals for massive diversion of Siberian rivers to the water-hungry south.[20]

Environmental critics in the West frequently seek the roots of the contemporary ecological crisis precisely in such a "mastery-of-nature" mentality, a mentality allegedly rooted deep in Western culture and religion. Lynn White, for example, has argued that, unlike some Eastern cultures that sanctify nature, Western culture regards nature as an instrument to serve human ends. This view is firmly rooted in a Judaeo-Christian tradition which places nature in opposition to humanity and sanctions human mastery over it. White sees this Western tradition as underlying the present ecological crisis and for this reason he calls for a radical transformation in society's attitude toward nature.[21] In this argument, Marxism, at least as practised in the Soviet Union, is firmly entrenched in this anthropocentric Western cultural tradition. John Passmore goes so far as to assert that whatever the claims of Marx and Engels, the Soviet Union "certainly has shared, in this respect, the ideology of capitalism," referring to its ignorant and greedy concern with short-term benefits at the expense of nature.[22]

Soviet theorists have shown considerable interest in the idea of a "classical Western attitude" toward nature and its alleged roots in the Judaeo-Christian tradition. Historically, Russian intellectuals have expressed deep ambivalence toward Western Europe, her philosophy and culture. Positioned between an industrializing Western Europe and the less developed Oriental East, many nineteenth-century Russian intellectuals bemoaned the "Asiatic" backwardness of Russia and sought to emulate the West. At the same time some intellectuals sought to affirm their identity by stressing Russia's unique cultural traits, superior to the secularism and greed of Western Europe. Lenin's adaptation of Marxism manifests this ambivalence about Russia's "backwardness" vis-a-vis Western Europe. Russia becomes, in Lenin's view, the spark to ignite the revolution in Western Europe, the trigger to push social progress onwards; by virtue of her backwardness, as the weakest link in the capitalist chain, Russia steps into the vanguard of the world revolutionary process.[23] Backwardness becomes an asset, an idea which Mao Zedong was to carry even further several decades later in his celebration of China as poor and blank.[24] Given this historical ambivalence to Western culture in Russia, it is not surprising that contemporary Soviet scholars should be intrigued by the thesis of a classical Western attitude underlying the ecological crisis. East German intellectuals, on the other hand, lie much more clearly in the mainstream of Western European cultural history and historically have not experienced the same degree of ambivalence to Western European culture. This may be one reason why the White thesis has not elicited such a visible response from mainstream scholars in the GDR. On the other hand, the responsibility of the church and Christianity for the ecological crisis is a theme that crops up frequently in GDR church writings. The emphasis in these writings on the importance of an altered value system and elevated moral responsibility in addressing the crisis has in turn elicited responses in GDR scholarly journals.

On the face of it, of course, a common cultural explanation for ecological problems in capitalist and socialist countries contradicts Soviet Marxist principles, and the overwhelming rejection of such an interpretation by Soviet commentators is predictable. Nonetheless several Soviet scholars have shown considerable interest in the suggestion and their answers are interesting. Most Soviet theorists emphasize that culture is an epiphenomenon, reflecting underlying material forces. Kantor, for example, points out that the different attitudes toward nature in Western Europe and Asia rest on differing natural conditions (e.g., climate and soil types), just as Marx explained the "Asiatic mode of production" partly on this basis.[25] The Soviet theorist Novik rejects the "cultural" hypothesis on somewhat different grounds. He characterizes it as the most extreme type of ecological pessimism and romanticism; it rests on a metaphysical assumption of an East-West split, which finally "means to absolutize the division of humanity, to immortalize backwardness." This view is romantic, Novik explains, because it misapprehends the real essence of the so-called "harmony with nature" of Eastern society, which was achieved by "ignoring the personality."[26] Novik rejects any resolution of ecological problems which is based on "ethical anti-knowledge"; a true resolution depends on a scientific synthesis, capable of unifying humankind through communist interrelationships.[27]

Similar logic leads Kantor and Shakhnovich to reject the notion that Western anthropocentrism is rooted in the Judaeo-Christian tradition. Shakhnovich argues that religion itself does not give birth to any special view of nature. Rather, religion expresses in its "language of fantastic images and ideas those attitudes toward nature which take shape in society under the influence of the socio-economic conditions of life of the people."[28] This effort to exonerate Christianity of responsibility for the ecological crisis is probably not intended to redeem Christianity, but rather to subordinate religion to the material forces of history. Although some Soviet scholars express sympathy for efforts of religious leaders to further humanistic values, they also accuse them of attempting to restrict science by placing it under religious-ethical control.[29]

In short, most Soviet theorists do not question the existence and worth of a Western anthropocentric tradition. What they do contest are (1) the origins of that tradition, and (2) its link to the ecological crisis. In his critique of capitalism Kantor openly acknowledges that Marxism (and socialist society) are the heirs to all the **human** values accumulated under capitalism: ". . . values of science, art, technics, material practice, progressive forms of social organization and management, **methods of mastery of the forces of nature and methods of the conservation of nature.**"[30] Kantor does not dissociate Marxism or Soviet society from these classical Western values, including mastery of nature, but only from the social forms which have given these values an alienated and exploitative expression.[31]

While this dominant interpretation is entirely consistent with Soviet Marxism-Leninism, a few Soviet scholars have provided a dissenting view. By suggesting a greater validity for the cultural thesis, these scholars imply both a common source for ecological problems in the capitalist and socialist world and a common need for a fundamental transformation of values. Iu. G. Rychkov, a biologist and geneticist, attributes more importance to culture than do other Soviet analysts. He identifies two attributes of the Western European or Euro-American cultural variant: (1) "a view of man as the crown of creation, the master of nature, the final stage of evolution"; and (2) development which "has always proceeded on the principle of denying equilibrium with the environment."[32] So far, this is in line with the dominant view. But for Rychkov these Western cultural traits have ambiguous consequences: The denial of equilibrium with nature has ensured "the surging of the economy, science, and culture, but it naturally led to the situation we are discussing as catastrophic" (i.e., ecological imbalance).[33] Rychkov is careful to refer to Western culture (presumably excluding the USSR). Nonetheless his discussion produces an ambiguous impression, for he ties the negative consequences not to capitalism, but to a cultural type. Furthermore, by never distinguishing the cultural type of the Soviet-bloc countries, he strongly implies that it is similar to that of the West: "However, developed countries with a socialist societal organization may find themselves facing a similar ecological crisis if the scientific basis and philosophical interpretation of the relationship between nature and society lag behind the rate of cultural-economic development."[34] Rychkov proceeds to point out that "in our country we have a somewhat larger margin of time" in which to analyze all aspects of the problem and he concludes that the experience of socialist countries affirms the possibility of a more harmonious society-nature relationship. Nonetheless, between the lines, he has implied the **lack** of a comprehensive approach in socialist countries ("trial and error"); a similarity in cultural type between socialist and capitalist countries; and the predominant role played by culture. In addition he clearly suggests the need for a new philosophical interpretation of the nature-society relationship. For "ecological optimism," in present circumstances, may "upon examination prove to be merely an instinctive consumerist attitude toward nature." What is required is the establishment of optimum boundaries and norms of development, based on philosophical, psychological, ethical-aesthetic, and natural scientific analysis. Further attempts at "conquest," "transformation" or "improvement of nature" must be abandoned, and replaced by a steady state economy.[35]

Rychkov is not the only scholar who emphasizes the cultural roots of ecological imbalance. A Soviet historian, S. A. Arutiunov, suggests that critical interaction of energy, demographic, food, and ecological problems could present humanity with very serious "shocks, which could lead to a sharp decline in its numbers, the quality of life, and the level of consumption, if not to death." According to Arutiunov, "ritual-prestige consumption" is responsible for large, potentially unlimited, expenditures of energy and material, "the rapid growth of which has become one of the direct causes of crisis phenomena in the environment."[36] Because prestige is a necessary component of a fulfilling human life, society must find ways of fulfilling this need which are labor-intensive and do not involve the "wasting" of nature which is so characteristic of virtually all past and present societies.[37] Arutiunov therefore advocates ethnic studies to discover which cultural patterns are ecologically sound. The future requires human self-knowledge, rather than continued concentration on rebuilding and adapting nature to human ends.[38]

These critiques take on added significance when placed alongside other evidence of disillusionment with the "mastery-of-nature" mentality in the USSR. Tsaregorodtsev, an environmental advocate, refers to frequent demands for a "return to nature" because of the deleterious effects of pollution on human health.[39] Such challenges to the "mastery-of-nature" mentality are no doubt politically weak in the USSR, but some commentators seem to view them as dangerously widespread. For example, the scientists Rutkevich and Shvarts complain that these "pessimists" appeal to emotion, they set their attention "only on the undesirable changes in living nature, and they have already drawn broad conclusions from them."[40] While such attacks may be directed against a wide range of environmental advocates, they reflect alarm over the extensive support for a more cautious attitude toward nature transformation.

While "back-to-nature" sentiments are criticized in the USSR, it is more difficult actually to find such sentiments revealed publicly. A few striking examples deserve mention. The famous writer Leonid Leonov, long concerned with forest conservation, published two important articles in **Literaturnaia gazeta** (Literary Gazette) and **Sovetskaia Rossiia** (Soviet Russia) in 1965 and 1970, respectively. The 1965 article addresses the waste of forest resources and calls for a "radical reassessment" of the "archaic view of the forest as an inexhaustible gratuitous bounty."[41] Not only does Leonov compare present-day bureaucrats to Gogol's tsarist officials, but he appeals to Russian history for an alternative ethical system:

> It is dangerous to hold ourselves aloof from the history of our forefathers, as though nothing ever happened then apart from feudalism. . . . in our country we often crush our native soil to bits with bulldozers and efface the memory of the warm and living. . . . The great tombs of history call for emulation of our fathers' deeds, stiffen the spine of weak conscience, sober our boastful brashness, inhibit excesses of administrative zeal.[42]

This general denunciation of modern arrogance is Leonov's answer to crude violation of the sanctity of the forest. The article is a masterful intertwining of external respect for the Soviet state (in his call, for example, for "urgent public intervention," for an all-Union conference, and for thrift), with fundamental and damning criticism. Concluding this discussion with a consideration of the problems of weaning children "from an arrogant, spoiled, consumer attitude toward nature," Leonov makes a double-edged attack on anthropocentrism and atheism:

> This kind of expression of ironic superiority to everything, even to what is most inviolate and even, one might think, to the existence of one's makers, has always repelled me in the image of our civilization. Much would be different if our children were to take for their commandment the idea that "I may be small, but I am infinitively stronger and needed because the world contains many creatures smaller and weaker than I, and I can do them good, if only by passing them by without noticing, touching, or perturbing them."[43]

By 1970, Leonov's attack is tempered, but still contains similar jibes at "over-zealous comrades," "overly taken with rationalism,"[44] and assertions that birds and animals "have just as much right to exist as we do!"[45]

Other voices resisting "progress" and its effects on nature are heard in rural and ethnic literature. As is well known, the orthodox Marxist-Leninist position sees the development of urban life as an indicator of social progress, a triumph over nature.[46] Much so-called rural literature, often closely associated with minority nationality literature, focuses on the value of country life, not on the triumph of the city. Soviet literary critics draw attention to the social implications of this preoccupation. For example, the Kazakh novelist Samtimzhan Sanbaev is criticized for expressing "opposition to social progress" in his idealization of the Kazakh village and in his rejection of urban life with its alienation from nature.[47] The literary critic Kozhinov takes a more positive attitude toward these tendencies. He describes art as turning "to the village [**derevne**], not for its own sake, but for deeper penetration into the meaning of the present, for precisely in the village, where social-technical progress directly and openly 'collides' with nature, the possibility arises of penetrating its meaning to the end."[48] Kozhinov is distressed by humanity's isolation from the natural environment. Out of this isolation grows an avid striving which "grabs millions of people," "a striving to escape from the ever more 'artificial' conditions of urban existence just for a time," a desire for the "least touched corner of nature."[49] Although he admits that it would be senseless, stupid, and anti-democratic to oppose scientific-technical progress, Kozhinov finds "intoxicating admiration" for "a technical paradise" unworthy of humanity. He calls upon writers to address the question of the costs of scientific-technical progress.[50]

Also in the GDR, the theme of environmental deterioration has been taken up in the belles lettres, particularly since the late 1970s. For example in his story "Swantow," Hanns Cibulka includes reflections on the release of radioactivity from nuclear power plants[51] and bemoans human alienation from nature. Ill health is directly connected to environmental pollution: "The majority of illnesses that we suffer from today are the result of our way of life and the environment." As nature loses its own regenerative capability, "man murders himself, to be sure, a murder on credit."[52] Not only is physical health damaged by a deteriorating environment, but spiritual health as well. Cibulka questions the meaning of scientific progress for "what we lack now is a science of human beings that finally asks, for once, how a person should live and work, what is actually necessary for a life worthy of a human being. . . . We live in an invisible trench; we lie under fire from all sides: water, air, food, radiation fields."[53] These sentiments are reflected in other literary works, along with descriptions of urban life: "the dirty old houses on the other side of the dirty hot street and dirty autos which let out dirty smoke behind them, and the pigeons, whose wings were so dirty that they couldn't fly. . . ."[54] Some literary commentaries are too sensitive and meet with the hand of the censor, as did Monika Maron's book **Flugasche** (Fly Ash), which was published in the West. Maron describes the frustrations of a young journalist, Josefa Nadler, as she seeks to draw public attention to the inhumane conditions resulting from pollution in the city of B. (presumably Bitterfeld). Nadler finally comes to question the meaning of her work and of progress: "'The worst is,' she said, 'they told us so much about revolution that life without revolution seems totally meaningless. . . . And I'm supposed to wash the revolution clean of 180 tons of fly ash. I should wash it and buff it with polish from a spray can and extol it on newspaper pages as a . . . powerful vehicle into the future. Afterwards I go to the hairdresser and have my hair dyed blond, because I crave a change.'"[55]

This type of scepticism regarding GDR-style progress, also reflected in the church writings which we will examine below, has elicited a response from GDR theorists in the official scholarly media. While little explicit attention has been devoted to the "mastery-of-nature" thesis or to the cultural origins of the ecological crisis in GDR scholarly journals, since 1979 a series of articles on the problem has appeared in **Deutsche Zeitschrift für Philosophie** (The German Journal for Philosophy), the leading philosophical journal in the GDR. These articles emphasize the importance of social democracy in addressing ecological problems (see page 60 above), the ecologicization of production (see page

94 below), and the "unalterable unity of man and nature," linking these themes to the writings of Marx and Engels, and to SED documents.[56] The richness of the Soviet discussion is still missing. For example, the philosopher Herbert Hörz affirms that "the essence of the cultural-function of science is the expansion of the domination of human beings over nature for the development of humanity. Freedom doesn't consist in the subordination to natural necessity, but rather in actions based on the competent decisions of human beings, which serve the purposeful formation of a benevolent nature."[57] In one of the more interesting pieces, Adolf Bauer, an environmentalist in his own right, cites Engels in emphasizing that "mastery of nature does not resemble the role of a conqueror, but rather consists in the correct application of natural laws."[58] But he also warns against illusionary efforts to solve the problem primarily through the propagation of a new attitude. In a footnote, presumably to alert only the initiated to the existence of the sentiments he critiques, he warns against the theological conception that "the contemporary environmental problem is finally the result of 'human arrogance vis-a-vis nature' – a view that is also propagated in socialist countries."[59] The editor of a leading literary journal clarifies the real role of social democracy, drawing a parallel between the "big" environmental questions and serious medical problems: "Should we put **these** to the normal consumer for a decision? . . . The serious treatment of such questions occurs among specialists, for good reason, in scientific journals whose language is hardly accessible to the simple reader."[60]

Now let us look at the more critical ideas which have emerged outside the official media in the GDR.

## The Theological Perspective: Voices from the GDR Evangelical Church

GDR church writings cautiously criticize the materialism implicit in the mainstream perspective. Although they do not embrace "Oriental" harmony with nature (à la White), these writings do call for spiritual renewal as the foundation of ecological responsibility. At the same time, they seek common points of orientation with the Marxist viewpoint and call for dialogue with Marxists. Elements of Christian asceticism (and occasional back-to-the-earth sentiments) are joined with a future-oriented perspective, in which science and religion work hand in hand to improve human life.[61]

An important report produced by a working group of young scientists, technicians and theologians at the Church Research Facility in Wittenberg identifies the search for security through material possessions and achievements as an important source of alienation between human beings and nature: "Nature increasingly becomes a thing which human beings want to secure and which is only seen in terms of its effective use. An understanding of its intrinsic value and of the human bond to nature gets lost." Although this viewpoint is an implicit critique of the ruling mentality in the GDR, the document invokes support in a quote from Marx himself: "With the exploitation (**Verwertung**) of the world of things, the devaluation (**Entwertung**) of the human world increases in direct proportion."[62] Despite such attempts to identify points of common reference with the regime, most church writings emphasize spiritual values and appeal to Biblical examples and insights. For example, the Biblical account of the flood and Noah's Ark is drawn upon to demonstrate the importance of Christian hope and action in confronting the environmental crisis.[63] Similarly, Christian teachings suggest that security should be sought in an atmosphere of mutual love and human acceptance, ultimately the love of Christ, rather than in material achievements. This would reduce the need for possessions produced at the expense of nature and afford the courage and insight to lead an ecologically responsible life-style (e.g., low meat intake, reduced use of automobiles, repair of old objects, resistance to faddish consumption).[64] "To reduce the threat to our lives from a devastated environment, in the rich lands of the earth (also the GDR), it is necessary to direct human work and needs more than previously to **spiritual-cultural affirmations [Bestätigungen], social commitment and a life-attitude of solidarity**, rather than giving continued priority to the further accumulation and accelerated consumption of material goods."[65] This, of course, would involve a fundamental change in cultural goals and values.

The direction of the needed value reorientation is suggested in other church writings as well. Heino Falcke, a church leader in Saxony, identifies three criteria – justice, ecological viability, and participation – which should govern human use of science and technology. In terms of responsibility, the relationship to poorer Third World countries takes an special meaning: "We [must] measure the level of our material consumption not by the higher level of Western society, but rather by the level of the developing countries. . . ."[66] The requirement of ecological viability demands, in Falcke's view, a critique of "technical reason" and a rejection of the reification of nature: "Living beings (**das Lebendes**) become a thing. The animal, the way of dealing with the animal, becomes an issue of meat production. . . . The workers become labor power. The issue is a critique of this approach of modern science and technology. It has to do with a new view of the reality with which we live."[67] Finally Falcke rejects decision-making by a technocracy or expertocracy. Rather, every individual must be ethically and morally mature, and able to participate in social decision-making. The Wittenberg document cited above supplements Falcke's analysis, identifying principles deriving from the scientific-technical method which must be **rejected** in the human relationship to nature: (1) the principle of "do-ability" (**Machbarkeit**), i.e., the belief that we can direct and compel all things to operate in our interest or that "we may do everything that we can do!"; (2) ownership and achievement as measures of evaluation; and (3) the delegation of responsibility for the environment to others.[68] Not only do church activists propose a new orientation, but they also take action to help bring it about. The Theological Study Division of the Federation of Evangelical Churches in the GDR (Berlin) operates a "reference service" which collects information and provides abstracts and analyses of official GDR, church, and Western publications relating to the ecological situation.[69] The Wittenberg Research Facility circulates "orientation letters on the man-nature conflict." The letters inform interested parties of local activities, provide book reviews, and carry reports on environmental damage, along with essays on the role of Christians and the church.[70]

A primary emphasis in almost all of these church writings is on individual responsibility and possibilities for meaningful, if not immediately efficacious, action. Indeed, much of the interest in the ecology issue was generated from below, by young Christians seeking to make a positive mark by planting trees in the local environment. Tree-planting actions (often carried out in cooperation with a local enterprise), environmental exhibits circulated in churches, "**mobil ohne Auto**" ("mobile without a car") week-ends, seminars for exchange of information, and bicycle "demonstrations" (sometimes interfered with by the state) are all examples of the so-called "small steps" which are deemed morally superior to resignation and inaction. Changes in personal life-style, such as reduced meat consumption (to help combat world hunger and reduce use of pesticides and herbicides) and reduced reliance on automobiles, may not produce noticeable improvements in the short term, but can help produce long-term changes in attitude.[71] Such life-style changes depend on a reassessment of needs. Götz Planer-Friedrich, head of the Theological Study Division, rejects the distinction, which he calls a Marxist distinction, between primary (material) and secondary (e.g., spiritual) needs. Similar to the dissident writer Rudolf Bahro, with his concept of compensatory interests, Planer-Friedrich suggests that "the pseudo-satisfaction of non-material needs through material, organizational, or institutional means leads to frustration. . . ." Furthermore, some material needs may be artificially created.[72] Although Planer-Friedrich does not directly link his critique to the ecology question, its relevance is clear. The Wittenberg document cited above asks Christians to reconsider their personal needs to improve their relationships with both nature and other human beings.[73]

Of course, a more modest definition of material needs is not necessarily antithetical to regime interests. In an unusual exchange of letters published in a GDR journal in 1973, Hermann Klare, then-President of the Academy of Sciences of the GDR, had raised the issue: "Is there a law that we must follow a capitalist pattern of need-creation? One can indeed create needs although no such needs are present."[74] The church writings, however, advocate not only a dampening of material desires but also an awakening of spiritual life, a suggestion perhaps not so amenable to regime priorities.

These church documents thus call for fundamental cultural change and implicitly reject Marxist assumptions about material determination. It may seem ironic that while mainstream Soviet theorists exonerate Christianity of responsibility for the present ecological situation, these GDR church writings accept Christian culpability in bringing about the present dilemma. This is, of course, because Marxism considers material forces to be the real motor of history (religion is only an epiphenomenon), while GDR Christians consider ultimate reality to be spiritual. Thus, some GDR theologians emphasize the church's responsibility for the environment and acknowledge that, historically, the church may indeed have been a contributing source of the problem. For example, Biblical teachings do provide a foundation for human domination of nature, which could easily be transformed into irresponsible exploitation, particularly as the long-term consequences could not be foreseen. In an error of omission the church also often failed to engage science in dialogue to assure the ethical application of its findings. Rather, the church accepted compartmentalization, concentrating on individual morality at the expense of social responsibility.[75] In acknowledging its own negative role in the past the church hopes to create a firmer base for future action: "A view of the culpable past of our own (church) tradition prevents us from self-righteously seeking fault in others and being over-bearing toward them. In this way solidarity with non-believers who are also actively involved with the environmental question becomes easier."[76] These church writings call on Christians to work with the state to reduce waste and to beautify the surroundings; Christians should also work with scientists to direct progress in a more humanistic manner.[77]

If church activists differ from Marxist thinkers in terms of some fundamentals, they do not really reject mastery of nature. For it is grounded in the Biblical charge ". . . to fill the earth and subdue it; and have dominion over the fish of the sea and over the birds of the air and over every living thing that moves on earth" (Genesis 1:28). Rather, these activists interpret this passage so as to soften the impact of human domination. According to the Wittenberg document, this should not be a justification "to exploit and treat nature as an external power treats a foreign people. . . ."[78] Nature's interests and intrinsic worth should be honored, since nature is also a creation of God, and human beings must care for nature as God cares for humans. Saint Francis of Assissi's attitude toward nature, as a subject to be esteemed, provides a model. The relationship to nature is described as a dialectic: humans are "dependent" on and "emancipated" from nature: "indebted to nature and asserting ourselves vis-a-vis her."[79]

## Some Comparative Speculation

While one finds sprinklings of scepticism about "progress" in the Soviet press, it is virtually invisible in the East German media, except in some literary works. Similar sentiments do exist, however, in the GDR; they are simply not allowed widespread public expression. Perhaps because official controls are tighter in the GDR, the "unofficial" church literature, as well as dissident writings on ecology, are richer. The **Zivilisationskritik** evident in recent GDR writings emanating from church organizations echoes themes found in Soviet scholarly debate, but in more radical form. As in the USSR, non-Marxist value systems appear to provide a key foundation for heightened ecological consciousness in the GDR. In part this reflects the fact that the church, at least in the GDR, provides an institutional alternative to the Party's closed attitude. The regime may be giving the church a greater role than it would otherwise enjoy by exiling the ecological critique from most official outlets.[80] If Soviet scholars provide a somewhat sterile refutation of Lynn White's thesis about the Judaeo-Christian roots of the ecological crisis, it is the ecological activism of GDR Christians which provides the practical critique of White's perspective. Whatever its historical role may have been, Christianity in the GDR is proving itself to be a springboard for reevaluation of the meaning of the mastery of nature. And in a backward sort of way this proves the Soviet theorists correct. Perhaps, as they suggest, religion does not give birth to any special view of nature; rather, in the GDR, the role of the church in the ecology debate is shaped by the socioeconomic and political realities confronting the population. For the church allows one of the few outlets for expression of their concern about deteriorating natural surroundings.

In summary, the anthropocentric approach to the society-nature relationship dominates in the USSR and the GDR. Most Soviet and GDR scholars justify this in terms of Marxism-Leninism, itself rooted in a broader Western cultural heritage. They make no attempt to dissociate themselves from this heritage. Those who do dissent from this view appeal to traditional, religious, or national values, and only marginally to Marxism-Leninism. They are on the fringes of environmental debate in the USSR. Nonetheless, the visibility of this "dissenting" view, both in the official Soviet media and in "unofficial" and dissident writings in the GDR, suggests that disillusionment with the prevailing version of "progress" may be widespread. Furthermore, if some scholars and writers question the **desirability** of unlimited exploitation of nature, natural scientists buttress this position by questioning the **possibility** of sustaining such an exploitative relationship.

## MASTERY OF NATURE: THE LIMITS

At the 25th Party Congress of the CPSU Leonid Brezhnev stated that "it is possible and necessary, comrades, to improve nature, to help nature reveal her living forces completely."[81] This statement fully echoes the official commitment to an interventionist approach to the natural environment. The ability of nature to absorb the onslaught of civilization is, however, an issue of contention among Soviet-bloc scientists. An increasingly vocal minority is alerting the population and leadership to the scientific limits of such mastery. Let us examine the areas of conflict and consensus in this debate.

Three general observations can be made about serious Soviet and East German scientific analyses of the evolving society-nature relationship. First, again in this area, academics in the GDR rarely articulate "meta-theories" in a public forum. Often they follow the Soviet lead, mimicking Soviet concepts and theories. Second, although Soviet scholars are broadly agreed on the desired direction of change in the biosphere, they employ a variety of conceptual frameworks to explicate it. This creates difficulties in distinguishing areas of genuine disagreement from differences in terminology and emphasis. Third, within this shared, but conceptually diffuse, consensus Soviet scholars address with seriousness and care the limits of desirable human influence on nature. We will examine these last two points in more detail, noting some contrasts with the more limited GDR debate.

Several important concepts for understanding ecological change have gained acceptance in the Soviet academic community. The most important of these is the idea of the "noosphere," a notion attributed to Teilhard de Chardin, but adapted to the materialist approach of Marxism-Leninism by the late Soviet geologist, V. I. Vernadskii.[82] According to Vernadskii, the noosphere is the final stage of evolution in the biosphere, in which "man, taken as a whole, becomes a powerful geological force. And before him, before his thought and labor, stands the question of the rebuilding of the biosphere in the interests of freely thinking humanity as a unified whole."[83] Soviet analysts regard the development of the noosphere as in accordance with the objective laws of natural and social development, but not as an automatic process.[84] As the noosphere evolves, increasing human knowledge and technical capabilities allow the conscious planning and direction of human destiny. In this sense, Vernadskii's view of the noosphere is deemed consistent with the Marxist view of human development;[85] furthermore, it underscores the **scientific** nature of the building of communism, by linking it to law-like geological evolution. GDR scholars also occasionally refer to the noosphere and frequently suggest that a healthy relationship to nature is part and parcel of the movement to communist society.[86]

A second, somewhat less common concept applied by some Soviet scholars is the "biotechnosphere," which the biologist Khil'mi proposed in his book, **Osnovy fiziki biosfery** (Leningrad, 1966).[87] Khil'mi describes it as a new condition of the environment in which the "initial biosphere and the urban surroundings created by man form a whole new system," a symbiosis of nature and technology.[88] The biotechnosphere includes "the physical surroundings, living organisms, and technical equipment, in particular — the large-scale structures, transforming the atmosphere, the hydrosphere, and the lithosphere of the Earth."[89] Subsequent discussion of this concept has revealed no important distinction between its meaning and that of the noosphere, Khil'mi's term being favored by some, Vernadskii's by others.[90] Khil'mi's concept appears, however, to provide a more fully developed framework for analyzing the society-nature relationship, but the connection to the building of communist society is not made so clear. The concept is therefore less ideological in function. Perhaps for this reason, the noosphere concept has gained wider currency in recent years, possibly with official encouragement.

Some Soviet analysts, especially social scientists, have developed yet other concepts to portray the movement toward more effective human mastery of the environment.[91] The diversity of concepts to elucidate similar ideas exemplifies the fragmentation of Soviet science, particularly the rift between natural and social scientists. (See chapter 4.) Nonetheless, most accept the basic goal implicit in the noosphere idea, namely, increased human control over natural processes. GDR scholars also ascribe to this view, even though they haven't developed an arsenal of concepts to describe it. In the late 1970s, a few voices in the USSR have introduced a differing emphasis. These analysts suggest that, contrary to conventional assumptions, humankind is becoming more, not less, dependent on nature, as the biosphere becomes destabilized and increasingly vulnerable to disequilibrium.[92] Such a conclusion implies that society is ever more subject to natural limits and implicitly challenges the reality of the noosphere. Freedom, to use the familiar Marxist phrase, becomes "recognition of necessity," not the assertive reshaping of the natural world.

Thus, even the most ardent environmentalists accept the inevitability of growing human intervention in nature, while all place some limits on the desirable scope of human intervention. Virtually all Soviet scholars reject the notion of the "technosphere," i.e., the replacement of the biosphere "by a whole complex of technical mechanisms to fulfil biological needs."[93] The clear consensus among most scientists, in the USSR and GDR alike, is that the self-regulating processes of nature must be protected and optimized, assuring a condition of dynamic equilibrium. In other words, temporary or minimal disequilibria are necessary for change, but not to the point where nature's cyclical processes of self-regulation are irreparably damaged. Human beings should not consider themselves above nature, but part of it, regulating it to increase its productivity, its capacity for self-regeneration, and its internally balanced biological cycles. The GDR philosopher Hegewald sets as a goal the ecologization of production, a process that will humanize industry in socialism and communism. This will involve the transfer "to the production process of the principles and mechanisms of ecological systems as they have developed and grown stronger over billions of years."[94] In fact, however, ever more Soviet scholars fear that damage done to the environment is already extensive and irreversible.[95]

## Ecological Reassurance and Ecological Caution

There is considerable disagreement in the USSR about the seriousness of the present level of environmental deterioration. In recent years, a growing number of analysts have dared to suggest that ecological **crisis** threatens the socialist world as well as the capitalist world.[96] Other scholars consider such warnings alarmist and spare no efforts to point out positive changes in the environment. Rutkevich and Shvarts, for example, are oracles of great hope. They argue that intervention in natural processes may ultimately enable humanity to regulate the productivity of living systems.[97] Kamshilov, on the other hand, points out that an "offensive" against certain parts of nature may be necessary for economic development, but it is undermining nature's own forces of self-regulation. In Kamshilov's view "we chop off the branch we are sitting on," for nature cannot adapt to the fast tempo of human intervention.[98] In the GDR scientists are reticent in making such grave predictions. Again, one must turn to GDR literature and church writings for expressions of alarm.

According to William Ophuls, in America "leading technologists . . . and leading environmentalists . . . agree that eventually we must learn to live in 'steady-state' or 'spaceship' societies characterized by great frugality in consumption of resources and by deliberate setting of limits to maintain the balance betwen man and nature." What distinguishes environmentalists (or "pessimists" as Ophuls calls them) is their belief that "time is short, and far-reaching action by the current generation is imperative to avoid overwhelming the earth's capacity to support us in dignity; failure to act soon and effectively could lead us into the apocalyptic collapse — wars, plague, and famine predicted by . . . Thomas Malthus. . . ."[99] Although Soviet environmentalists dare not appeal to Malthus (whose theories Marx specifically refuted), the line between technologists (or optimists) and environmentalists (or pessimists) is essentially the same in the USSR. Even among those who seriously address environmental problems, some are more ecologically cautious, that is, less willing to entertain the risks involved in present approaches. And others, like their counterparts in the West, de-emphasize the risk involved in continued technological intrusion into nature's processes. We should recall here that our analysis can deal only with those who think (and write) about environmental problems. Most of these scholars reflect a considerably higher level of "ecological caution" than the silent ones, who are simply not attuned to the problem.

Among ecological optimists one might include the Soviet scientists Shvarts, Kalesnik and Davitaia. Shvarts gives particular importance to a high level of productivity in nature and asserts that human influence on the environment can already be positively assessed, since an increased level of energy exchange in the ecosystem may accelerate the evolutionary process.[100] Likewise, Kalesnik and Davitaia argue that pollution of the environment (specifically, Lake Baikal) is necessary up to a point, for it "helps insure the flowering of life and, through it, helps purify the environment itself."[101] Similarly, Soviet "optimists," like their counterparts in the West, tend to argue that the risk of death from a nuclear accident is minuscule compared to risks of everyday life, like driving a car or travelling by air. One might say that Shvarts, Davitaia, and Kalesnik offer "ecological reassurance" and so provide the regime with professional approval.

Other Soviet scholars like Novik, Rychkov, Kamshilov, and

Zharikov are markedly more "ecologically cautious." Zharikov explicitly warns that "in these conditions it is better to be overcautious (**perestrakhovat'sia**) in determining objects of nature protection and to include as great an amount as possible."[102] Likewise, Novik advocates the exploitation of energy sources such as solar power, wind, and ocean waves that do not disturb the ecosystem, and he warns against activities with unknown long-term consequences; for example, he questions the advantages of electric cars, pointing out that their impact on the electric field of the city has not been assessed.[103] Like Rychkov and Tsaregorodtsev, Novik also questions the genetic adaptability of human beings. The danger is imminent for "man, by virtue of the complexity of his body structure and chiefly as a result of the delicacy of his genetic apparatus, will, under the influence of the noted change in the parameters of his surroundings, either die out or leave posterity fraught with the danger of the degeneration of homo sapiens."[104] Echoing these sentiments, Medunin proceeds to an unusually critical analysis of the potential dangers of nuclear power.[105]

In the GDR these nuances are hard to detect.[106] Only occasionally are macro-issues discussed publicly in depth. It is the exceptional scholar who refrains from the usual criticism of capitalism and praise of socialism. In one exceptional analysis Horst Paucke and Adolf Bauer advocate closer examination of those changes which produce ecological crises and ecological catastrophe, as opposed to "ecological problems." They imply that ecological crisis could, from a theoretical viewpoint, affect socialist society as well.[107] In the official GDR media, no forum has been provided for a dialogue **between** disciplines, such as exists in **Voprosy filosofii** and **Priroda** in the USSR. The closest equivalents of these journals in the GDR, **Deutsche Zeitschrift für Philosophie**, **Urania**, and **Wissenschaft und Fortschritt** (Science and Progress) are theoretical wastelands by comparison (at least in the environmental area). Advocates of ecological caution must again turn to dissident writings or the church as an outlet. The writings of the philosopher and ecological critic, Wolfgang Harich, provide an example of a dissident view which had to seek a Western publication outlet. In his book **Kommunismus ohne Wachstum** (Communism without Growth), published in West Germany in 1975, Harich suggested that environmental deterioration and resource depletion are grave problems; their solution is possible only in a centrally-directed no-growth economy in the context of communist society.[108] The book was ideologically orthodox in its critique of capitalism, but Harich questioned the conventional GDR wisdom about the relationship between communism and material abundance. He surely must be considered ecologically cautious. Harich sensed some resonance for his views in the public positions taken by Kapitsa, Rychkov, Medunin, Kamshilov, and others in the USSR.[109] In the GDR, however, Harich found that neither his more radical assessment nor views similar to those of these Soviet writers found public airing.[110]

By the late 1970s, however, ecological caution was finding expression within the church and in literary works published in the GDR. The Wittenberg document cited above is entitled "the earth is to be saved," implying it needs saving. The environmental problem is called a "survival question." Like the Soviet philosopher Novik, the document suggests that "one may not wait for the proof" that a given technology is safe (or harmful); one must embrace an "ethic of future responsibility" now, by "taking account of such effects, even if it is not sure they'll come about."[111] Church documents also report that some young people suffer from depression and resignation as they confront the dimensions and apparent hopelessness of environmental decay.[112] This disillusionment also appears in literary works: "Here, for example, the masses are beginning to stir, so that something is getting under way. Only too slowly: the pollution is faster than the regeneration."[113] Likewise Günter Kunert, in the literary journal **Sinn und Form**, considers the harmful effects of chemicalized agriculture and their link to world hunger:

> Starve or poison ourselves, there doesn't seem to be a third possibility of an escape. And this is only one aspect of the creeping catastrophe to which we, in general and in particular, are closing our eyes. . . . One believes that one is doing something decisive to maintain or restore the environmental ecology and doesn't notice that in important respects we've long since lost the ground under our feet.[114]

Like the Soviet biologist Medunin, Kunert proceeds to express doubts about nuclear power safety: "So long as nuclear fusion remains in the stage of theory, this will never become 'clean.'"[115] Kunert concludes that survival itself "is not guaranteed."[116] Clearly the pessimists are present in the GDR too; but the regime demands ecological reassurance from its own scientists and experts. Thus the editor of **Sinn und Form** replies to Kunert: "The only really deadly serious danger for all of humanity is atomic war. All others are petty. . . ."[117]

## CONCLUSION

Official theoretical discussion of the environmental problem is limited, narrow in scope, and clichéd in the GDR, while Soviet theorists have confronted the implications of the crisis more boldly. It is unlikely that this contrast derives from cultural differences between the two countries, since it applies across the board in scholarly publications, and yet there is a healthy wave of "ecological caution" expressed in "unofficial" writings in the GDR. The East German regime exercises stricter control over theoretical innovation; for the authorities must balance the response to the West German challenge with loyalty to Soviet ideological leadership. The continued stability of the regime depends on both. East German authorities prefer to muzzle debate which may raise uncomfortable questions about the future of GDR society (as compared to the FRG) or which may contradict interpretations preferred by the Soviets. It is safer to stress analysis of capitalist society and to focus on concrete solutions of technical-economic problems. Therefore, East German challenges to predominant ideological and practical values – anthropocentrism, ecological optimism, and mastery of nature – are pursued almost exclusively in unofficial or dissident outlets. In these outlets we find reflection on topics which are touched only marginally in the Soviet debate; for example, the reassessment of needs in these GDR writings in some ways resembles similar preoccupations in advanced Western societies. This GDR discussion is not marked by national revivalism or glorification of the past, as are some Soviet writings; rather, these GDR writers seek to redefine the shape of a scientific and modern society. Because the church-related discussion in the GDR serves to reinforce regime initiatives for recycling, material savings, and local beautification (e.g., tree planting), the political authorities may tolerate some autonomous activities, so long as they remain within the confines of the church. The sacrifice of material luxuries called for in church writings may make more palatable the austerity measures demanded by the regime in times of economic stringency.[118] Implicit criticism of the West as a standard of reference also serves regime interests. To be sure, some undertakings of these church activists have been suppressed, for example, bicycle tours to view environmental damage.[119] Nonetheless, the ecological engagement of young people has not threatened the regime as much as the independent peace initiatives in the church. The latter more clearly implicate the GDR and Soviet regimes and thus challenge fundamental policy commitments. One wonders whether the ecology activists will come under increasing control as they have begun to join environmental and peace issues in their statements.[120]

In the USSR there is a substantial challenge to the predominant tone of ecological optimism emerging in the official scholarly journals as well as in some wider circulation media. As Seweryn Bialer has recently suggested, in this area as in others "the future looks less like a unilinear, unbridled progression upwards."[121] While pessimistic voices remain a minority, their growing visibility in the official media probably indicates that elements in the Soviet leadership itself recognize the prophetic truth in their warnings. As in the GDR, the strength of these attitudes is impossible to measure and sentiments of these "loyal dissenters" influence the tone of official ideology and policy only gradually.[122] But even now the chorus of optimism, anthropocentrism, and mastery of nature is more cautious and nuanced than it was twenty, or even ten, years ago.

How have Marxist values affected the assessment of the environmental situation? Basic assumptions underlying mainstream approaches (e.g., material determinism, anthropocentrism, scientific-technical domination of nature, ecological optimism) are consistent with Marxist values and are explicitly attributed to Marx and Engels. To this extent, Marxism seems to help shape the response. One should be cautious in ascribing these values exclusively to Marxist influences, however, since they reinforce other regime priorities as well. To compete economically with the West, a utilitarian relationship to nature, utilizing methods of scientific and technical control, is necessary. It is striking that all of the above-mentioned assumptions, with the possible exception of material determinism, also predominate in Western non-Marxist industrial societies. These values may reflect the Soviets' "imitative" path of development[123] more than their Marxist commitments. Furthermore, ecological optimism and faith in technological solutions reinforce regime legitimacy by placing a positive light on future prospects. Almost all ruling parties tend to downgrade threats resulting from policy failures and focus on external threats which can forge social solidarity. This has little to do with Marxism. Thus it is not possible to say that the predominate values derive from Marxism as such. More likely, those parts of Marxism which are supportive of other regime priorities are emphasized. Of course this begs the question of the origin of regime priorities and here the reader is referred back to chapter 1 where we explored the (dare one say dialectical) interaction between short-term goals and ideological precepts. Marxism contains diverse strains, which writers like Gorz, Bahro, and Harich have developed to produce vastly different conclusions about the ecological crisis. Perhaps the most we can say, then, is that the ideas underlying most Soviet/GDR analyses of the problem lie clearly within the mainstream interpretation of Marxism-Leninism, as it has developed in the USSR since the 1920s. From **this** Marxism, Soviet and GDR theorists have not deviated in any significant sense, nor have they derived any stunning insights. On the other hand, those who dissent from an anthropocentric and utilitarian attitude toward nature appeal explicitly to non-Marxist values, e.g., to religious, national, or traditional ideals. It seems clear that in the USSR and GDR the belief prevails that Marxism-Leninism implies those anthropocentric and utilitarian values.

In the USSR, as in the GDR, theorizing continues to be inextricably bound to political legitimation, despite the fact that the political leadership has loosened the reigns of control in certain areas. Nonetheless, natural scientists and philosophers inject a subtle political commentary into their expert judgments. There is considerable strain on these scholars to emphasize ecological reassurance, that is, the adaptability of nature to human influence. For this reason, the numerous and frank calls for ecological caution coming from Soviet scholars is even more significant. They indicate a high degree of concern for the future of our planet earth. By the same token, in allowing these debates, the Soviet leaders demonstrate some confidence that the existing system of power can survive a measured dose of ecological pessimism. This may be precisely because the leadership underestimates the potential ramifications of the emerging ecological crisis. Or, on the other hand, it may reflect their growing sensitivity to the problem and their belief that the system can (or must) allow at least some small degree of honesty in confronting it.

## NOTES

1. Also in East Germany Adolf Bauer and Horst Paucke refer to numerous open philosophical questions: "whether our theory of the scientific-revolution already correctly apprehends the ecological side of the process . . .; whether, and if so how socialism should create new social forms for mastering ecological demands or whether forms of consciousness formation and institutional control of an economic and non-economic sort which came into being in a time of less aggravating ecological problems are adequate to meet the newly emerging problem situation"; and so forth. See Adolf Bauer and Horst Paucke, "Einheit und Kampf zwischen Natur und Gesellschaft," **Deutsche Zeitschrift für Philosophie** (hereafter **DZfPh**) 27 (1979): 600-601.

2. See, e.g., V. Zh. Kelle, "The Leninist Conception of Scientific Ideology and Its Critics," **Voprosy filosofii** (hereafter **VFil**), no. 4 (1970), trans. in **Soviet Studies in Philosophy** 9 (Fall 1970): 114; and "Vernost' printsipam Leninizma," **VFil**, no. 9 (1968), p. 4.

3. For a discussion of the Lysenko affair see Zhores A. Medvedev, **The Rise and Fall of T. D. Lysenko** (New York:

Columbia University Press, 1969); and David Joravsky, **The Lysenko Affair** (Cambridge, MA: Harvard University Press, 1970).

4. See Loren Graham, **Science and Philosophy in the Soviet Union** (New York: Alfred A. Knopf, 1972).

5. See George Holliday and other articles in the same collection for general assessments of Soviet scientific contributions and effects of the political system on them. "Summary of Discussion" in John R. Thomas and Ursula M. Kruse-Vaucienne, eds., **Soviet Science and Technology: Domestic and Foreign Perspectives** (Washington, DC: National Science Foundation, 1977), pp. 142-46.

6. The meetings were actually held in November 1972, but were reported in **Voprosy filosofii** in the following months. See "Chelovek i sreda ego obitaniia," **VFil**, nos. 1-4 (1973).

7. Numerous definitions of global problems are offered. D. M. Gvishiani, deputy head of the State Committee for Science and Technology of the USSR (GNKT), and director of the All-Union Scientific Research Institute on Systems Research (VNIISI) of GNKT, distingishes global problems from universal problems. The former have vital importance to all humanity and require international cooperation for their constructive resolution. Universal problems, on the other hand, are simply those which may face many nations simultaneously, but can be addressed at the national or local level. Most scholars consider the arms race also to be a global problem. See D. M. Gvishiani, "Nauka i global'nye problemy sovremennosti," **VFil**, no. 3 (1981), pp. 99-101. Other recent articles in the series on global problems include I. T. Frolov, "Filosofiia global'nykh problem," **VFil**, no. 2 (1980), pp. 29-44; N. M. Mamedov, "Global'nye aspekty ekologicheskogo vzaimodeistviia," **VFil**, no. 11 (1981), pp. 74-83; and V. V. Zagladin, "Metodologicheskie problemy issledovaniia global'nykh protsessov mirovogo razvitiia," **VFil**, no. 9 (1981), pp. 17-31. Already in 1974 **Voprosy filosofii** played an important role in initiating the dialogue by sponsoring a round table discussion on global problems. See "Nauka i global'nye problemy sovremennosti," **VFil**, no. 8 (1974), pp. 102-24; no. 9 (1974), pp. 67-83; no. 10 (1974), pp. 46-60; and no. 11 (1974), pp. 60-75.

8. I. T. Frolov, "Nauka i gumanisticheskie idealy v reshenii global'nykh problem," **VFil**, no. 6 (1979), p. 118.

9. Ibid.

10. Many Soviet articles are translated in **Sowjetwissenschaft Gesellschaftswissenschaftliche Beiträge** (Berlin: Verlag Volk und Welt), monthly. On global problems see, e.g., Herbert Hörz, "Globale Probleme der Menschheitsentwicklung," **DZfPh** 30 (1982): 1301-22; and Horst Paucke, "Globale Probleme und der Menschheitsfortschritt," **DZfPh** 30 (1982): 1497-1508.

11. See Martin McCauley, **Marxism-Leninism in the German**

**Democratic Republic** (London: Macmillan Press, 1979), pp. 172-73.

12. I am indebted to Hubertus Knabe for his insightful treatment of this in " 'Der Mensch mordet sich selbst'. Ökologiekritik in der erzählenden DDR-Literatur," **Deutschland Archiv** 16 (1983): 954-73.

13. For brief historical background see Hubertus Knabe, "Gesellschaftlicher Dissens im Wandel. Ökologische Diskussionen und Umweltengagement in der DDR," pp. 2-3, presented at the XVII. DDR-Forscher-Tagung at the Europäische Akademie Lerbach, Bergisch-Gladbach, FRG, June 12-15, 1984. The conference papers will be published by the Verlag Wissenschaft und Politik (Cologne).

14. On these activities see also chapter 7, p. 220; and Wolfgang Büscher and Peter Wensierski, **Null Bock auf DDR: Aussteigerjugend im anderen Deutschland** (Reinbek bei Hamburg: Spiegel Verlag, 1984), chap. 3.

15. The five newspapers put out by the Evangelical Landeskirchen are the **Mecklenburgische Kirchenzeitung**, circulation 15,000; the **Potsdamer Kirche**, 13,500; **Der Sonntag** (Saxony), 40,000; **Die Kirche** (Berlin), 42,000; and **Glaube und Heimat** (Thüringen), 36,500. **Die Kirche** also publishes editions with special inserts for Görlitz, Griefswäld, and Magdeburg. See Detlef Urban, "Kirchen treten an die Öffentlichkeit," in Reinhard Henkys, ed., **Die evangelischen Kirchen in der DDR** (Munich: Chr. Kaiser Verlag, 1982), pp. 341-49.

16. See, for example, Peter Franz, "Eine Sache aller Bürger," **Neue Zeit**, Mar. 15, 1980; and Olaf Bastian, "Schützen wie nutzen zu aller Wohl," **Neue Zeit**, Apr. 27, 1983, p. 3.

17. See Urban in Henkys, pp. 341-42.

18. See below, p. 90; and chapter 7, p. 220.

19. See Boris Komarov (pseud.), **Destruction of Nature in the Soviet Union** (White Plains, NY: M.E. Sharpe, 1980), pp. 58-61; and Thane Gustafson, **Reform in Soviet Politics** (Cambridge, ENG: Cambridge University Press, 1981), pp. 53-54.

20. Ibid., chapter 5. On the present status of the Siberian river diversion proposals see the comments and references cited on page 33, note 58.

21. Lynn White, "The Historical Roots of our Ecological Crisis," **Science** 155 (March 10, 1967): 1203-7. See also Christian Mayer [Carl Amery], **Das Ende der Vorsehung: die gnadenlosen Folgen des Christentums** (Reinbek bei Hamburg: Rowohlt, 1972).

22. E.g., John Passmore, **Man's Responsibility for Nature: Ecological Problems and Western Traditions** (New York: Charles Scribner's Sons, 1974), p. 25.

23. See Alfred Meyer's excellent explication of this point in **Leninism** (Cambridge, MA: Harvard University Press, 1957), Part

4.

24. See Stuart Schramm, ed., **The Political Thought of Mao Tse-tung** (New York: Praeger, 1969), documents VII, C2 and C3.

25. K. M. Kantor, "Ekologiia i progress," **VFil**, no. 8 (1977), p. 114. The editors of **Voprosy filosofii** make a similar point in "Ekologiia, nauka, obshchestvo," no. 9 (1974), pp. 99-100. M. I. Shakhnovich calls the Western fascination with pantheism and other Eastern religions reactionary and utopian, in "Kritika teologicheskoi interpretatsii problem ekologii," in V. G. Marakhov, ed., **Nauchno-tekhnicheskaia revoliutsiia, chelovek, ego prirodnaia i sotsial'naia sreda** (Leningrad: Iz. Leningradskogo universiteta, 1977), p. 153.

26. I. B. Novik, "Budushchee nauki i ekologicheskaia problema," **VFil**, no. 1 (1975), p. 114.

27. Ibid., pp. 114-15.

28. Shakhnovich, in Marakhov, **Nauchno-tekhnicheskaia revoliutsiia, chelovek**, p. 151; and Kantor, pp. 109-11. S. S. Shvedov, however, accepts White's thesis about the influence of the Judaeo-Christian tradition on Western (specifically American) attitudes toward nature. See his interesting discussion of "American Social Philosophy and the Sources of the Ecological Crisis," **VFil**, no. 8 (1978), translated in **Soviet Studies in Philosophy** 17, no. 4 (Spring 1979): 76.

29. See Iu. B. Krianev, "Ekumenicheskie kontseptsii global'nykh problem," **VFil**, no. 5 (1981), pp. 144-46.

30. Kantor, p. 116. Emphasis added.

31. Ibid., p. 114.

32. Iu. G. Rychkov in "Chelovek i sreda," **VFil**, no. 4 (1973), p. 72. Other analysts like Gudozhnik specifically reject the idea of equilibrium in the environment, since "local ecological crises" are the basis of social progress. Gudozhnik advocates "progressive ecological equilibrium." See G. S. Gudozhnik, **Nauchno-tekhnicheskaia revoliutsiia i ekologicheskii krizis** (Moscow: Mezhdunarodnye otnosheniia, 1975), pp. 18-19, 23-27, 42, 71.

33. Rychkov, in "Chelovek i sreda," **VFil**, no. 4 (1973), p. 72.

34. Ibid.

35. Ibid., pp. 70-73.

36. S. A. Arutiunov, "Kul'turologicheskie issledovaniia i global'naia ekologiia," **Vestnik Akademii Nauk SSSR**, no. 12 (1980), p. 95. Shvedov also emphasizes the importance of cultural values in shaping attitudes toward nature, this time in the American context. Philosopher E. S. Markarian discusses the process of socio-cultural adaptation to ecological conditions in "Lokal'nye istoricheskie osobennosti kul'tury i protsessy ekologicheskoi adaptatsii," **Vestnik Akademii Nauk SSSR**, no. 1 (1981), pp. 111-17.

37. Arutiunov uses this word which he calls an Anglicism ("**veisting**," presumably a transliteration of "wasting"), apparently

in the sense of conspicuous consumption, p. 95.

38. Ibid., pp. 96-98.

39. G. I. Tsaregorodtsev, " 'Tekhnizatsiia' sredy i zdorov'e cheloveka," **VFil**, no. 10 (1972), p. 59. See also "Marksistsko-leninskaia teoriia nauchno-tekhnicheskoi revoliutsii," **VFil**, no. 2 (1974), p. 12; and E. S. Zharikov, "Filosofskii podkhod k postroeniiu sotsial'no-ekologicheskoi kontseptsii," in Marakhov, **Nauchno-tekhnicheskaia revoliutsiia, chelovek**, p. 82.

40. M. N. Rutkevich and S. S. Shvarts, "Filosofskie problemy upravleniia biosfery," **VFil**, no. 10 (1971), p. 55. See also S. S. Shvarts, "Problemy ekologii cheloveka," **VFil**, no. 9 (1974), p. 102; and O. K. Gusev in "Chelovek i sreda," **VFil**, no. 3 (1973), p. 62. Similar sentiments are expressed in the GDR by Wilhelm Girnus, in Günter Kunert and Wilhelm Girnus, "Anlässlich Ritsos: Ein Briefwechsel," **Sinn und Form** 31 (1979): 856-58.

41. Leonid Leonov, "Concerning Large Chips of Wood," **Literaturnaia gazeta**, March 30, 1965, trans. in **Current Digest of the Soviet Press (CDSP)**, vol. 17, no. 20, pp. 13-14.

42. Ibid., pp. 14-15.

43. Ibid., p. 15.

44. Leonid Leonov and Boris Riabinin, " 'Preserve' Means Inviolable," **Sovetskaia Rossiia**, May 24, 1970, trans. (condensed) in **CDSP**, vol. 22, no. 23, p. 5.

45. Ibid., p. 7.

46. See V. M. Gokhman, G. M. Lappo, and Iu. L. Pivovarov, "Urbanization and its Impact on the Environment," in I. P. Gerasimov, ed., **Man, Society, and the Environment**, USSR Academy of Sciences, Institute of Geography (Moscow: Progress Publishers, 1975), p. 292; and Siegfried Grundmann, "Mensch und Umwelt," **DZfPh** 21 (1973): 194-95. In another article Siegfried Grundmann in fact criticizes idealization of city life in literature, in "Soziologie des Territoriums," **Wirtschaftswissenschaft** 23 (January 1975): 50. The extent to which the orthodox position is questioned is indicated by N. P. Namurov's statement: "I don't consider that urbanization is necessarily harmful." But, he continues, "it is inevitably accompanied by a number of undesirable phenomena." In "Chelovek i sreda," **VFil**, no. 3 (1973), p. 56.

47. Vadim Kovskii, " 'Romantika Kochev'ia' i real'nost' narodnoi zhizhni," **Literaturnoe obozrenie**, no. 3 (March 1974), pp. 39-40. Other authors also claim that rural literature idealizes the past and that it addresses the society-nature relationship in this context. See, for example, L. Iakimenko, "Height of Criteria," **Pravda**, Jan. 27, 1974, trans. in **CDSP**, vol. 26, no. 4, p. 33; and A. Mikhailov, "The Rhythms of the Seventies," **Novyi mir**, no. 3 (March 1976), trans. (abstract) in **CDSP**, vol. 28, no. 22, p. 17.

48. Vadim Kozhinov, "Avtoritet istorii," **Literaturnoe obozrenie**, no. 3 (March 1977), p. 66.

49. Ibid., pp. 64, 67.

50. Ibid., p. 67. For a surprisingly gentle review of Kozhinov's article see Leonid Timofeev, "Etalon chelovechnosti," **Literaturnoe obozrenie**, no. 6 (June 1977), pp. 66-68.

51. Hanns Cibulka, "Swantow," **Neue Deutsche Literatur** 29 (April 1981): 39-40, 46, 48-49. In 1982 essentially the same piece was published as a book in the GDR, but with the passages relating to nuclear energy considerably weakened (Halle, Leipzig: Mitteldeutscher Verlag).

52. Ibid., p. 49.

53. Ibid.

54. Edith Anderson, **Der Klappwald** (Berlin: Kinderbuchverlag, 1978), p. 5, as quoted in Knabe, "'Der Mensch mordet sich selbst,'" p. 962.

55. Monika Maron, **Flugasche** (Frankfurt am Main: S. Fischer Verlag, 1981), p. 102.

56. On the unity of man and nature see Adolf Bauer, "Probleme der Naturaneignung," **DZfPh** 30 (1982): 1324. On social democracy see ibid., p. 1325 ff.; and Helmar Hegewald, "Moralische Triebkräfte bei der rationellen Nutzung der Naturressourcen im Sozialismus," **DZfPh** 30 (1982): 1465. On the ecologization of production see, esp., Hegewald, ibid. For references to Marx, Engels, and SED documents see Ellenor Oehler, "Aktuelle Fragen der Lehren von Karl Marx zur Gestaltung der Umweltbeziehungen," **Staat und Recht** 32 (1983): 305-14; Siegfried Stoof, "Fragen des Verhältnisses von sozialistischer Gesellschaft und Natur," **DZfPh** 27 (1979): 587-592; A. Bauer and Paucke, "Einheit und Kampf"; and most of the other articles cited below from **DZfPh**.

57. Herbert Hörz, "Natur, Naturwissenschaften, Kultur," **DZfPh** 31 (1983): 788.

58. A. Bauer, "Probleme der Naturaneignung," p. 1326. Other GDR and Soviet scholars also frequently refer to Engels' writings, particularly to parts of the **Dialectics of Nature**, in which Engels both justifies human mastery of nature and considers limits to it. Most specifically Engels writes: "For each such victory, nature takes its revenge on us. Each victory, it is true, in the first place brings about the results we expected, but in the second and third places it has quite different, unforeseen effects which only too often cancel the first. . . . Thus at every step we are reminded that we by no means rule over nature like a conqueror over a foreign people, like someone standing outside nature – but that we, with flesh, blood, and brain belong to nature, and exist in its midst, and that all our mastery of it consists in the fact that we have the advantage over all other creatures of being able to learn its laws and apply them correctly." Frederick Engels, **Dialectics of Nature** (Moscow: Progress,

1954), p. 180.

59. Ibid., p. 1329.

60. Girnus, in Kunert and Girnus, p. 855.

61. On asceticism see, e.g., Michael Kinze, "Ein Gebot unserer Zeit: Zurückhaltung," **Glaube und Heimat** (Thüringen), Oct. 1, 1978. For back-to-the-earth sentiments see the report from a living community in Saxony, "Auf der Suche nach neuen Werten," in Peter Wensierski and Wolfgang Büscher, eds., **Beton ist Beton: Zivilisationskritik aus der DDR** (Hattingen, FRG: Scandica-Verlag, 1981), pp. 93-98.

62. Forschungsheim, das Kirchliche (Wittenberg), **Die Erde ist zu retten: Umweltkrise, christliche Glaube, Handlungsmöglichkeiten** (hectographic document, 1982), p. 34. An earlier version of this report was published in Wensierski and Büscher, eds., **Beton ist Beton,** under the title "Kein Sonnenschein ohn' Unterlass," pp. 51-84. Where possible, I quote from the latter because of easier accessibility, but some changes introduced in the 1982 version are worthy of note. Another important report, written by a working group for the Commission "Church and Society" of the Federation of Evangelical Churches in the GDR, deals with the impact of agriculture on the environment. The report, entitled "Agrarwirtschaft und Umwelt," appeared in 1982. This and other "inner-church" documents are available in West Berlin at the Berliner Arbeitsgemeinschaft für Kirchliche Publizistik. The latter also publishes the serial, **Kirche im Sozialismus**, which reports on GDR church activities and publishes occasional excerpts from church reports.

63. Forschungsheim (Wittenberg), **Die Erde**, pp. 22-25. See also Peter Gensichen (head of the Wittenberg Research Facility), "Vor uns die Sinflut," **Glaube und Heimat**, Feb. 27, 1980; and Bernhard Lange, "Von klugen Menschen — oder Wie kann man leben?" **Die Kirche** (Berlin), Sept. 30, 1979.

64. Forschungsheim (Wittenberg), **Die Erde**, pp. 35-37.

65. Ibid., p. 27.

66. Heino Falcke, "Kritische Partizipation im realen Sozialismus," in Wensierski and Büscher, eds., **Beton ist Beton**, p. 198.

67. Heino Falcke, "Gerechtigkeit, ökologische Lebensfähigkeit und Participation," in ibid., p. 213.

68. Forschungsheim (Wittenberg), "Kein Sonnenschein," in ibid., pp. 59-61.

69. For example, in June 1979 the Theological Study Division (Theologische Studienabteilung) put out a hectographically-reproduced report with extensive documentation and commentary on issues relating to "The Environment — the Future — Faith."

70. The first of the "Briefe zur Orientierung im Konflikt Mensch-Natur" was produced in January 1980. By February 1983 seven letters had been issued. These hectographically-reproduced

documents are distributed by mail, although available materials and facilities apparently make it necessary to limit the circulation. Thus the letters are circulated from hand to hand. Contributions from readers are printed in the letters. An excerpt from the third letter is published as "Jahrzehnte zu spät . . .?" in Wensierski and Büscher, eds., **Beton ist Beton**, pp. 99-101.

71. See, esp., Forschungsheim (Wittenberg), "Kein Sonnenschein," in ibid., pp. 55-57, 74-84.

72. Götz Planer-Friedrich, "Schranken instrumenteller Vernunft," in ibid., pp. 123-28. See also the document from a working circle in Greifswald, "Bedürfnisbefriedigung als Basis der Macht," in ibid., pp. 129-37. See also Rudolf Bahro, **The Alternative in Eastern Europe** (London: Verso, 1978), pp. 272 ff.

73. Forschungsheim (Wittenberg), **Die Erde**, p. 39.

74. Jürgen Kuczynski and Hermann Klare, "Briefe," **Spektrum** (June 1973), p. 3. The Soviet analyst Zharikov agrees. He calls the Western ideal of a consumer society an "advertisement for the suicide of humanity" due to its ecological consequences. He believes that regulation of needs offers some hope for mollification of the situation, although the fundamental solution involves the restructuring of production. See his piece in Marakhov, **Nauchno-tekhnicheskaia revoliutsiia, chelovek**, p. 86.

75. See Forschungsheim (Wittenberg), "Kein Sonnenschein," in Wensierski and Büscher, eds., **Beton ist Beton**, pp. 64-66. See also Günter Krusche, "Gute Nachricht für die ganze Schöpfung," **Potsdamer Kirche**, Dec. 27/28, 1981; and "Ohne Rücksicht geht es nicht! " **Glaube und Heimat**, Oct. 3, 1982.

76. Forschungsheim (Wittenberg), **Die Erde**, p. 21.

77. See ibid., pp. 40, 43, where cooperation with the newly formed Society for Nature and the Environment is called for as well; and Falcke, "Gerechtigkeit," in Wensierski and Büscher, eds., **Beton ist Beton**, pp. 205-9.

78. Forschungsheim (Wittenberg), **Die Erde**, p. 32.

79. Forschungsheim (Wittenberg), "Kein Sonnenschein," in Wensierski and Büscher, eds., **Beton ist Beton**, pp. 68-69. See also Hans-Peter Gensichen, "Die Natur – nicht nur Dienerin unseres Magens," **Potsdamer Kirche**, Oct. 5, 1980, p. 1; and B. Seite, "Mach in mir deinem Geiste Raum," **Mecklenburgische Kirchenzeitung**, May 30, 1982.

80. Some proponents of such critiques have quite literally been forced into exile. The best example is Rudolf Bahro, who now is an outspoken activist in the Green Party in the Federal Republic of Germany.

81. Quoted by E. M. Sergeev, "Ratsional'noe ispol'zovanie geologicheskoi sredy," **Priroda**, no. 1 (1977), pp. 89-90.

82. Rutkevich and Shvarts, pp. 61-62. For a general discussion of the noosphere concept, see I. V. Kuznetsov, "Estestvo-

vanie, filosofiia, i stanovlenie noosfery," **VFil**, no. 12 (1974), pp. 128-38; and V. P. Kaznacheev, "Voprosy dialektiki prirody v rabotakh V. I. Vernadskogo," **VFil**, no. 8 (1981), pp. 100-109.

83. Rutkevich and Shvarts, p. 62.

84. I. D. Laptev, "Soznatel'noe i stikhiinoe vo vzaimodeistvii obshchestva i prirody," **VFil**, no. 7 (1976), p. 90. A. V. Bykhovskii and V. A. Los' note that the noosphere concept was not fully developed by Vernadskii and thus it does not have a unified meaning for those who use it. See their review of **Razmyshleniia naturalista. Prostranstvo i vremia v zhivoi i nezhivoi prirode** and of **Razmyshleniia naturalista. Nauchnaia mysl' kak planetnoe iavlenie** by V. I. Vernadskii, in **VFil**, no. 1 (1978), p. 174.

85. See, e.g., Laptev, pp. 88-89; Rutkevich and Shvarts, pp. 62-63; and A. G. Doskach, Iu. P. Trusov, and E. T. Faddeev, "Problema vzaimodeistviia prirody i obshchestva i sovremennaia geografiia," **VFil**, no. 4 (1965), p. 113.

86. See, e.g., Jutta and Günter Haase, "Die Mensch-Umwelt Problematik," **Geographische Berichte: Mitteilungen der Geographischen Gesellschaft der DDR** 16, no. 61 (1971): 254-55; and A. Bauer and Paucke, "Einheit und Kampf," p. 600. On the link to building communism see also Oehler, "Aktuelle Fragen," pp. 306-7.

87. G. F. Khil'mi, **Foundations of the Physics of the Biosphere** (Leningrad: Hydrometeorological Publishing House, 1967), p. 281. The book was originally published in Russian in 1966 as **Osnovy fiziki biosfery**.

88. G. F. Khil'mi, in "Global'nye problemy," **VFil**, no. 9 (1974), p. 82.

89. M. M. Kamshilov, "Chelovek i zhivaia priroda," **Priroda**, no. 3 (1969), p. 33.

90. M. M. Kamshilov sees the two terms as virtually synonymous. See ibid., p. 33. E. K. Fedorov and I. B. Novik, on the other hand, refer only to biotechnosphere in "Problemy vzaimodeistviia cheloveka s prirodnoi sredoi," **VFil**, no. 12 (1976), p. 57.

91. A. K. Astaf'ev, for example, uses systems analysis and does not refer to the noosphere or technosphere. See his "Sistemnyi podkhod i problema reguliatsii prirodnoi sredy," **VFil**, no. 2 (1977), pp. 71-72. The economist P. Oldak suggests a bioeconomics approach, applying systems analysis concepts, in "Chelovek i sreda" (round table), **VFil**, no. 2 (1973), p. 50 and "Novoe mesto prirody v ekonomicheskikh issledovaniiakh," **Priroda**, no. 2 (1973), pp. 36-41. B. P. Vysotskii criticizes the noosphere concept for inadequately reflecting the unity of interaction of social and natural elements in L. S. Abramov, "Problema vzaimodeistviia prirody i obshchestva i mesto geograficheskikh nauk v ee razrabotke," **VFil**, no. 2 (1965), p. 146.

92. E. V. Girusov notes that "in contemporary conditions the dependence of society on nature is not declining, but growing." "Metodologicheskie voprosy teorii vzaimodeistviia prirody i obshchestva," in Marakhov, **Nauchno-tekhnicheskaia revoliutsiia, chelovek**, p. 91. Siegfried Stoof reaches a similar conclusion, p. 590.

93. Kamshilov, "Chelovek i zhivaia priroda," pp. 30-33. For rejections of the idea see, in addition, I. K. Pletnikov, "Die Wechselwirkung von Gesellschaft und Natur und Probleme des Noosphäre," **DZfPh** 24 (1976): 152; and N. P. Fedorenko and N. F. Reimers, "Priroda, ekonomika, nauka," **Priroda**, no. 3 (1974), pp. 8-9.

94. Hegewald, p. 1459. See also Oehler, "Aktuelle Fragen," p. 306; and A. Bauer, "Probleme der Naturaneignung," p. 1328.

95. This view is not uncommon. See, e.g., Horst Paucke and Adolf Bauer, "Zum Verhältnis von Natur- und Produktionskreisläufen," **DZfPh** 28 (1980): 906; and the views of philosophers and biologists reported in M. B. Zykov et al., "Problemy global'noi ekologii," **VFil**, no. 1 (1981), pp. 165-68.

96. See, e.g., Zharikov in Marakhov, **Nauchno-tekhnicheskaia revoliutsiia**, p. 80; Zykov et al., p. 168; V. A. Anuchin, "Sootnoshenie obshchestva i prirody v geograficheskoi srede i filosofskie problemy geografii," **VFil**, no. 4 (1975), p. 87; G. Tsaregorodtsev, "Obshchestvo, okruzhaiushchaia sreda i meditsina," **VFil**, no. 2 (1975), p. 63; and A. E. Medunin, "Vliianie nauchno-tekhnicheskoi revoliutsii na prirody Zemli," **VFil**, no. 3 (1969), p. 24.

97. Rutkevich and Shvarts, "Filosofskie problemy," p. 61.

98. Kamshilov, "Chelovek i zhivaia priroda," p. 29.

99. William Ophuls, **Ecology and the Politics of Scarcity** (San Francisco: W. H. Freeman & Co., 1977), p. 2.

100. Shvarts, "Problemy ekologii cheloveka," pp. 107-9. Ophuls points out that in some instances increased productivity necessarily reduces ecosystem stability or raises energy consumption, pp. 58-59.

101. S. V. Kalesnik and F. F. Davitaia, "The Task of Soviet Geography in Supporting Economic Development During the 10th Five-Year Plan," **Materialy VI S"ezda Geograficheskogo Obshchestva SSSR**, translated in **Soviet Geography: Review and Translation (SGRT)** 17 (April 1976): 226. Iu. A. Izrael', the head of the recently created State Committee on Hydrometeorology and Control of the Natural Environment, has also noted the positive effects of human activity on the environment. For example, in 1979 he pointed to increased carbon dioxide as potentially allowing the formation of biological systems of high productivity. It is, however, difficult to say how ecologically cautious he is, since his public role may require him to make ecologically reassuring

statements. See Iu. A. Izrael', "K strategii okhrany prirodnoi sredy i ratsional'nogo prirodopol'zovaniia v SSSR," **VFil**, no. 6 (1979), pp. 119-20.

102. Zharikov, in Marakhov, **Nauchno-tekhnicheskaia revoliutsiia, chelovek**, p. 84.

103. I. B. Novik, "Metodologicheskie aspekty optimizatsii biosfery," **Priroda**, no. 9 (1972), p. 8.

104. Novik, "Budushchee nauki," p. 110. See also Rychkov, in "Chelovek i sreda" (round table), **VFil**, no. 4 (1973), p. 73; and G. I. Tsaregorodtsev in ibid., no. 3 (1973), p. 66.

105. Medunin, "Vliianie," pp. 28, 32.

106. A notable exception is the exchange of letters between Günter Kunert (a pessimist) and Wilhelm Girnus (an optimist). See Kunert and Girnus, "Anlässlich Ritsos." A more typical optimistic position is found in Herbert Hörz, "Globale Probleme," pp. 1308-9.

107. Adolf Bauer and Horst Paucke, "Umweltprobleme in der Sicht des historischen Materialismus und bürgerlicher Weltmodelltheorien," **DZfPh** 24 (1976): 783-84. The more recent article by Bauer and Paucke is equally interesting ("Einheit und Kampf"). See also Haase and Haase, "Die Mensch-Umwelt Problematik," pp. 254-55.

108. Wolfgang Harich, **Kommunismus ohne Wachstum? Babeuf und der 'Club of Rome'** (Reinbek bei Hamburg: Rowohlt, 1975). See also the interview, "Ich hatte alle gegen mich," **Der Spiegel**, June 11, 1979, pp. 70-76. Harich emigrated from the GDR in 1979 but reportedly maintains friendly relations with the SED and is free to return.

109. Harich, **Kommunismus**, pp. 49, 70.

110. Ibid., p. 52.

111. Forschungsheim (Wittenberg), **Die Erde**, pp. 29-30.

112. See, e.g., the readers' letters in the seventh "Briefe zur Orientierung im Konflict Mensch-Natur," put out by the Forschungsheim (Wittenberg) (Feb. 1983), p. 19.

113. From Marianne Bruns, **Der grüne Zweig** (Halle, Leipzig: Mitteldeutscher Verlag, 1979), p. 81, quoted by Knabe, "'Der Mensch mordet sich selbst,'" p. 963.

114. Kunert, in Kunert and Girnus, p. 851.

115. Ibid.

116. Ibid., p. 853.

117. Ibid., p. 864.

118. Compare government measures to control use of gasoline with church-supported "Mobil ohne Auto" actions. On government controls see Michael Mara, "Renaissance des Pferdefuhrwerks," **Tagesspiegel** (West Berlin), June 28, 1983. On measures initiated by church leaders to reduce use of gasoline by church workers see "Bischöfe wollen weniger Benzin verbrauchen," **Evangelischer Pressedienst** (Landesdienst Berlin), no. 86

(May 8, 1984).

119. E.g., see Büscher and Wensierski, **Null Bock auf DDR**, pp. 41-42, 57.

120. Compare the 1980 document put out by the working group at the Forschungsheim (Wittenberg) (published as "Kein Sonnenschein" in Wensierski and Büscher, eds., **Beton ist Beton**, pp. 51-84) with the 1982 version, **Die Erde**, in which more prominence is given to the link between the ecology and peace issues, esp. pp. 16-17. See also Hans-Dietrich Spengler, "Friedensdienst und Umweltverantwortung," **Die Kirche** (Magdeburg), Nov. 11, 1981. In "Menschen über Bäume," **Mecklenbergische Kirchenzeitung**, Nov. 21, 1982, there is reference to a lecture by Gensichen linking the issues, with a note that October 1982 was the first time this theme was addressed at a tree-planting action.

121. Seweryn Bialer, **Stalin's Successors: Leadership, Stability, and Change in the Soviet Union** (Cambridge, ENG: Cambridge University Press, 1980), p. 55.

122. On loyal dissenters see David Lane, **Politics and Society in the USSR** (New York: New York University Press, 1978), pp. 247-51.

123. See Bahro, **The Alternative**, pp. 134-35. See also Ithiel de Sola Pool, "Communication in Totalitarian Societies," in Ithiel de Sola Pool et al., eds., **Handbook of Communications** (Chicago: Rand McNally College Publishing Co., 1973), pp. 474-75.

# 4
# Science, Scientists, and Environmental Research

Environmental problems challenge not only basic assumptions about the nature of progress and the "good life." They also reveal deficiencies in our knowledge of the world and in the structure of scientific research. Many Soviet scientists agree that Soviet science is ill-prepared to meet the ecological challenge. The inadequate level of scientific knowledge is explained in part, these critics argue, by social factors: high military expenditures drain resources both from environmental research and from practical implementation of scientific advances; the seriousness of environmental deterioration is not recognized; and social priorities emphasizing short-term production goals rather than long-term effects of production are mirrored in scientific research programs.[1] Other problems are, however, rooted in the very structure of the scientific community; the fragmentation of science and its disciplinary specialization are inadequate to address complex problems of environmental protection which impinge on many fields of knowledge.[2] As ecological proponents say, "everything is connected to everything." If this is so, science itself must also become better integrated if it is to help stave off ecological crisis.

The ecological problem has thus contributed to a rethinking of the organization of scientific research, and this rethinking has been reflected in statements by top party leaders. In 1976 at the 25th Party Congress of the CPSU, Leonid Brezhnev pointed out that "new possibilities for fruitful research of both a theoretical, basic, and applied character are being opened up at the boundary between sciences, specifically between natural and social sciences. These possibilities ought to be utilized to the fullest extent."[3] Behind this seemingly innocuous statement, however, lies a long-standing conflict within Soviet Marxist circles about the proper relationship between social processes and natural processes, and thus between the social and natural sciences.

Indeed, the organization of scientific research may reflect an understanding of the relationship between different types of knowledge, and thus is a reflection of important elements of the predominant worldview. Probably the clearest example of this is the existence of the discipline of political-economy in the Soviet bloc, whereas in the West we generally have two separate disciplines, namely, political science and economics. The definition of disciplinary boundaries and relationships in each society reflects the prevailing interpretation of the relationship between politics and economics. The way in which scientific disciplines are perceived is particularly important in the Soviet bloc because the authorities emphasize the scientific nature of their worldview and the potential which science, joined with the insights of Marxism-Leninism, offers in suggesting solutions to pressing policy problems.

Many Soviet scientists are concerned that, despite gains made in particular disciplines, fragmented knowledge affords only limited guidance for environmental policy-making. "Only separate elements of the process" can be influenced and "consequently the results are rather limited."[4] V. A. Anuchin, a controversial geographer whom we will encounter later in this chapter, captures well the sentiment of many Soviet scientists. In a tone familiar to Western academics, he bemoans the absence of synthesizers among an abundance of specialists: "We have lived to see the end of university education. Universities exist only formally."[5] Such statements have both practical and theoretical implications, suggesting a need to readjust both traditional conceptions and the practical jurisdictions of existing scientific disciplines. In this chapter we are, therefore, concerned with the manner in which the environmental dilemma has affected the Soviet and GDR understanding of the relationship between different types of knowledge and the proper organization of research. As scholars have sought to produce an appropriate solution, has Marxism-Leninism offered guidance? As we shall see in the first part of this chapter, disciplinary training and institutional structures have proven more influential here, while Marxism-Leninism seems to offer little practical or theoretical guidance. Only in the geography discipline, which we examine at the end of the chapter, have issues of scientific integration been joined with the theoretical concerns of Marxism-Leninism.

## THE SEARCH FOR A FRAMEWORK: DISCIPLINARY COMPETITION AND COOPERATION

Scientists in the USSR disagree about the appropriate disciplinary and conceptual framework for synthetic environmental research. (This disciplinary competition is much less visible in the GDR.) The Soviet philosopher Novik, for example, has argued that an "ecologized physics" should provide the conceptual basis for study of environmental problems. To accomplish this a thermodynamic conception of the biosphere and of physical climatology must be developed, according to Novik. "Physicalization of research on the biosphere" will involve "finding out the tendencies of development of an ecological situation from the fundamental laws of physics."[6] Furthermore, "the conceptual structure of physics will also change in the direction of a more organic integration of the laws of conservation of physics and a systems-evolutionary approach. . . ."[7] Within this scheme, Novik feels that ecology can fill the gap until physical-mathematical sciences are adequately "ecologized" to take on the role they more naturally (**zakonomerno**) fulfil.[8]

Other scientists argue that ecology itself must become the leading scientific discipline in the near future. According to Medunin, in the late 19th and early 20th centuries physics was the leading science, since it provided the basic methods to serve as the foundation for other sciences. Presently, he argues, the social sciences play the leading role, since they can address the most critical contemporary issues, such as the arms race. In the future, however, the most important tasks will relate to complex (interdisciplinary) environmental problems. Ecology should form the core of this complex research.[9] Medunin's understanding of ecology apparently involves a broadening of the traditional Soviet understanding of the field, which referred to that branch of the biological sciences dealing with "the study of the relations of individual plants to their environment."[10] As in the West, the term "ecology" (**ekologiia**, **Ökologie**, and particularly the adjectival forms of these words) has come to take on an ever-expanding meaning in the USSR and GDR. Giliarov, Vinberg, and Chernov, for example, define ecology as "a system of scientific disciplines studying life on the level of organization above that of the organism."[11] Terms such as social ecology, human ecology, and global ecology, widely used in the USSR, reflect a further broadening, involving ecology in the study of interconnections between society and natural surroundings. As Gerasimov points out, in the USSR it is already impossible to limit the term "ecology" to its narrow biological sense, since the word is widely applied in the broader sense.[12]

Some Soviet scientists see ecology, in this broader sense, emerging as an independent scientific discipline, a new science capable of producing methods to guide environmental research in other disciplines as well.[13] Other scientists, most notably the head of the Institute of Geography of the AN SSSR, I. P. Gerasimov, reject the notion that ecology (in the broad sense) is an independent science capable of integrating and guiding environmental research. Rather, ecology is a general scientific approach like cybernetics or systems analysis. As such, the ecological approach should, according to Gerasimov, provide an orientation for the scientific disciplines proper (physics, biological sciences, chemistry, economics, etc.), revealing "the links existing between the object of study of this and that science, and its environment."[14] Gerasimov and other geographers insist that geography should play the leading role in the environmental area. Not only is study of the environment the subject of geographical research, but geography has experience with interdisciplinary research, standing as it does at the junction of the social and natural sciences. In 1976 Gerasimov and Preobrazhenskii even suggested that "it is obvious that the problem of the protection and improvement of the environment ought to be, and also in the short and long-term perspective is, the main subject of fundamental development in the geographical science."[15] Gerasimov does not deny the role of other disciplines, but suggests that geographic research must be supplemented by: (1) geological research, dealing, for example, with mineral resources and earthquakes; (2) biological research on genetic problems and human health; (3) technological studies to produce low-waste technology; and (4) social and economic scientific investigations to consider policy issues.[16] Soviet geographers are insistent, however, that geography maintain the leading role in providing the synthetic framework to integrate these various perspectives. In addressing "the danger of ecological crisis and the gravity of optimal resource use and environmental protection" T. V. Zvonkova and the late Iu. G. Saushkin, then chairpersons of the Departments of Physical Geography and of Economic Geography, respectively, at Moscow University, exemplify this sentiment. They write: "Virtually all scientific disciplines have now become involved in this problem. This is all very well, but it seems to us that the International Geographical Congress should make it plain that this is essentially a geographical problem. . . ."[17]

Other scientists are sceptical of the ability of geography to tackle the task of environmental synthesis. The Soviet economist Loiter suggests that geographers over-dramatize environmental problems.[18] Conversely, some geographers express alarm that it should have been "economists, of all people" who sponsored the 1978 Moscow conference on the Man-Nature Relationship.[19] Some economists and other social scientists suggest the establishment of the field of bioeconomics to "study the productive environment – to study the relationship between rates of growth, level of technology, and the quality of the environment."[20] Using systems analysis, Oldak tries to demonstrate the applicability of this type of modelling to analysis of the optimal use of Lake Baikal.[21]

Oldak rejects the simple coordination of scientific disciplines as inadequate, because each scientific field would continue to conduct research based "on various points of view, with various concrete goals, with various methods."[22] This would inhibit scientific development in a situation where time and resources are limited and problems are urgent. Oldak favors the creation of a new department, specifically established to analyze problems of nature utilization from a systemic bioeconomic perspective. Such a department would include highly trained economists, ecologists, and systems analysts, and consult with other experts as well. And while "bioeconomics is interdisciplinary, formed at the junction of a number of sciences, we think it possible to consider it an immediate development out of one of the directions of economic research."[23]

Los' presents yet another perspective, namely that it is no longer scientific disciplines which set the tone of research, but problem areas. The problem of the biosphere, he suggests, is undoubtedly assuming a leading role; thus the "ecologizing" of the various scientific disciplines is increasingly important.[24] Although not explicitly stated, this seems to be the perspective of some GDR scholars. The extensive disagreement over the disciplinary division of labor is not visible in the GDR. Indicative of this is the emphasis placed by some East German geographers on the interdisciplinary nature of the problem.[25] Haase and Haase do attribute a potentially unique position to geography, since it contains both social and natural science perspectives.[26] But, like other GDR geographers, they emphasize that a focus on the "territorial aspect" of the problem is appropriate for the geographical sciences. L. Bauer suggests that the geography discipline must establish limiting values for the capacity of the natural environment to tolerate stress.[27] All of these geographers are careful to respect the complex, interdisciplinary nature of the environmental problem. One should not overemphasize the contrast between the GDR and USSR here. Most Soviet scholars also want to retain the specific insights of particular disciplines. But there is much more visible controversy over the theoretical framework which should unify the complex research, and this controversy grows out of differing theoretical perspectives, in part arising from differences between disciplines. In the GDR this conflict may also exist, but it has not surfaced as an important issue of open theoretical dispute.

The diversity of approaches to the study of environmental problems may seem a bit surprising, given the unifying rubric which Marxist-Leninist theory is supposed to provide. It is only in the Soviet geography discipline, however, that debates over disciplinary jurisdiction are conducted in terms of Marxist-Leninist categories. And here, as we shall see later in this chapter, geographers seem to have had an important influence in shaping the Party's articulation of Marxist-Leninist guidelines. The fragmentation of Soviet science and limited intrusion of Marxist-Leninist doctrine into disciplinary competition is a clear indicator of the increasing independence of science from direct political intrusion. The weakness of attempts to develop an integrated conceptual framework for environmental research has the merit of permitting a diversity of approaches, a first step if science and scientists are to become forces furthering the move toward "institutional pluralism."[29]

## PRACTICAL COORDINATION: FRAGMENTATION AND SCIENTIFIC LOBBIES

Scientists can agree on some practical measures to address scientific fragmentation. Many call for better coordination of research, most typically through the Academy of Sciences, and propose the establishment of central or specialized research institutes to facilitate synthetic research related to environmental problems.[30] Most also support expanded training of specialists to handle environmental aspects of production at all levels of the economy, as well as inclusion of related courses for post-secondary students in technical, economic, and biological fields.[31] In recent years some educational programs in both the GDR and USSR have been adapted or initiated to reflect the ecological dimension. For example, a department (**kafedra**) for the economics of nature-use (**prirodopol'zovanie**) has been developed at Moscow State University and at the Moscow Institute of the National Economy (MINKh).[32] In the GDR "Environmental Protection (selected problems)" has become an area of post-graduate study, and at the Technical University of Dresden engineering and natural science programs familiarize students with the foundations of waste-free and low-waste technology. In the same area a new discipline, ecology engineering, is developing in the GDR.[33]

Coordination of research is, however, more difficult than educational reform. In actual practice interdisciplinary research is scattered and unsystematically integrated. It is estimated that over three hundred research establishments and university departments in the USSR are conducting research dealing with biological aspects of nature conservation, yet this work remains uncoordinated, plagued by departmental barriers.[34] In the late 1970s an All-Union Institute on Nature Conservation and Nature Preserves was opened under the jurisdiction of the Ministry of Agriculture in the USSR. Its tasks involve improvement of the system of nature preserves, provision of information on the preserves, and publication of the updated Red Book on endangered fish and fauna.[35]

There is no interdisciplinary institute in the USSR which deals comprehensively with environmental problems. Some individual scholars with diverse expertise collaborate regularly. A notable example is the joint efforts of N. F. Reimers, a biologist, and economists associated with the Central Mathematical-Economics Institute (TsEMI) in the USSR.[36] Multi-disciplinary conferences and symposia offer opportunities for interdisciplinary interaction as well. In January 1978, for example, the first Soviet Conference on the Man-Nature Relationship was held at Moscow University and other interdisciplinary conferences are occasionally reported in scholarly journals or in the popular media.[37] The Academy of Sciences is a key force in environmental research in both the GDR and USSR. Numerous individual institutes stand out (e.g., in the USSR, the Institute for Geography of Siberia and the Far East, the Central Mathematical-Economics Institute, and union republic bodies such as the Institute of the Economics of Industry of the Voroshilov filial of the Ukrainian Academy of Sciences;[38] in the GDR, e.g., the Research Agency for Chemical Toxicology, the Institute of Economics, and the Institute of Geography and Geoecology[39]). These Academy of Sciences' institutes have a specialized disciplinary focus, however. Thus in both the GDR and USSR coordinating bodies have been established.

In the USSR coordination of interdisciplinary research efforts is facilitated by organizations such as the Council of Ministers' State Committee on Science and Technology (GKNT), the Academy of Sciences' Scientific Council on Problems of the Biosphere, and the State Committee on Hydrometeorology and Control of the Environment (Goskomgidromet, formerly the Hydrometeorological Service). In 1972 these three organizations, along with others, were charged with developing long-term prognoses on the condition of the environment.[40] The State Committee on Science and Technology (GKNT), created in 1965, is primarily concerned with applied research and not with basic scientific investigation. In May 1973 the Interdepartmental Council on Complex Problems of the Environment and Rational Use of Nature, attached to GKNT, was established to coordinate research,[41] and in 1974 the GKNT's Interdepartmental Council on Science and Technology was charged with coordination of scientific research and design work, involving utilization of natural resources and protection of the environment.[42] Although GKNT is not a part of the Academy of Sciences the membership of the Interdepartmental Council on Science and Technology is approved by the Academy as well as by the GKNT. It includes biologists, foresters, medical workers, technical specialists, and city planners. It is not clear how important a role these interdepartmental councils have played in coordinating research in institutions not under their direct jurisdiction, nor what the link between the various councils is. Another organization of GKNT and the Soviet Academy of Sciences, the All-Union Scientific Research Institute for Systems Research (VNIISI), has been given responsibility for coordinating a whole complex of research on modelling of global development and on systems approaches. It seems that several recent articles in **Voprosy filosofii** have been generated by this program, under the leadership of D. M. Gvishiani.[43] In the late 1970s GKNT and the Presidium of the Academy of Sciences of the USSR, with the participation of the State Planning Commission (Gosplan) and the State Construction Committee, also developed a "Comprehensive Program of Scientific and Technical Progress and its Social and Economic Consequences." The program provided the basis for identifying research topics to be pursued by economics institutes in the 1981 to 1985 period, including treatment of environmental issues.[44]

The Soviet Academy of Sciences' Scientific Council on Problems of the Biosphere, headed by A. V. Sidorenko, is drawing up programs of research for the coming years dealing with such diverse problem areas as the complex use of water resources, the genetic consequences of pollution, prognosis of climatic changes, and development of mathematical modelling methodologies.[45] The Council also has sub-units, including one dealing specifically with socio-philosophical problems of the biosphere, headed by I. T. Frolov. Among the tasks set out for this section are examination of the methodological bases of ecological modelling; the place of ecological problems in contemporary global problems; the integration of the scientific, natural and technical sciences; and critical analyses of bourgeois socio-ecological conceptions. The membership of this particular section of the Council includes philosophers, biologists, at least one systems theorist, a geographer, and others.[46] Iu. A. Izrael', head of Goskomgidromet, identifies the Scientific Council on Problems of the Biosphere as the coordinating body for research carried out by Academy institutes, Goskomgidromet, and other establishments. The Council must, however, apparently rely on existing institutes and organizations to realize its programs. Some scientists point out the inadequacy of such an arrangement (although not specifically referring to this Scientific Council); they note that the success of synthetic research depends on adequate financial resources and personnel, in effect, on an independent institutional setting.[47]

Soviet coordination problems are illustrated well in the Siberian context. The extreme climatic conditions of Siberia produce a highly sensitive ecosystem susceptible to disruption by even minimal human intervention. Massive development projects such as the Baikal-Amur Mainline (BAM) and the creation of territorial production complexes for the exploitation of natural resources threaten to induce severe ecological imbalances. The

Siberian Division of the Soviet Academy of Sciences (SO AN SSSR), an early and outspoken defender of Lake Baikal, has a long-standing record of ecological awareness. In recent years the Siberian Division has developed a long-term scientific program on the "Complex Development of the Natural Resources of Siberia" (Sibir').[48] The Sibir' program focuses on use of mineral, land and biological resources, as well as on the development of particular geographic regions such as the areas surrounding BAM, the Kuzbass, Lake Baikal, and the Kansk-Achinsk complex. One of more than thirty specialized programs in the Sibir' project involves "Protection of the Environment and Rational Use of Natural Resources in Siberia," and the environmental theme runs through other specialized programs as well (many of which, for example, focus on particular regions). The environmental program has succeeded in involving 32 scientific sub-units of the SO AN SSSR, 25 organizations of other departments and 1400 individuals, including 60 scientists with doctoral degrees and 300 candidates of sciences. Some scientists complain, however, that there is not one specialized nature-protecting subdivision in the SO AN SSSR. Although the Presidium of the Iakutsk filial of the SO AN SSSR has created a division on nature-protection, it can fulfil mainly scientific-organizational functions since it consists of only seven individuals. Nonetheless the division did succeed in becoming a coordinating point for work on "Complex Schemes for Nature Protection in the Iakutsk ASSR." On July 7, 1979, the Presidium of the SO AN SSSR passed a resolution requiring all scientific sub-units to take measures for the organization of fundamental research on the environment and on rational utilization of nature. Nonetheless, additional financing, cadres, and organizational support are needed to make the effort successful. In an attempt to marshal such support, the Presidium forwarded its resolution to the Council of Ministers of the USSR, Gosplan, GKNT, and the Presidium of the Academy of Sciences of the USSR.

East German scientists, particularly geographers, also emphasize the need for interdisciplinary research.[49] Here again the Academy of Sciences of the GDR (AdW DDR) plays an important role. In April 1971 a special section (**Klasse**) on "Optimal Development of Environmental Conditions (Humans and the Environment)" was set up under the leadership of Edgar Lehmann, a prominent geographer. The section involves regular meetings between scholars of various disciplinary backgrounds, mainly natural scientists, in which diverse topics related to the environment are discussed.[50] In addition, in April 1974 a Scientific Council for Environmental Research was formed with Hans Mottek, a prominent economist, as chair. The Council was charged with development of a unified research strategy.[51] In the early 1970s coordination of geographical research on **Landeskultur** was centered in the Geography Institute of AdW DDR in Leipzig, involving participation of representatives from all significant geography departments in higher educational institutions.[52] In 1976 the Institute of Geography and Geo-ecology (AdW DDR) was formed by uniting the environmental research efforts of the former geography institute with those of other Academy bodies. The Institute is now a major research agency on environmental topics (Forschungstelle Umweltgestaltung) specifically focusing on the geo-ecological basis for measures related to **Landeskultur** and on socially determined aspects of territorial structures.[53] Outside of the Academy of Sciences, in March 1982 a Center for Environmental Development (Zentrum für Umweltgestaltung) was formed out of the Institute for Water-Economy, to serve as a coordinating point for research and development work on environmental protection, with an emphasis on the application of scientific advances in the production process.[54] Since 1973, the Technical University in Dresden has also furthered interdisciplinary coordination of environmental research carried on in its various sections. The result has been numerous publications integrating natural scientific insights with economic analysis.[55] Other research projects in the GDR are commissioned by the Ministry for Environmental Protection and Water Management. The existence of a single government ministry with responsibility for environmental protection would seem an important advantage in the GDR.[56] Furthermore, research integration is undoubtedly much easier in the GDR than in the USSR simply because the country is geographically more compact and the number of institutes and scientists involved is much smaller. Interdisciplinary conferences also further contact between scholars in the GDR, as in the USSR.[57]

Despite the inadequacy of specialized research institutes in both the USSR and GDR, the above discussion should suggest that environmental proponents have developed quasi-institutional bases for environmental lobbying. In 1976, Donald Kelley identified "informal ad hoc coalitions" as the main means through which scientists (other than those working in state-related environmental protection agencies) could attempt to influence environmental policy.[58] As Thane Gustafson has pointed out, however, in the 1960s environmentalists "began to form institutes and training programs, mobilize data, publish journals, and gain established channels of participation."[59] As a result, the environmental lobby has become progressively less ad hoc (focusing on particular cases like Lake Baikal) and more institutionally based. Certain research institutes in the USSR and GDR (including those discussed above) have consistently furthered environmental research and represented an environmentalist perspective. These institutes are not strictly speaking institutional lobbies for environmental protection, in that generally only particular groups of scholars devote their attention to ecological problems and the organization as such may or may not take an official stand. Nonetheless, scholars in these institutions can often use their organizational positions to support these efforts and to lend credence to their findings. They can also use organizational publications to disseminate their views more broadly in the scientific community. In this sense they may be considered quasi-institutional interest groups. A notable recent example of an environmentally sensitive publication emerging from such an institutional base was the founding in 1980 of the journal **Geography and Natural Resources (Geografiia i prirodnye resursy)** by the Institute of Geography of Siberia and the Far East of the SO AN SSSR. Although fundamentally directed toward "stimulating the development of geographical works having paramount theoretical, methodological and economic meaning," a clear focus of the journal is the ecological issue. The editors indicate that

> . . . the processes of exploitation of natural resources are engendering ever more wide and far-reaching changes in the environment of man. They have a universal character, appearing on the local, regional, and global levels. Therefore the development of scientific programs, not only to eliminate the possibility of future ecological crises, but also to assure improvement of socialist nature-use, is becoming a basic matter for geographical research.[60]

The journal has an all-union orientation and has already attracted articles from diverse parts of the USSR. Other Soviet journals inspired by rising ecological concern include **Ecology (Ekologiia)** (initiated in 1970), which has a largely biological-technical focus, and **Nature and Man (Priroda i chelovek)**, a popular science magazine. Unfortunately it is not possible to point to even one new East German publication which has emerged from rising ecological consciousness in the past two decades. Indeed, since 1983 publication of some outlets for discussion (e.g., **Forum, Die Wirtschaft**) has ceased, reportedly due to a paper shortage. Since direct contact is a much more viable means of interchange of information in the GDR than in the USSR, this need not indicate fewer possibilities for scientific lobbying in the GDR. On the other hand, lobbying efforts in both countries would almost surely be more effective if environmental research had a unified institutional base.

## INTERNATIONAL COOPERATION

East German and especially Soviet scholars express enthusiastic support for international research in the environmental field. On an official level, the Soviet leadership also expresses readiness "to participate, together with other interested states, in the solution of such problems as the conservation of the natural environment, the development of power engineering and other natural resources. . . ."[61] Because of the GDR's exclusion from the 1972 UN conference on the environment (held in Stockholm) other East European countries, including the USSR, boycotted that gathering. Since the diplomatic recognition of the GDR by the FRG that obstacle has been removed and both the GDR and USSR have participated in international conferences and concluded bilateral agreements with capitalist nations in the environmental area. It is well-known that the Soviet and GDR leaders desire scientific exchange with and technology transfer from the West.[62] However the leadership is also wary of the ideological impact of free intellectual dialogue between East and West and of free travel to the West on the part of their scientists. Heavy restrictions remain on that type of free interchange.[63] Nonetheless both Soviet and East German scholars participate in international seminars and research projects such as those sponsored by the International Institute for Applied Systems Analysis in Laxenburg, Austria, and the Institute of Life in Paris.[64] In August and September of 1977 representatives of the Soviet scientific community also had the opportunity to consult with members of the Club of Rome.[65] The UNESCO project on "Man and his Biosphere," SCOPE (the Scientific Committee for Problems of the Environment, part of the International Council of Scientific Unions) and projects of the United Nations' Economic Commission for Europe (ECE) receive support in both the GDR and USSR.[66] Cooperation within COMECON has also been formalized in special environmental programs at least since 1974.[67]

Soviet scientists appear to use official support for international cooperation as a springboard for demands for expanded international contacts. By contrast, East German scholars' calls for East-West cooperation are more restrained. Joint research within the Soviet bloc and praise of Soviet initiatives receive more attention.[68] East German scholars are more careful than Soviet scholars to include caveats that environmental agreements with capitalist countries are not "apolitical," but rather represent a way of furthering peaceful coexistence and detente.[69] Soviet scholars demonstrate an impressive familiarity with Western environmental writings and exhibit an intense interest in the approaches taken in Western environmental research.[70] In discussing the need for increased international cooperation, some

Soviet scholars appear temporarily to forget their ideological scruples and imply the possibility of research which is system-neutral, in the social as well as the natural scientific and technical fields. A striking example is Ananichev's suggestion that international global modelling, along the lines of the Club of Rome studies, be carried out with Soviet participation, and not representing specific interests.[71] Khozin endorses a proposal of American scientists for establishment of a world center on environmental cooperation:

> The existence of such a center would allow the development of the basis for a new political ideology, which would take account of the latest changes in the system of social values called forth by the growing concern among ever wider circles of the public in many countries with the problem of the worsening quality of the environment.[72]

This is indeed a radical political statement, by anyone's account, which is tempered only moderately by Khozin's observation that the USSR should provide the guiding principles for such a center, based on its experience with central planning. He urges a business-like and practical approach.[73]

The call for international cooperation extends also to such prominent Soviet figures as the late world-renowned physicist Petr Kapitsa and I. T. Frolov (former editor-in-chief of **Voprosy filosofii**). Kapitsa acknowledged that some measures may come into contradiction with the national interests of particular countries, but urged them nonetheless – a suggestion implying a movement toward binding international law. He urged scientists to encourage the diversion of military expenditures to the solution of global problems.[74] Frolov likewise considers that international cooperation in this area may further "a general rapprochement of peoples and the strengthening of peace in the world," based on a concrete, scientific, and practical approach.[75] Oldak urges the development of an international division of labor and cooperation "in the stream of knowledge which is created by world science."[76] Soviet scientists strongly identify themselves as part of an international scientific community and show more than scientifically indifferent support for increased contact with the West.

## "UNIFIED GEOGRAPHY" AND THE ENVIRONMENT

As noted above, geographers in the Soviet Union claim a central role for their discipline in the emerging field of environmental studies. Many Soviet geographers fear, however, that the geography discipline is ill-prepared to seize the moment. Long-standing debate among geographers about the interrelationship between sub-divisions within geography have come to center stage as geographers seek to clarify their role as synthesizers of environmental research.. As A. A. Mints and V. S. Preobrazhenskii pointed out in 1975, geography is in a good position to become a leader in the environmental field since geography is, by tradition, "an interdisciplinary science with high potential for integrated research."[77] They caution, however, that as topics become of widespread concern, they tend to slip away from geography as it falters in developing the necessary "conceptual field." This is because geography has not generated a framework for synthesis of analytical data, reflected in the fact that the very definition of "geographicness" remains unresolved.[78]

Our purpose in this section is to demonstrate the interconnection between theoretical issues of long-standing dispute in the Soviet geography discipline and approaches to the study of environmental problems. Despite some evidence of similar conflicts in the geographic discipline in the GDR, their relevance to the environmental issue is not so clear.[79] East German geographers, including Günter and Jutta Haase, Gerhard Schmidt-Renner, Heinz Lüdemann, and Edgar Lehmann have been actively involved in environmental research and consultation. It seems, however, as noted above, that there is greater unanimity on the centrality of territorial studies in East German geographic approaches to environmental problems.[80]

Examination of the Soviet geography discipline is important because this field has historically played a significant role in policy deliberation in the USSR. The great territorial expanse and diversity of the Soviet Union make geography particularly important for economic development policy. In the present period adequate raw material and energy supplies for Soviet European industrial centers can only be assured through exploitation of the rich resource deposits of Siberia. Similarly, water supply problems in the Soviet Union stem from unequal distribution of water resources rather than from an absolute shortage.[81] The spatial dimension of resource allocation thus commands considerable attention in the Soviet Union, particularly as easily accessible resources in the more heavily populated western part of the country approach depletion. Knowledge of the peculiarities of different geographical regions is essential for optimal solution of these locational aspects of natural resource use. For this reason Soviet economic and political leaders may turn to geographers for practical advice. Some Soviet geographers complain that their skills are too infrequently called upon by expert commissions dealing with projects involving transformation of nature.[82] Nonetheless geographers are a vocal and potentially influential group in terms of the evolution of attitudes toward environmental problems in the Soviet Union. Individual geographers, for example, I. P. Gerasimov, head of the Institute of Geography of the Academy of Sciences, have been highly visible figures in the environmental field.[83]

Conflict within the Soviet geography discipline is also interesting because of its relationship to Marxism-Leninism. Questions regarding the causal interrelationship between changes in the natural and social environments have been self-consciously addressed by geographers, since their discipline stands at the crossroads of the social and natural sciences. The role of natural factors in social development is an issue of established concern to Soviet Marxist theoreticians and politicians. In the Stalinist period official doctrine affirmed the non-identity of natural laws and social laws, and asserted the primary impact of productive relations (as opposed to natural factors) on social development. This tenet was in turn reflected in the organizational and theoretical separation of physical geography from economic geography in the USSR. For, according to Stalinist doctrine, the physical world is governed by one set of laws and the economic world by another. Therefore it would be inappropriate to mix the two spheres in one scientific discipline. With the publication of a controversial book, **The Theoretical Problems of Geography**, by geographer V. A. Anuchin in 1960, dispute over the relationship between economic and physical geography was reignited, but this time in the context of increasing awareness of the threat of ecological crisis.[84]

Anuchin's book argued for a "unified geography," in contrast to the prevailing position of the leadership of the Institute of Geography of the Soviet Academy of Sciences which continued to support the separation between physical and economic geography. Physical geography was viewed by the leadership of the Institute as a natural science and economic geography as a social science, different laws governing each branch. Anuchin's "unified geography" was rejected in the early 1960s by leading figures in the Institute, including its head, Gerasimov. Only after considerable travail were Anuchin's works approved as fulfilling requirements for his doctoral degree. Controversy over Anuchin's position continued into the late 1970s, as Anuchin offered new support for his position in another book, **Fundamentals of Nature-Use**, published in 1978.[85] Although the dispute over Anuchin's writings is by no means primarily focused on treatment of environmental deterioration, nevertheless, it has affected the attitude of geographers to the study of environmental problems. It also offers an example of the role of Marxist-Leninist concepts in such disciplinary conflicts.

Anuchin's argument centered around his definition of the geographical environment to include both nature and human society.[86] With the increasing impact of human economic activity on the natural environment, natural and societal factors become increasingly intertwined, creating a "humanized nature." According to Anuchin, any attempt to separate study of the physical world from the effects of human activity on nature is bound to produce unsatisfactory results.[87] In 1965, **Literaturnaia gazeta** published several articles in which geographers discussed the view of geography fostered within the Academy of Sciences' Institute of Geography. Here Anuchin repeated his argument for the concept of a unified geographical environment and discussed the practical importance of "synthetic geographical studies":

> Progress in the development of mathematics, physics, and chemistry has armed mankind with powerful means for influencing nature. Through the use of modern technology people change the course of rivers, radically transform the soil and vegetation cover, improve climate, and level the contours of the land. But these powerful tools must be correctly applied: the consequences of our influencing nature must be foreseen.[88]

Other economic geographers, in particular Iu. G. Saushkin of Moscow University, supported the core of Anuchin's position, as did the editors of **Literaturnaia gazeta.**[89] In his 1965 article, Anuchin charged his opponents with misunderstanding Marx, also a key argument in his earlier book. As Anuchin was to elaborate in his 1978 book, Marxism affirms the unity of the material world and sees human beings as a part of nature, governed in important respects by the same material laws.[90] In his 1965 article Anuchin associated the present leadership of the Institute with Stalin's so-called "voluntaristic" attitude toward nature which had led to well-known debacles, attempts at "transformations of nature" which had no scientific basis. Under Stalin, "the development of society on earth was regarded as isolated from the development of nature, and vice versa."[91] With this statement, by implication, Anuchin linked those wishing to maintain the split in Soviet geography with the Stalinist approach. This was a somewhat powerful indictment. For the Stalinist view of the geographical environment had been officially rejected by a prominent ideological spokesperson, L. F. Il'ichev, in 1963. Il'ichev also gave support to the integration of natural and social scientific approaches, specifically referring to the geography discipline, while rejecting Anuchin's formulation of a "unified geography" in some points. The rejection of the Stalinist view, frequently referred to in the Soviet press as "voluntarism," demanded recognition of the impact of natural laws on societal development and the impact of societal development on nature.[92] This official recognition of limits on the malleability of nature has formed an important foundation for arguments of environmental defenders in the Soviet Union.[93] The American analyst Ian Matley suggests that Il'ichev's proclamation was the result of "the gradual evolution of these ideas among a group of geographers," namely Anuchin, Saushkin, and Baranskii, the dean of Soviet economic geographers. Matley supports this conclusion by pointing out formulations in Il'ichev's proclamation taken virtually verbatim from Anuchin's earlier writings.[94] While Il'ichev expressed disagreement with Anuchin's view on some points, overall his pronouncement lent strength to Anuchin's case, vis-a-vis the leadership of the Geography Institute.

Early supporters and opponents of Anuchin recognized the importance of environmental problems for the future of the geography discipline. Already in the early 1960s Gerasimov, an opponent of Anuchin's ideas, had expressed strong support for the protection of Lake Baikal.[95] In 1966 he noted "a crisis in the science of geography." Refuting those who saw its real function as exhausted, Gerasimov pointed to new tasks, namely, providing a "full-fledged general science theory" for forging a "fundamentally new constructive-transformative attitude" towards the environment.[96] In 1965 Anuchin's distinguished supporter, Iu. Saushkin, also identified geography as the key discipline in this area, despite its lack of preparedness. He specifically noted the importance of Anuchin's book, explaining that "it is sometimes said that ideas are generated by social necessity," presumably referring to the developing ecological situation. He went on to relate it to geography's role.

> The natural and the engineering sciences, which open the way for new methods of influencing the geographical environment, have been developing at lightning pace. There has also been great progress in the social sciences, which deal with problems of the future course of mankind and the prospects of the economy. At the boundary between the natural and social sciences lies geography, which focuses on problems of development and the future use of the environment. What other branch of knowledge, if not geography, should be concerned with the study of the geographical environment of human society and the key question: What will happen to the environment in the immediate and more distant future?[97]

At about the same time that Saushkin and Gerasimov were stressing the role of geography in environmental studies, a conference was held in Moscow on "the interaction of nature and society and some problems of contemporary geography."[98] In addition to discussion of substantive topics, the participants became engaged in a heated conflict over the "unified geography" concept. Some geographers emphasized the need for a general theory, but rejected both the attempt to tear the physical and economic sciences apart, as well as the attempt to merge them.[99] Doskach suggested a toning down of the controversy over unified geography; instead, study of the nature-society interaction could provide a methodological pivot to clarify the relationship between both geographical sciences.[100] The larger theoretical conflict surrounding Anuchin's thesis had become an important source of disagreement over the appropriate approach to the study of environmental problems. The controversy was, however, viewed by some geographers as exaggerated in proportion to the importance of the substantive issues at stake. One element fuelling the controversy may well have been long-standing rivalry between the Academy's Institute of Geography (headed by Gerasimov) and the geography division at Moscow University, where Saushkin headed the Department of Economic Geography, and with which V. A. Anuchin was also associated.[101]

The controversy continued into the late 1960s and 1970s. While not endorsing the idea of a unified geographical science, by 1967 Gerasimov acknowledged the "overall methodological problem of properly combining and interrelating a progressive specialization or differentiation of knowledge with the need for synthesis or integration."[102] He affirmed the need for

> a proper proportioning of geographic research between the primary characteristics of the physical environment and the changes produced by man as a result of his activities, and, on the other hand, . . . a well grounded evaluation of the role of natural factors in any particular aspect of population settlement, man's economic activities or various aspects of the economy.[103]

Throughout this period, Saushkin and Anuchin continued to support a unified geographical approach to environmental problems. They attributed particular importance to the development of "territorial-production complexes" and to the reuse of waste products in national economic planning.[104]

In 1971, an opponent of "unified geography," Isachenko, tried to associate the Anuchin-Saushkin position with the bourgeois approach of American geographers.[105] Isachenko stressed the interdisciplinary character of the issue, defining geography's appropriate concern as follows:

> One aspect of the relationship – the impact of human society on natural complexes (geosystems) – would fall within the province of physical geography, and the other –

> the impact of natural complexes on the formation and evolution of spatial economic systems – would be investigated by economic, or to be more exact, socio-economic geography. This would cover all the geographical aspects of the problem. And since the man-nature interplay does not result in a new "mixed" material object ("man-nature"), the investigation of the interplay would not lead to the merging of physical and economic geography.[106]

Anuchin replied in an article in **Priroda** (Nature), a journal, like **Literaturnaia gazeta**, known for its outspoken environmental advocacy. He suggested that without a general geographic approach "technological progress may degenerate into technological rapacity, and man will inflict irreparable harm to the environment. . . ."[107]

In 1975 the Geography Institute of the Academy of Sciences, with the full participation of Gerasimov, published **Man, Society, and the Environment.**[108] This was a significant step by the Institute in developing its role as synthesizer in the environmental sphere. The book took no stand on "unified geography" and, as a collection of articles, did not attempt to synthesize approaches within the geography discipline itself. The participation of economic geographers indicated an attempt to produce some sort of integrative work, however. **Man, Society, and the Environment** focused in various chapters on a detailed analysis of issues such as natural disasters; air, water, and thermal pollution; resource cycles; urbanization; and an unimaginative socio-political analysis of the roots of the present environmental situation. Indeed the book was ideologically "orthodox" enough to be translated into foreign languages and put out as part of a Progress Publisher's series on current problems. Nevertheless the book stirred unusual controversy. A review published in the journal of the Geographical Society of the USSR charged that the authors had failed to provide a clear picture "of the specific role that geography can play as an organizer of research in this area."[109] Furthermore, emphasis on "the transformation of nature on a global scale" is a throw-back to the "nature-transforming zeal of a well-known period," in other words, the Stalinist era. The "usurping approach to nature" is evidenced, the review continued, by the neglect of appropriate discussion of nature conservation and of cultural landscaping; specifically, the works of S. V. Kalesnik and A. G. Isachenko (among others) had been overlooked.[110] These, of course, were the outspoken opponents of the "unified geography" concept. The tables had been turned, and Gerasimov's efforts were described from the opposite side as resembling the earlier Stalinist approach. The intensity of debate over the book indicated that older issues were resurfacing as geography sought the correct approach to the society-nature problem.[111]

The conflict over a "unified geography," although somewhat tempered, continued to occupy Soviet geographers in the late 1970s.[112] Although Gerasimov and even some of Anuchin's most ardent critics acknowledge the need for cooperation between the economic and natural sciences, their opposition to merging economic and physical geography remains strong. In 1978 Isachenko called "unified geography" "simply a myth in the heads of some theorists" which "rests on shaky ground." Indeed, "no one has thus far been able to offer a sound theoretical basis for it or to perform a concrete piece of research in its spirit."[113] On the other end, Saushkin and Zvonkova acknowledge that physical and economic geography each have "stable cores" which come into interaction most effectively in the resolution of applied problems. Such interaction, they assert, assumes "particular importance in light of the danger of ecological crisis." Furthermore "the differentiation of geography . . . is beginning to lead into a dead end, both in theoretical and in practical terms. What is needed now is a greater interplay between physical and economic geography."[114] This statement differs little in tone from Saushkin's earlier position in support of Anuchin's ideas.[115] Other geographers are still disturbed that the geography discipline exhibits a kind of "theoretical and organizational helplessness . . . which prevents it from making full use of the favourable circumstances."[116] The difficulties confronting the geography discipline in attempting to relate natural and social scientific perspectives may well express, in microcosm, the larger problems of interaction within Soviet science as a whole.

We will not go into further detail here as to the dimensions and implications of this controversy, for the issue extends beyond our range of concern. Two points are of importance to us. First, the controversy provides an example of how broader disagreements over scientific theory may have important implications for the study of the society-nature interrelationship. Indeed the interconnections are so complicated that it is difficult to unravel them. We will try to summarize some of them briefly. Regarding official party policy, Anuchin's influence was important in shaping the terms of the rejection of the Stalinist approach to the geographical environment. This ideological shift, as noted, has in itself been a buttress for environmental proponents in the USSR. Il'ichev's official proclamation and additional pressure placed on Gerasimov to alter the approach within the geographical discipline derived in part from Anuchin's writings. The Presidium of the Academy of Sciences (presumably at the behest or at least with the approval of the USSR Council of Ministers) urged that more attention be given to economic geography.[117] As the American geographer Hoosen suggests, the interest shown by

Anuchin and Saushkin in territorial-production complexes and in the economic effects of geographical study may have been one factor behind this pressure and behind the favorable reception offered the unified geographical approach by the Soviet leadership.[118] The adaptability of Gerasimov, from a staunch opponent of Anuchin to an advocate of a "constructive" integrated (if not "unified") geography, indicates his responsiveness to these cues from the leadership. To be sure, in the early 1960s Gerasimov was already voicing concern over damage to Lake Baikal. Nonetheless, most likely the Geography Institute's increased emphasis on its integrative role in the study of environmental problems was in part a result of the theoretical discussion initiated by Anuchin. This inference is supported by the fact that Anuchin and Saushkin themselves linked the need for a unified geography to the threat of ecological crisis. The support offered Anuchin by **Literaturnaia gazeta** and his use of **Priroda** as a mouthpiece are consistent with this conclusion. In any case, it is clear that what was at stake in the theoretical conflict was, at a minimum, the **manner** in which the society-nature relationship should be studied, if not also the call for a more explicit focus on that issue by geographers.

The controversy between Anuchin and his opponents was phrased in terms of their different understandings of Marxist methodology, and in particular, of the nature of determinism. The parties disagreed over the relationship between social and natural laws, and between social influences and natural influences. Anuchin's branding of his opponents as "indeterminists" and "voluntarists"[119] and the attempts of his opponents to associate him with "bourgeois" American scientific approaches ("bourgeois determinism") signal the centrality of Soviet Marxist categories in establishing the legitimacy of one's position. Likewise, both sides endeavoured to associate their opponents with the officially-repudiated Stalinist understanding of the geographical environment ("voluntarism"). Anuchin implied that his adversaries underestimated the mutual interconnections between society and nature, a view which potentially might lead to projects which are not based on a scientific understanding of their effects. Anuchin's opponents, on the other hand, charged that an integrated approach might overestimate human ability to transform the environment, when actually more attention should be given to issues such as nature conservation. These opposing assessments of society's ability **scientifically** to transform nature are explained by both sides as manifestations of differing theoretical understandings of social and economic causality. Each then labels the other's as ideologically incorrect.

## CONCLUSION

Examination of conflict between natural and social scientists over the appropriate method for studying environmental problems demonstrates that disagreement over approaches to environmental research is substantial and that this disagreement cannot be separated from larger theoretical controversies within the academic community. These disagreements stem in part from differences in the background and training of the various analysts. Within the geography discipline, as a specific example, the larger theoretical conflict was closely related to differences in perspective between physical and economic geographers.

Only in the geography discipline does correct interpretation of Marxism-Leninism appear to affect controversies over the appropriate scientific approach to environmental problems. It is difficult to assess whether geographers marshal Marxist-Leninist tenets primarily to legitimize their positions or whether a genuine dispute over Marxist-Leninist theory is at stake. The highly polemical nature of the argumentation suggests the importance of the former, but most likely the participants confuse the two motivations themselves. For scholars outside the geography discipline Marxist-Leninist categories seem to be of little relevance as they seek to find an integrative framework for environmental research.

## NOTES

1. See P. G. Oldak, in "Nauka i global'nye problemy sovremennosti" (round table), **Voprosy filosofii** (hereafter **VFil**), no. 8 (1974), pp. 110-14; Hans Mottek, "Zu einigen Grundfragen der Mensch-Umwelt Beziehung," **Wirtschaftswissenschaft** 20 (January 1979): 43; V. A. Anuchin, in "Nauka i global'nye problemy," **VFil**, no. 8 (1974), p. 121; M. M. Kamshilov, "Printsipy organizatsii zhivoi prirody," **Priroda**, no. 2 (1969), p. 28; and G. S. Gudozhnik, **Nauchno-tekhnicheskaia revoliutsiia i ekologicheskii krizis** (Moscow: Mezhdunarodnye otnosheniia, 1975), pp. 185-86.

2. See, e.g., L. S. Abramov, "Problema vzaimodeistviia prirody i obshchestva i mesto geograficheskikh nauk v ee razrabotke," **VFil**, no. 2 (1965), p. 149; Ludwig Bauer, "Zu einigen Problemen und Aufgaben der Landeskultur und Umweltgestaltung," **Geographische Berichte: Mitteilungen der Geographischen Gesellschaft der DDR** (hereafter **Geographische Berichte**) 17, nos. 64-65 (1972): 178; Hans Mohs et al., "Forschung und Wissenschaftsentwicklungen an der Sektion Geographie der Martin-

Luther Universität Halle-Wittenberg," **Geographische Berichte** 19, no. 72 (1974): 192; Horst Paucke and Adolf Bauer, "Zu einigen Prinzipien der Naturnutzung im Sozialismus," **Deutsche Zeitschrift für Philosophie** (hereafter **DZfPh**) 26 (1978): 1033; and G. A. Aksenenok, in "Vzaimodeistvie estestvennykh i obshchestvennykh nauk na sovremennom etape" (round table), **VFil**, no. 10 (1973), p. 47.

3. Quoted by S. B. Lavrov and B. S. Khorev, review of **Osnovy prirodopol'zovaniia: Teoreticheskii aspekt**, by V. A. Anuchin, in **Izvestiia Vsesoiuznogo Geograficheskogo Obshchestva**, no. 6 (1979). Translated in **Soviet Geography: Review and Translation** (hereafter **SGRT**) 21 (April 1980): 238. Originally from **Materialy XXV S"ezda KPSS** (Moscow, 1976), p. 72.

4. A. G. Doskach, Iu. P. Trusov, and E. T. Faddeev, "Problema vzaimodeistviia prirody i obshchestva i sovremennaia geografiia," **VFil**, no. 4 (1965), p. 105.

5. V. A. Anuchin, in "Nauka i global'nye problemy," **VFil**, no. 8 (1974), p. 123.

6. I. B. Novik, "Budushchee nauki i ekologicheskaia problema," **VFil**, no. 1 (1975), pp. 112-13.

7. Ibid., p. 113.

8. I. B. Novik and V. Los', review of **Chelovek, obshchestvo i okruzhaiushchaia sreda**, by I. P. Gerasimov, ed., in **VFil**, no. 4 (1974), p. 173.

9. A. E. Medunin, in "Nauka i global'nye problemy," **VFil**, no. 9 (1974), pp. 76-77.

10. V. Sukachev and N. Dylis, **Fundamentals of Forest Biogeocoenology** (Edinburgh: Oliver & Boyd, 1964), p. 16.

11. M. S. Guliarov, G. G. Vinberg, and Iu. I. Chernov, "Ekologiia-zadachi i perspektivy," **Priroda**, no. 5 (1977), esp. pp. 3-6, 10-11. They clearly relate ecology to the solution of environmental problems. Another term, biogeocoenology, generally refers to the study of the interaction of non-living and living units within material and energy cascades. See M. N. Rutkevich and S. S. Shvarts, "Filosofskie problemy upravleniia biosfery," **VFil**, no. 10 (1971), p. 60 for a definition of **biogeotsenoz**.

12. I. P. Gerasimov, "Metodologicheskie problemy ekologizatsii sovremennoi nauki," **VFil**, no. 11 (1978), pp. 66-67. In the same article Gerasimov discusses the evolution of the idea of ecology, pp. 62-76. N. M. Mamedov also discusses social ecology and global ecology in "Global'nye aspekty ekologicheskogo vzaimodeistviia," **VFil**, no. 11 (1981), pp. 74-77.

13. See the views of A. P. Markevich and N. N. Kiseev in the report by V. A. Los' and E. T. Faddeev, "Chelovek, obshchestvo i priroda v vek NTR," **VFil**, no. 12 (1981), p. 53.

14. Gerasimov, "Metodologicheskie problemy," pp. 67, 72; and G. I. Tsaregorodtsev and V. G. Epokhin, "Ekologiia cheloveka

v sisteme meditsinskogo znaniia," **VFil**, no. 9 (1980), pp. 74-86. The latter provide an elaborate analysis of the problem, concluding that the ecological approach provides a synthetic systemic viewpoint which can supplement existing perspectives within traditional scientific disciplines.

15. I. P. Gerasimov and V. S. Preobrazhenskii, "Osnovnye zadachi sovetskoi geograficheskoi nauki v svete reshenii XXV S"ezda KPSS," **Izvestiia Akademii Nauk, seriia geograficheskaia**, no. 4 (1976), p. 10.

16. Gerasimov, "Metodologicheskie problemy," pp. 70-72; and Gerasimov and Preobrazhenskii, p. 14.

17. T. V. Zvonkova and Iu. G. Saushkin, "Interaction Between Physical and Economic Geography," **Vestnik Moskovskogo Universiteta, geografiia**, no. 3 (1976), trans. in **SGRT** 18 (April 1977): 250. See B. S. Khorev, "Ekologicheskaia politika i rol' geograficheskoi nauki," **Geografiia i prirodnye resursy**, no. 1 (1981), pp. 177-79, for a view of the roles of geography, economics, and ecology.

18. M. N. Loiter, **Prirodnye resursy i effektivnost' kapital'nykh vlozhenii** (Moscow: Nauka, 1974), p. 13. See also Novik and Los' for a criticism of emphasis on the central organizing role of the geographical sciences in the environmental sciences, pp. 172-73.

19. Lavrov and Khorev, p. 238.

20. P. G. Oldak, in "Chelovek i sreda" (round table), **VFil**, no. 2 (1973), p. 50.

21. P. G. Oldak, "Novoe mesto prirody v ekonomicheskikh issledovaniiakh," **Priroda**, no. 2 (1973), pp. 36-41. See also Gudozhnik, **Nauchno-tekhnicheskaia revoliutsiia**, pp. 199 and 202, for a discussion of the approach. N. P. Fedorenko and N. F. Reimers also make use of the bioeconomics approach. See their "Priroda, ekonomika, nauka," **Priroda**, no. 3 (1974), pp. 3, 10, 13. Others favor application of a systems approach as a general synthetic method to integrate insights of the specific sciences. See, e.g., G. M. Gvishiani, "Nauka i global'nye problemy sovremennosti," **VFil**, no. 3 (1981), p. 106; and N. N. Moiseev, "Aktual'nye voprosy ekologicheskoi evoliutsii i 'mirovaia dinamika' Dzheia Forrestera," **VFil**, no. 7 (1978), p. 179. For many of these scientists the systems approach is closely associated with development of mathematical models of ecological processes.

22. Oldak, "Novoe mesto," p. 41.

23. Ibid., pp. 34, 40-41. See also his discussion in **Sovremennoe proizvodstvo**, pp. 6-7.

24. V. A. Los', in "Chelovek i sreda," **VFil**, no. 4 (1973), pp. 75-76.

25. This view, of course, resembles that of Gerasimov, although GDR geographers don't emphasize the leading role of the

geography discipline.

26. Günter Haase attempts to clarify the respective roles of the natural and social sciences in the study of natural resource use. See "Zur Ableitung und Kennzeichnung von Naturpotentialen," **Petermanns Geographische Mitteilungen** 122 (1978): 113-25. See also Dieter Graf's comments in "Naturpotentiale und Naturressourcen – Bemerkungen aus ökonomischer Sicht," **Petermanns Geographische Mitteilungen** 124 (1980): 53-57. See also Jutta Haase and Günter Haase, "Die Mensch-Umwelt Problematik: Gedanken zum Ausgangspunkt und zum Beitrag der geographischen Forschung," **Geographische Berichte** 16, no. 61 (1971): 243-70.

27. Ibid., p. 249. See also Ludwig Bauer, "Zu einigen Problemen und Aufgaben der Landeskultur," pp. 178-79; Mohs et al., "Forschung und Wissenschaftsentwicklung," pp. 191-92; and Heinz Lüdemann, "Perspektiven der geographischen Forschung," **Spectrum**, no. 9 (1981), pp. 11-13.

28. L. Bauer, "Problemen und Aufgaben der Landeskultur," p. 179.

29. The term is used here in the sense discussed by Jerry Hough in **The Soviet Union and Social Science Theory** (Cambridge, MA: Harvard University Press, 1977), pp. 23-24

30. See, e.g., Aksenenok, "Priroda i chelovek," p. 62; and K. Sytnik, "The City's Breath," **Izvestiia**, March 23, 1976, trans. in **Current Digest of the Soviet Press (CDSP)**, vol. 28, no. 12, p. 7. In the GDR see Adolf Bauer, "Probleme der Naturaneignung," **DZfPh** 30 (1982): 1325; and Horst Paucke, "Globale Probleme und der Menschheitsfortschritt," **DZfPh** 30 (1982): 1507.

31. A. G. Bannikov, in "Nauka i global'nye problemy," **VFil**, no. 8 (1974), p. 112; Gudozhnik, **Nauchno-tekhnicheskaia revoliutsiia**, p. 186; and N. P. Namurov, in "Chelovek i sreda," **VFil**, no. 3 (1973), p. 56.

32. See P. M. Nesterov, in "Aktual'nye voprosy ekonomiki prirodopol'zovaniia" (round table), **Voprosy ekonomiki**, no. 2 (1982), p. 92. For a discussion of the term "prirodopol'zovanie" see Iu. P. Mikhailov, "Geograficheskie grani protsessa prirodopol'zovaniia," **Geografiia i prirodnye resursy**, no. 3 (1980), pp. 159-65; and Ian Matley, review of **Theoretical Problems of Geography**, by V. A. Anuchin, in **SGRT** 19 (October 1978): 578.

33. Helmar Hegewald, "Moralische Triebkräfte bei der rationellen Nutzung der Naturressourcen im Sozialismus," **DZfPh** 30 (1982): 1467; Manfred Schubert, **Zur Entwicklung abproduktarmer bzw. -freier Verfahren in Einheit von Umweltschutz, Material- und Energieökonomie**, Sitzungsberichte der Akademie der Wissenschaften der DDR, Mathematik-Naturwissenschaften-Technik, no. 21N (Berlin: Akademie Verlag, 1982), p. 23; and Karl-Franz Busch, "Probleme der sozialistischen Umweltgestal-

tung," **Wissenschaftliche Zeitschrift der Technischen Universität Dresden** 32 (1983): 99. See also Ilka Norhara-Schnabel, "Zur Entwicklung der Umweltpolitik in der DDR," **Deutschland Archiv** 9 (1976): 823-24.

34. A. Iablokov, "Every Facet is Priceless," **Pravda**, Sept. 24, 1983, trans. (excerpts) in **CDSP**, vol. 35, no. 38, p. 23.

35. Interview with E. Syroechkovskii, "Knowing How to Protect Nature," **Izvestiia**, March 30, 1979, trans. (condensed) in **CDSP**, vol. 31, no. 13, pp. 26-27; and D. Brezhnev, "The Eternal Values of Flora," **Pravda**, June 5, 1980, trans. (condensed) in **CDSP**, vol. 32, no. 23, pp. 7, 14.

36. See, e.g., N. P. Fedorenko, M. Ia. Lemeshev, and N. F. Reimers, "Sotsial'no-ekonomicheskaia effektivnost' okhrany prirody," **Priroda**, no. 10 (1980), pp. 2-13; and N. P. Fedorenko and N. F. Reimers, "Sblizhenie ekonomicheskikh i ekologicheskikh tselei v okhrane prirody," **Priroda**, no. 9 (1981), pp. 3-12. Both of these articles are in a popular science journal. It is not clear how much actual research is carried on jointly by these individuals.

37. For reference to the Moscow conference see Lavrov and Khorev, p. 237. For a report on a conference involving biologists and philosophers see M. B. Zykov et al., "Problemy global'noi ekologii," **VFil**, no. 1 (1981), pp. 165-68. For a brief report on a symposium on "Natural Scientific Environmental Problems" held in Leipzig in January 1981 and involving about 200 scientists, doctors, and practitioners see **Neues Deutschland**, Jan. 15, 1981, p. 2.

38. This is the institute which put out the book edited by N. G. Chumachenko and L. A. Belashova, **Aktual'nye problemy okhrany okruzhaiushchei sredy** (Kiev: Naukova Dumka, 1979). Its reviewer, E. E. Drozbovskii, refers to the institute as the leading organization in the Ukrainian Republic conducting socio-economic research on the problem of environmental protection. See his review in **Geografiia i prirodnye resursy**, no. 3 (1981), pp. 146-48. The other Soviet institutes mentioned here will be dealt with in the pages that follow or in chapter 5. Additional examples could be offered. For example in 1970 a new Institute of Animal and Plant Ecology of the USSR Academy of Sciences was opened. See V. Sintsov, "Laboratory of the Far North," **Izvestiia**, May 29, 1970, trans. in **CDSP**, vol. 22, no. 22, p. 18.

39. H. G. Riedel and K. L. Enders refer to the Forschungstelle für Chemische Toxikologie der AdW der DDR in "Ökonomische und betriebswirtschaftliche Aspekte technischer Umweltschutzmassnahmen im industriellen Bereich Energie- und Chemiewirtschaft," in **Ausgewählte Probleme der Leitung und Planung des Umweltschutzes**, Freiberger Forschungshefte D102 (Leipzig: VEB Deutscher Verlag für Grundstoffindustrie, 1976), p. 39. Norhara-Schnabel indicates that in 1976 over 50 research agencies

were involved in some aspect of environmental research, in addition to research carried on in large enterprises, pp. 823-24. Other institutes mentioned in the text will be dealt with more specifically in the following pages or in chapter 5.

40. See Iu. A. Izrael', "Fundamental'nye issledovaniia po okhrane i uluchsheniiu prirodnoi sredy i ratsional'nomu prirodopol'zovaniiu," **Vestnik Akademii Nauk SSSR**, no. 7 (1980), p. 108. Gudozhnik also refers to joint work between government organs and the Soviet Academy of Sciences to develop a scientific-technical prognosis of possible changes in the biosphere and their effects on the national economy in the next 20 to 30 years. He does not designate the specific organizations involved (**Nauchno-tekhnicheskaia revoliutsiia**, p. 189). Likewise Efremov refers to the role of GNKT's Scientific Council on Science and Technology in the development of such prognoses. L. Efremov, "Planet's Health," **Izvestiia**, June 5, 1974, trans. (excerpts) in **CDSP**, vol. 26, no. 23, pp. 30-31. The dissident writer Boris Komarov (pseud.) refers to prognoses ("Nature 1980," "Nature 1990") developed by special commissions involving prominent figures such as Iu. A. Izrael', I. P. Gerasimov, and N. P. Fedorenko. These reports, according to Komarov, have not been made public. I have also never found a public report issued by the groups which Izrael' and Gudozhnik discuss. It is certainly possible, although at this point impossible to verify, that the forecasts Izrael' and Gudozhnik refer to are Komarov's "Nature 1980" and "Nature 1990." If this were so it would be evidence to support the authenticity of Komarov's work. See Komarov, **Destruction of Nature in the Soviet Union** (White Plains, NY: M. E. Sharpe, 1980), p. 19 and other pages listed in his index under "Nature 1980" and "Nature 1990."

41. See Loren R. Graham, "The Place of the Academy of Sciences in the Overall Organization of Soviet Sciences," in **Soviet Science and Technology: Domestic and Foreign Perspectives**, edited by John R. Thomas and Ursula M. Kruse-Vaucienne (Washington, DC: National Science Foundation, 1977), pp. 45-47 for a discussion of the State Committee on Science and Technology and its relationship to the Academy of Sciences. Chumachenko and Belashova, eds., pp. 242-43, and P. Oldak refer to this Interdepartmental Council. See P. G. Oldak, **Sovremennoe proizvodstvo i okruzhaiushchaia sreda** (Novosibirsk: Nauka, sibirskoe otdelenie, 1979), p. 125.

42. L. Efremov, "Planet's Health," p. 30.

43. See I. T. Frolov, "Filosofiia global'nykh problem," **VFil**, no. 2 (1980), p. 28; and D. M. Gvishiani, "Nauka i global'nye problemy sovremennosti," **VFil**, no. 3 (1981), p. 97. See also chapter 3, page 101, note 7.

44. N. Fedorenko, "Long-Range and Current Problems in

Economics," **Voprosy ekonomiki**, no. 7 (1979), trans. (abstract) in **CDSP**, vol. 31, no. 33, pp. 1-2.

45. Izrael', "Fundamental'nye problemy," pp. 105-8.

46. For a report on the first meeting of the group see V. A. Los', "Sotsial'no-filosofskie problemy biosfery," **VFil**, no. 9 (1980), pp. 168-69. Participants at that meeting included Frolov, I. B. Novik, A. V. Bykhovskii, G. S. Gudozhnik, G. S. Khozin, A. G. Doskach, N. F. Reimers, and E. V. Girusov.

47. E.g., I. V. Luchitskii, V. V. Vorob'ev, and V. D. Ermikov, "Kompleksnaia programma 'Sibir'' i okhrana okruzhaiushchei sredy," **Geografiia i prirodnye resursy**, no. 2 (1980), pp. 8, 13.

48. The discussion below is based on ibid., pp. 3-13.

49. E.g., Günter Jacob, "25 Jahre Geographische Gesellschaft der Deutschen Demokratischen Republik," **Petermanns Geographische Mitteilungen** 123 (1979): 148.

50. Norhara-Schnabel, p. 824. More specific information on the nature of the **Klasse** was gained from an interview with Edgar Lehmann in Leipzig, July 1980. Norhara-Schnabel refers to another interdisciplinary commission for environmental protection set up at the end of 1972. She does not, however, provide detailed information on its nature or purpose, nor have I found subsequent reference to it elsewhere. See p. 824.

51. **Neues Deutschland**, April 27, 1974.

52. Mohs et al., "Forschung und Wissenschaftsentwicklungen," pp. 190-91. The Academy's Institute of Geography was formed only in the late 1960s or in 1970. See Heinz Lüdemann, "Geographische Forschung in der DDR — Entwicklung und Perspektiven," **Petermanns Geographische Mitteilungen** 124 (1980): 97-103, for a discussion of the institute's development and focus.

53. Ibid. Lüdemann points out that the leading research agency for territorial planning (Forschungsleitstelle für Territorialplanung) is attached to the State Planning Commission, but it is in close contact with nearly all geographical institutes in the GDR, p. 100. Information on the Institute of Geography and Geoecology is also based on the author's interview with Heinz Lüdemann in Leipzig in July 1980. Dr. Lüdemann is the institute's director.

54. Interview with Eberhard Stief, "Zentrum für Umweltgestaltung," Berliner Rundfunk (Aug. 10, 1982), reported in **RIAS Monitor** (Aug. 11, 1982), pp. 6-11.

55. On the program at the Technical University Dresden see Karl-Franz Busch, pp. 99-100. See also the numerous articles on ecological issues in **Wissenschaftliche Zeitschrift der Technischen Universität Dresden** 32 (1983): 101 ff.

56. Norhara-Schnabel, p. 823. For example, the ministry concluded an agreement with the Bergakademie Freiberg dealing with research and education in the area of environmental protec-

tion and water economy. See **Neues Deutschland**, April 15, 1981, p. 16.

57. E.g., in December 1983 the fourth Central Scientific Conference of students and young scientists on "Environmental Protection and Environmental Development" was held at the Bergakademie Freiberg, with an emphasis on interdisciplinary work. See **Das Hochschulwesen**, no. 5 (1984), pp. 136-37.

58. Donald R. Kelley, "Environmental Policy-Making in the USSR: The Role of Industrial and Environmental Interest Groups," **Soviet Studies** 28 (October 1976): 578-79.

59. Thane Gustafson, **Reform in Soviet Politics** (Cambridge, ENG: Cambridge University Press, 1981), p. 50.

60. "Ot redaktsionnoi kollegii," **Geografiia i prirodnye resursy**, no. 1 (1980), p. 3. On the Institute of Geography of Siberia and the Far East see L. D. Eliseeva, "Institutu geografii Sibiri i Dal'nego Vostoka 20 let (1959-1979)," **Geografiia i prirodnye resursy**, no. 2 (1980), pp. 184-85.

61. L. I. Brezhnev, "The Report of the CPSU Central Committee to the 24th Party Congress of the CPSU," **Pravda**, Mar. 31, 1971, trans. in **CDSP**, vol. 23, no. 12, p. 13. For the GDR position see "Umweltschutz zum Nutzen der Völker," **Leipziger Volkszeitung**, Nov. 23, 1973, p. 5.

62. E.g., on cooperation between the FRG and GDR see Wilhelm Bruns, "Deutsch-deutscher Umweltschutz. Notwendig und möglich! " **DDR Report** 17, no. 2 (1984): 70-73.

63. See John R. Thomas and Ursula M. Kruse-Vaucienne, "Toward Understanding Soviet Science and Technology: An Overview," in Thomas and Kruse-Vaucienne, pp. xxvi-xxxii. For a vivid description of restrictions on this type of scientific interchange see Zhores A. Medvedev, **The Medvedev Papers** (London: Macmillan, St. Martin's Press, 1971), especially pp. 116-54.

64. On the Institute of Life see N. N. Moiseev, "Koevoliutsiia cheloveka i biosfery v vek comp'iuterov," **Vestnik Akademii Nauk SSSR**, no. 3 (1982), pp. 90-97. The Soviet and East German Academies of Sciences are members of the International Institute for Applied Systems Analysis (IIASA). IIASA describes itself as a non-government, multi-disciplinary research institute and sponsors studies of "global" issues such as energy and food supplies. IIASA's office is located at A-2361 Laxenburg, Austria. A list of participants in a recent IIASA forum on "Global Modelling at the Service of the Decision-Maker" (September 14-18, 1981) included several participants from the GDR and USSR. Scholars frequently praise IIASA. See, e.g., D. M. Gvishiani, "Methodological Problems in Modeling Global Development," **VFil**, no. 2 (1978), trans. in **Soviet Studies in Philosophy** 17 (Fall 1978): 14-15.

65. A brief report is contained in A. I. Prigozhin, "Rimskii Klub pered sotsial'nymi problemami," **VFil**, no. 5 (1978), pp. 164-

71. The report does not indicate which Soviet scientists were present other than D. M. Gvishiani. Komarov writes in his book: "The leadership of the Committee for Science and Technology and the Academy of Sciences may invite members of the Club of Rome to unannounced meetings in Moscow and have affable conversations with them. This changes nothing in the course of our ecological policy and most likely will continue to change nothing" (p. 108). Gvishiani is deputy head of GKNT and director of VNIISI.

66. The Soviet popular science magazine **Priroda** has given consistent attention to the "Man and His Biosphere" project and in fact devoted an entire issue (no. 9, 1981) to articles related to the project. See also D. M. Gvishiani and V. E. Sokoleva, eds., **Programma 'chelovek i biosfera' v stranakh sotsializma** (Moscow: VNIISI, 1979). For reference to SCOPE see Moiseev, "Koevoliutsiia cheloveka i biosfery," p. 95; and Lüdemann, "Geographische Forschung," p. 102. On the ECE see Schubert, **Zur Entwicklung abproduktarmer Verfahren**, pp. 27-28.

67. See B. Gorizontov and V. Prokudin, "Environmental Protection in COMECON Member Nations," **Voprosy ekonomiki**, no. 4 (1978), trans. in **Problems of Economics** 21 (December 1978): 24-37.

68. For calls for international cooperation see, e.g., Bernd Bittighöfer, Herbert Edeling, and Hans Kulow, "Theoretische und politisch-ideologische Fragen der Beziehung von Mensch und Umwelt," **DZfPh** 20 (1972): 61; and Alfred Kosing, "Mensch-Gesellschaft-Umwelt," **Einheit** 31 (1976): 801. For a less ideological approach see Erhard Schuster, "Charakter und Stellung der Umweltprobleme und einige sich daraus ergebende Konsequenzen," **Wissenschaftliche Zeitschrift der Technischen Universität Dresden** 31 (1982): 212. For references to Soviet examples see Adolf Bauer and Horst Paucke, "Umweltprobleme in der Sicht des historischen Materialismus und bürgerliche Weltmodelltheorien," **DZfPh** 24 (1976): 791, 798; and Klaus Müller, "Subjektiver Faktor und Zusammenhang von Gesellschaft und Natur," **DZfPh** 22 (1974): 592, 596.

69. See, for example, Alfred Bönisch and Dieter Graf, "Mensch am Wendepunkt," **Deutsche Aussenpolitik** 20 (1975): 1025-26.

70. The discussion of the Club of Rome publications provides numerous examples of this. See chapter 6. Other reviews of Western literature include I. P. Gerasimov and I. V. Komar, "Future Environments of America," **Izvestiia Akademii Nauk SSSR, seriia geograficheskaia**, no. 3 (1968), pp. 141-46, trans. in **SGRT** 9 (November 1968): 797-805; S. A. Parshenkov and S. N. Solomina, "Dve storony 'krizisa okruzhaiushchei sredy,'" **Priroda**, no. 2 (1975), pp. 119-23; Gudozhnik, **Nauchno-tekhnicheskaia revo-**

**liutsiia**, chapter 5; K. I. Shilin, "'Ekologicheskaia revoliutsiia' v sovremennoi zapadnoi nauke," **VFil**, no. 11 (1972), pp. 122-28; and S. S. Shvedov, "American Social Philosophy and the Sources of the Ecological Crisis," **VFil**, no. 8 (1978), trans. in **Soviet Studies in Philosophy** 17 (Spring 1979): 69-87. A. E. Medunin refers to "a considerable number of progressive authors who study the problem objectively" in the US, in "Chelovek i sreda," **VFil**, no. 4 (1973), pp. 68-69. See also T. S. Khachaturov, **Ekonomika prirodopol'zovaniia** (Moscow: Ekonomika, 1982), chap. 2.

71. K. V. Ananichev, **Problemy okruzhaiushchei sredy, energii, i prirodnykh resursov: mezhdunarodnyi aspekt** (Moscow: Progress, 1975), pp. 157-58.

72. G. S. Khozin, in "Chelovek i sreda," **VFil**, no. 3 (1973), pp. 72-73. Oddly enough in another context Khozin suggests the possibility that "environmental aggression" will be an international problem in 10-15 years (presumably if international cooperation fails). See his comments in another round table discussion, "Nauka i global'nye problemy," **VFil**, no. 8 (1974), p. 109.

73. Khozin, in "Chelovek i sreda," **VFil**, no. 3 (1973), p. 73.

74. See P. L. Kapitsa, "Nauchnyi i sotsial'nyi podkhod k reshenii global'nykh problem," **VFil**, no. 1 (1977), pp. 58-59; and Kapitsa, in "Chelovek i sreda," **VFil**, no. 2 (1973), p. 41.

75. I. T. Frolov, in "Chelovek i sreda," **VFil**, no. 1 (1973), p. 50.

76. Oldak, in "Nauka i global'nye problemy," **VFil**, no. 8 (1974), p. 115.

77. A. A. Mints and V. S. Preobrazhenskii, "Timely and Controversial Problems in the Systems Orientation of Geography," **Izvestiia Akademii Nauk SSSR, seriia geograficheskaia**, no. 6 (1973), trans. in **SGRT** 16 (February 1975): 65.

78. Ibid., p. 69. S. B. Lavrov and B. S. Khorev express similar sentiments, p. 238.

79. As in the USSR, the geography discipline in the GDR is divided into subdisciplines for economic geography and physical geography. According to Konrad Scherf, economic geography has come to focus on practically-oriented interdisciplinary territorial research, while physical geography has been more involved in environmental protection, water management, and landscape planning. See his "Zum wissenschaftlichen Beitrag der Geographie bei der Erarbeitung eines Generalschemas für die Standortverteilung der Produktivkräfte in der DDR," **Wissenschaftliche Zeitschrift der Universität Halle. Mathematisch-Naturwissenschaftliche Reihe** 31 (1982): 133-34, 138-39.

80. Gerhard Schmidt-Renner writes not only on issues relating to economic policy, as discussed in chapter 7, but also deals with more technical questions of the spatial aspects of environmental pollution. See, e.g., "Zum räumlichen Aspekt der Umwelt-

verschmutzung," **Geographische Berichte** 23, no. 88 (1978): 161-76. Other individuals mentioned in the text are referred to in this chapter. See also H. Richter, **Geographische Aspekte der Sozialistischen Landeskultur**, Studienbücherei Geographie für Lehrer, vol. 17 (Gotha/Leipzig: VEB Hermann Haack, 1979).

81. For the importance of geographers in the debate over re-routing of the flow of Siberian rivers see Gustafson, **Reform**, chapter 5.

82. See A. M. Alpat'ev and A. M. Riabchikov, "Rol' geografii v reshenii problemy sokhraneniia i uluchsheniia prirodnoi sredy," **Vestnik Moskovskogo Universiteta, seriia geograficheskaia**, no. 3 (1981), pp. 13, 16.

83. References to specific writings will be included later in the chapter. According to Komarov, Gerasimov chaired a commission set up by the Academy of Sciences to study the Lake Baikal problem in the 1960s. Komarov describes the disapproval which met the commission's report because it described the consequences of construction of the lake as "ruinous." See pp. 8-9.

84. Anuchin's book was published in English in 1977, edited by Ronald J. Fuchs and George J. Demko (Columbus: Ohio State University, 1977). For background to the Anuchin controversy and reactions of American geographers see Ian Matley, "The Marxist Approach to the Geographical Environment," **Annals of the Association of American Geographers** 56, no. 1 (March 1966): 97-111; Ian Matley, review of **Theoretical Problems of Geography**, pp. 576-79; David J. M. Hoosen, "Methodological Clashes in Moscow," **Annals of the Association of American Geographers** 52 (1962): 469-75; David J. M. Hoosen, "Introduction" to **Theoretical Problems of Geography** by V. A. Anuchin, edited by Ronald J. Fuchs and George J. Demko (Columbus: Ohio State University, 1977); John E. Chappell, Jr., "Marxism and Geography," **Problems of Communism** 14 (Nov.-Dec. 1965): 12-22; and John E. Chappell, Jr., "The Ecological Dimension: Russian and American Views," **Annals of the Association of American Geographers** 65 (March 1975): 144-65.

85. V. A. Anuchin, **Osnovy prirodopol'zovaniia. Teoreticheskii aspekt** (Moscow: Mysl', 1978).

86. Anuchin, **Theoretical Problems**, pp. 147-53.

87. Ibid., p. 154, 173-76.

88. V. A. Anuchin, "A Sad Tale about Geography," **Literaturnaia gazeta**, Feb. 18, 1965, trans. in **SGRT** 6 (September 1965): 28.

89. Iu. G. Saushkin, "The Today and Tomorrow of Geography," **Literaturnaia gazeta**, June 17, 1965, and editors' note which follows, trans. in **SGRT**, ibid., pp. 52-56.

90. Anuchin, **Osnovy prirodopol'zovaniia**, pp. 39-48.

91. Anuchin, "A Sad Tale," p. 29. On Stalinist projects for the "transformation of nature" and hydropower projects see also Komarov, pp. 58-62.

92. L. F. Il'ichev, "Metodologicheskie problemy estestvovaniia i obshchestvennykh nauk," **Vestnik Akademii Nauk SSSR**, no. 11 (November 1963), pp. 14-15.

93. See, for example, the discussion by Saushkin of the harmful effects of the Stalinist view on the interaction of society and nature. "A History of Soviet Economic Geography," in N. P. Nikitin et al., eds., **Ekonomicheskaia geografiia v SSSR** (Moscow: Prosveshchenie, 1965), trans. in **SGRT** 7 (October 1966): 43. Leonid Leonov and Boris Riabinin also refer to destruction of wildlife preserves "in the never to be forgotten days of rampant voluntarism," contrasting this with the present demand of the scientific community to open new ones. "'Preserve' Means Inviolable," **Sovetskaia Rossiia**, May 24, 1970, trans. in **CDSP**, vol. 22, no. 23, p. 6. The editors of **Voprosy filosofii** criticize vast "offensives against nature" which "cause nothing but damage," in "Obshchestvo i priroda," **VFil**, no. 4 (1973), p. 55. Likewise N. Melnikov attributes past troubles to "people who attempted to do good deeds in 'transforming nature.'" "Before Reversing the Rivers," **Literaturnaia gazeta**, July 12, 1967, trans. in **CDSP** 19, no. 32, p. 12. B. P. Vysotskii suggests that voluntarism is still often observed regarding transformation of nature, in his comments on E. K. Fedorov's "Metodologicheskie problemy nauk o zemle," **VFil**, no. 7 (1966), p. 101. Gusev suggests the same, in "Chelovek i sreda," **VFil**, no. 3 (1973), p. 62.

94. Matley, "The Marxist Approach," p. 108.

95. See, for example, I. P. Gerasimov and A. A. Trofimuk, "Sokhranit' chistotu vod ozera Baikala," **Priroda**, no. 11 (1965), p. 50; and I. P. Gerasimov, D. L. Armand, and K. M. Efron, **Natural Resources of the Soviet Union: Their Use and Renewal** (San Francisco: W. H. Freeman & Co., 1971), a translation of **Prirodnye resursy Sovetskogo Soiuza: ikh ispol'zovanie i vosproizvodstvo**, published in 1963 in Moscow.

96. I. P. Gerasimov, "Constructive Geography: Aims, Methods, and Results," **Izvestiia Vsesoiuznogo Geograficheskogo Obshchestva**, no. 5 (1966), trans. in **SGRT** 9 (November 1968): 739-41, 749-50. Iu. Efremov provides evidence of Gerasimov's long-standing interest in problems of the development and rational use of nature. He also provides a survey of geographical works on this topic in the 1960s. See his "Preobrazovanie prirody kak sostavnaia chast' prirodopol'zovaniia," in **Voprosy geografii** 108 ("Prirodopol'zovanie: geograficheskii aspekt"), edited by S. A. Kovalev, pp. 15-18.

97. Iu. Saushkin, "A History of Economic Geography," in Nikitin et al., p. 85. See also T. V. Zvonkova and Iu. G. Saushkin,

"Problems of Long-Term Geographic Prediction," **Vestnik Moskovskogo Universiteta, geografiia**, no. 4 (1968), trans. in **SGRT** 9 (November 1968): 757-59.

98. See Abramov for a summary of the conference.

99. Doskach, Trusov, and Faddeev, p. 115. This article is the published report given by Doskach, Trusov, and Faddeev at the conference.

100. Doskach, in Abramov, pp. 147-48.

101. I am grateful to Professor L. A. Kosinski for this insight. Chauncy D. Harris provides information on institutional affiliations of many of the geographers who play a role in this controversy. See "Geographers and Geographic Institutions in the Soviet Union Visited by a Delegation of American Geographers August 17 - September 13, 1961," n.d. Saushkin was head of the Department of Economic Geography at Moscow State University (MGU) at the time and V. A. Anuchin was in the Department of Economic Geography of the Peoples' Democracies, according to Harris. Gerasimov, as indicated, was head of the Academy's Institute of Geography. Two other outspoken opponents of Anuchin, S. B. Kalesnik and A. G. Isachenko, were on the faculty of Leningrad State University. Both are physical geographers. V. A. Anuchin's book, **Theoretical Problems of Geography**, had been rejected for his doctoral degree at Leningrad State University. Following this, Anuchin attempted to have it accepted at MGU where it was rejected by a small margin. Anuchin was later granted his doctorate on the basis of several of his publications, not his book alone. See sources in note 84 for elaboration of the controversy.

102. I. P. Gerasimov, "Fifty Years of Development of Soviet Geographic Thought," **Izvestiia Akademii Nauk SSSR, seriia geograficheskaia,** no. 5 (1967), trans. in **SGRT** 9 (April 1968): 247.

103. Ibid., p. 248.

104. Anuchin, **Theoretical Problems,** chapter 6, especially pp. 209-17. Also see the review of Saushkin's book on economic geography by N. T. Agafonov and S. B. Lavrov, **Vestnik Moskovskogo Universiteta**, no. 2 (1974), pp. 125-28, trans. in **SGRT** 15 (November 1974): 585.

105. A. G. Isachenko, "On the Unity of Geography," **Izvestiia Vsesoiuznogo Geograficheskogo Obshchestva**, no. 4 (1971), trans. in **SGRT** 13 (April 1972): 195. Anuchin had been subjected to this attack previously and addressed the issue in **Theoretical Problems**, pp. 139-42.

106. Isachenko, "On the Unity of Geography," pp. 208-9.

107. V. A. Anuchin, "Straddling the Boundaries Between Sciences," **Priroda**, no. 4 (1972), trans. in **SGRT** 13 (September 1972): 435.

108. I. P. Gerasimov, ed., **Man, Society, and the Environ-**

**ment**, USSR Academy of Sciences (Moscow: Progress Publishers, 1975).

109. B. A. Popov, review of **Man, Society, and the Environment**, I. P. Gerasimov, ed., in **Izvestiia Vsesoiuznogo Geograficheskogo Obshchestva**, no. 2 (1975), trans. in **SGRT** 17 (December 1976): 701.

110. Ibid., pp. 702-3.

111. A. G. Doskach et al., "Rejoinders," **Izvestiia Vsesoiuznogo Geograficheskogo Obshchestva**, no. 4 (1975), trans. in **SGRT** 17 (December 1976): 705-8.

112. See, e.g., V. A. Zubakov, "On the Content and the Tasks of Historical Geography (as the History of Nature Management)," **Izvestiia Vsesoiuznogo Geograficheskogo Obshchestva**, no. 6 (1976), trans. in **SGRT** 19 (March 1978): 170-80; A. G. Isachenko, "Geography and Historical Geography," **Izvestiia Vsesoiuznogo Geograficheskogo Obshchestva**, no. 6 (1976), trans. in **SGRT** 19 (March 1978): 180-85; Iu. P. Mikhailov, "Geograficheskie grani protsessa prirodopol'zovaniia," **Geografiia i prirodnye resursy**, no. 3 (1980), p. 161; and S. B. Lavrov and N. T. Agafonov, "Conceptual Issues in Economic Geography at the 23rd International Geographical Congress," **Izvestiia Vsesoiuznogo Geograficheskogo Obshchestva**, no. 1 (1977), trans. in **SGRT** 19 (May 1978): 326-27.

113. Isachenko, "Geography and Historical Geography," p. 181.

114. T. V. Zvonkova and Iu. G. Saushkin, "Interaction Between Physical and Economic Geography," **Vestnik Moskovskogo Universiteta, geografiia**, no. 3 (1976), trans. in **SGRT** 18 (April 1977): 250.

115. Iu. G. Saushkin, "Geograficheskoe issledovanie 'ochelovechennoi,' ili 'istoricheskoi' prirody," **Voprosy geografii** 106 ("Vliianie cheloveka na landshaft"), p. 29.

116. Lavrov and Khorev, p. 238.

117. Hoosen, "Introduction," p. 10. Gerasimov refers to this in 1970 in "Materialy V S"ezda Geograficheskogo Obshchestva," trans. in **SGRT** 12 (April 1971): 210-11.

118. Hoosen, "Introduction," p. 15.

119. Anuchin uses the term "indeterminism" in **Theoretical Problems**, pp. 186-88.

# Part 2

# Economics and the Environment

The second section of this book focuses on economic issues relating to environmental protection. Examination of the economic debates is particularly revealing because of the importance attributed to the economic base of society in Marxist theory. One would expect Marxist-Leninist concepts to lay the foundation for the Soviet and GDR understanding of the economics-ecology interface. This is, however, only sometimes the case. Furthermore, Marxist concepts have been molded by economists to support diverse positions.

In chapter 6 we explore the goals of economic production in the USSR and GDR as they relate to environmental protection. Particular attention is paid to the relationship between economic growth and environmental protection. Here we consider the important influence of Western environmental writings proceeding from the Club of Rome. In chapter 7 we examine elements of macro-economic theory and policy. We observe how the concepts and proposals of environmental advocates are influenced by Marxist-Leninist precepts, and how these advocates base their claims on officially-sanctioned values. In chapter 8 we assess the impact of the Marxist labor theory of value on policies for economic evaluation of natural resources, the most widely acclaimed tool for changing micro-economic incentives for enterprise managers.

Before examining the ideological and theoretical debate, it behooves us to establish the context for these considerations by providing more concrete background on present policies, on the role of economists in policy-making, and on the political-economic obstacles to effective environmental protection. We turn to these questions in chapter 5.

# 5
# The Political-Economic Context

From a Marxist-Leninist viewpoint, political economy is at the root of socialism's superior potential in responding to problems like environmental deterioration. For in the absence of antagonistic class relations, central authorities can set goals in the long-range interests of the whole population and implement them in a rational, efficient, coordinated planning system. Such an economic structure presumably provides the foundation for a policy of rational use and protection of nature on a societal scale. Not only Soviet officials, but also some Western environmentalists assert that a centrally planned economy has at least potential advantages over profit-oriented capitalism in dealing with collective problems like the environment.[1]

It may, therefore, seem ironic that in the final analysis the greatest obstacles to effective environmental protection in the USSR and GDR are almost certainly political-economic in character.[2] In the Soviet Union and to a lesser extent in the GDR the incentive structure makes environmental protection a low priority for enterprises and supra-enterprise organizations, despite some recent innovations. As a leading Soviet economist explains: "Since the economic mechanism is above all oriented toward the fulfillment of traditional plan indicators, the enterprises are not interested in the realization of nature-protecting legislative acts and plan targets for the regulation of nature-protection."[3] Changes in the incentive structure more favorable to environmental protection meet with various types of resistance. First, the bureaucratic conservatism of enterprise and ministry officials is reinforced by their wariness of increased and perhaps contradictory demands which binding environmental "indicators" or "stimuli" might imply. For example, application of charges for natural resource use or penalties for pollution may raise the cost of production per unit,[4] making it more difficult for enterprises to meet profitability and other planning norms and also reducing the flexibility of enterprises in assuring supplies of raw materials. Previous economic reforms have, unwittingly, imposed such contradictory demands on enterprises, due in part to the simple technical difficulty of developing indicators which simultaneously stimulate the achievement of multiple goals.[5] Furthermore, forceful application of environmental standards might imply a shift in the power balance between, for example, branch and territorial organizations, or between now influential ministries (like the extractive and industrial branches) and insurgent enforcement and inspection services. Such bureaucratic resistance complicates the difficulty of formulating and enforcing policies which can deal with complex and regionally-differentiated problems.

Second, consistent with historical tradition, the present political leadership of both countries is committed to maximum growth of material production.[6] This commitment is reinforced by real economic pressures associated with international competition and consumer demands. But it collides with the perceived burden of costly environmental protection. In a Soviet popular science article three environmentally-conscious scientists refer to experts, including "industrialists abroad and some of our economic executives" who consider that "environmental measures are not only economically ineffective, but also reduce the rate of economic development by 10-11%."[7] Most likely such a viewpoint is widespread among economic practitioners, not only at the enterprise level, but in planning organs as well.[8] This bias against increased environmental expenditures is further buttressed by certain ideological assumptions deriving from Soviet Marxism-Leninism such as the primary importance of material production in social progress and emphasis on rising labor productivity (as opposed to natural resource use) as the fundamental source of economic growth and advancement.[9]

All of these factors impede a heavy commitment of resources to environmental protection. But, simultaneously, wasteful use of natural resources and environmental deterioration impose their own costs, ever more apparent to the political leadership. Resource shortages, rising extraction costs for mineral and fuel resources, and skyrocketing prices for raw material imports pose problems for the political leadership of both countries. These factors make improved utilization of domestic raw materials and energy resources a high priority; this in turn requires more complete and effective extraction from mines, multiple component use of mined ore, reduced raw material inputs in production, development of long-lasting products, and reduction of waste emissions through extraction of secondary raw materials from waste products. These themes pervade the Soviet and GDR media. To some extent, these goals overlap with ecological concerns; therefore, an opening is provided for environmental advocates to press for simultaneous solution of productive and ecological problems. Evidence also mounts that environmental pollution introduces less easily measurable, but nonetheless significant economic costs, for example increased health costs, absenteeism, corrosion of equipment, reduced agricultural output, and harm to the fishing and lumber industries. Other costs are long-term and/or less apparent such as climatic changes or genetic damage. Some Western analysts have speculated that health problems resulting from pollution were a major factor motivating increased attention to environmental protection in the mid- to late-1960s in both countries.[10]

These contradictory economic pressures have assured economists an important role in the environmental debate in the Soviet bloc. In his study of Soviet environmental advocacy from 1965 to 1972, John Kramer concluded that "scientists and economists are well-represented among Soviet environmentalists."[11] This conclusion has become even more valid in the 1980s, with environmentalists forming an important opinion group within the economics discipline.[12] This is evidenced in the increasing attention granted to environmental and nature-use issues in the two major scholarly journals in the economic field in the USSR: **Voprosy ekonomiki** (Questions of Economics), the publication of the Economics Division of the Academy of Sciences of the USSR; and **Ekonomika i matematicheskie metody** (Economics and Mathematical Methods), published by the Central Mathematical-Economics Institute (TsEMI) of the Academy of Sciences of the USSR and representing the increasingly important mathematical-economics approach (see table 5.1). Discussion of natural resource pricing was a focal point of articles in **Voprosy ekonomiki** in the late 1960s. Since then articles dealing with the environment cover a broader range of issues. While more recent articles focus on ecological problems as such, many of the early articles on natural resource pricing probably arose as much out of the effort to perfect the cost-accounting system for enterprises as they did out of concern for the environment. **Ekonomika i matematicheskie metody** was established only in 1965. From 1965 through 1972 only an occasional article dealt with ecological or natural resource issues, these few articles addressing almost exclusively the problem of economic evaluation of land and differential land rent.[13] Since the passage of the Council of Ministers' resolution "On Strengthening the Protection of Nature and Improvement of the Use of Natural Resources" in 1972, the attention granted to problems of ecology and natural resource use has increased markedly.[14] In the previous chapter we observed the gradual emergence of institutional bases for environmental research. The attention devoted to environmental issues in

**TABLE 5.1 The Number of Articles on Environmental Protection and Rational Use of Nature in Major Soviet Economics Journals 1965–1981[a]**

| | Total number of articles on environmental or nature-use topics | | | Of these, with a specific focus on the economic evaluation of natural resources | | |
|---|---|---|---|---|---|---|
| Year | Questions of Economics[b] | Economics and Mathematical Methods[c] | Total in Both Journals | Questions of Economics | Economics and Mathematical Methods | Total in Both Journals |
| 1965 | 3 | 0 | 3 | 1 | 0 | 1 |
| 1966 | 4 | 0 | 4 | 1 | 0 | 1 |
| 1967 | 4 | 1 | 5 | 3 | 0 | 3 |
| 1968 | 6 | 1 | 7 | 6 | 1 | 7 |
| 1969 | 2 | 1 | 3 | 2 | 0 | 2 |
| 1970 | 4 | 1 | 5 | 0 | 1 | 1 |
| 1971 | 3 | 0 | 3 | 0 | 0 | 0 |
| 1972 | 2 | 0 | 2 | 0 | 0 | 0 |
| 1973 | 7 | 2 | 9 | 0 | 0 | 0 |
| 1974 | 6 | 3 | 9 | 2 | 1 | 3 |
| 1975 | 4 | 5 | 9 | 0 | 2 | 2 |
| 1976 | 5 | 4 | 9 | 0 | 1 | 1 |
| 1977 | 4 | 4 | 8 | 0 | 0 | 0 |
| 1978 | 4 | 4 | 8 | 0 | 2 | 2 |
| 1979 | 4 | 6 | 10 | 0 | 0 | 0 |
| 1980 | 5 | 4 | 9 | 2 | 1 | 3 |
| 1981 | 5 | 5 | 10 | 0 | 1 | 1 |

a. Articles, reports on conferences, letters, book reviews, and other pieces were included only if they exceeded two pages in length. Pieces focussing primarily on the efficiency of the productive process have been excluded, even though they may deal with natural resource use, most commonly land or mineral raw materials. For example, articles on the agricultural efficiency of irrigated land were excluded if the main concern was with returns on economic investment, but included if the main concern was with maintenance of soil quality. Likewise, articles dealing with the energy-fuel balance were generally included only if they gave significant attention to the availability of raw materials. Inclusion or exclusion involved an element of judgment, as any concern with raw material use can be construed to involve issues of rational use of nature. In most cases of serious doubt, the article was included. Because of this, the chart may, to a certain extent, mask the growing concern over the years with explicit issues of environmental protection and natural resource scarcity.

b. **Voprosy ekonomiki**, put out by the Institute of Economics of the Academy of Sciences of the USSR.

c. **Ekonomika i matematicheskie metody**, put out by the Central Mathematical-Economics Institute of the Academy of Sciences of the USSR. The journal commenced publication in 1965.

**Ekonomika i matematicheskie metody** reflects the increasing importance of the TsEMI in the development of policy proposals in this area.[15] Leading economists in the Academy of Sciences of the USSR have undoubtedly facilitated the important public role of economists in the environmental debate. T. S. Khachaturov, editor of **Voprosy ekonomiki** since 1966, has devoted his own scholarly energies to advocacy of natural resource pricing as a means of improving rational utilization and husbanding of natural resources.[16] Likewise, N. P. Fedorenko, editor of **Ekonomika i matematicheskie metody** and head of the Economics Division of the Academy of Sciences, has consistently supported assertive environmental policies and has frequently authored articles in such pro-environmental popular magazines as **Priroda** (Nature).[17]

In the GDR as well economists concern themselves with environmental problems. Publications like **Wirtschaftswissenschaft** (Economic Science), the major journal of economics in the GDR, and scientific journals of higher educational institutions, particularly the Higher School for Economics (Hochschule für Ökonomie) in Berlin-Karlhorst and the Technical University in Dresden, devote consistent attention to the economic problems associated with environmental protection.[18] Hans Mottek, member of the Academy of Sciences and long-time director of the Institute of Economic History at the Hochschule für Ökonomie in Berlin, has played an important role in stimulating the debate. Mottek, for example, has been active on the Academy of Sciences' Council for Environmental Research and was its first chairman at its inception in 1974. Several other individuals at the Hochschule für Ökonomie have been outspoken on environmental issues as well.[19] The Central Institute for Economic Sciences (Zentralinstitut für Wirtschaftswissenschaften) of the Academy of Sciences is also a focal point for environmental research in the GDR.[20] As in the USSR the environment has become a consistent concern of several economists in the GDR. Because the country is much smaller, the interaction and cohesiveness of this opinion group is undoubtedly higher than in the USSR.

Not only economists, but also legal scholars have devoted attention to economic aspects of environmental policy. Leading figures here are O. S. Kolbasov, a section head at the Institute of State and Law of the Academy of Sciences of the USSR and Ellenor Oehler at the Academy for State and Legal Sciences of the Academy of Sciences of the GDR. The attention of jurists has been directed more toward formal definitions of economic relations, while economists focus on optimal efficiency and productivity. While our main concern in the next two chapters is with the latter type of issues, in chapter 8 we will examine the interaction of jurists and economists regarding natural resource pricing as a stimulus to economical use of resources.

## ECONOMIC POLICY AND THE ENVIRONMENT

While economists form an important component of the Soviet and East German environmental lobby, their efficacy is more difficult to establish. Several important changes in economic policy affecting the natural environment have occurred in both countries in the last fifteen years. But as one Soviet economist has noted, "the fact is, that the number of proposals from the field of economics that remain unimplemented is quite large and continues to grow."[21] Another scholar notes that there is usually a ten to fifteen year time lag between the articulation of theoretical principles and their fusion "into practical methods and normative documents accessible to any plant economist."[22] Where policy changes have occurred the direct influence of particular scholars or institutes is often hard to establish, although there are some exceptions. For example, many Soviet economists have proposed methods for evaluating the economic effectiveness of expenditures on protection of the environment.[23] In August of 1980 adoption of a temporary methodology for this purpose was announced in **Ekonomicheskaia gazeta.**[24] The methodology, according to this report, was developed by the Academy of Sciences' Scientific Council on the Economic Effectiveness of Fixed Funds (**osnovnykh fondov**), Capital Investment, and New Technology (the council is headed by T. S. Khachaturov), and by the Institute of Economics of the Academy. How much effect this change will have remains to be seen, and the exact process of political persuasion which led the Council of Ministers to encourage its development and Gosplan (State Planning Commission) to adopt it is shrouded in mystery. As another example of this type of institutional connection, the draft law on air quality was developed in the Soviet Academy of Sciences' Institute of State and Law, according to Kolbasov.[25] For other policy changes which economists or jurists have long supported, it is impossible to establish such direct results.

Although the impact of economists on policy cannot be directly established, some of the most important innovations in the economics of environmental protection in the GDR and USSR are noteworthy. Since 1973 in the GDR and since 1975 in the USSR environmental protection has been included as a component of the national economic plan, the plans of ministries, territorial sub-units, and productive associations or enterprises.[26] In both countries productive enterprises and associations must file detailed forms indicating planned measures for protection and rational use of water, air, and mineral resources, and including information about past and projected use of water, emissions of pollutants, and extraction of useful components from wastes.[27] In the USSR a special section on environmental protection now exists in Gosplan.[28] Centrally funded investments for environmental protection increased from 6.3 billion rubles in the Ninth Five Year Plan period (1971-1975) to 9.2 or 9.3 billion rubles in the Tenth Five Year Plan period (1976-1980).[29] Actual expenditures in the Tenth Five Year Plan period, however, represented only 85 percent of those planned, although these central investments were supplemented by 16.8 billion rubles of branch and enterprise outlays.[30] Planned state investments for the Eleventh Five Year Plan period (1981-1985) are about 10.2 billion rubles (slightly less than the **planned** state expenditures for the previous five year period), to be supplemented by 19 or 20 billion rubles more from other levels.[31] As the experience of the Tenth Five Year Plan indicates, productive organizations sometimes fail to spend centrally allocated amounts. This is often because suppliers of pollution control equipment cannot be located, since this branch of the economy remains underdeveloped.[32] Rather than attempt to produce the equipment itself and thus expend some of its own resources, the enterprise or branch may simply leave the funds unspent. As we shall see in the next chapter, installation of pollution control equipment may cause cost-accounting indicators of some enterprises to suffer. Thus minimal penalties for environmental pollution may be more attractive than non-fulfillment of material output or other planning norms. Of the approximately 9.3 billion rubles invested in environmental protection between 1975 and 1980, 7.1 billion were committed to protection and rational use of water; 1.12 billion to land, forest resources, nature preserves, reproduction of fish stocks, and protection of animal life; only .84 billion to protection of air purity; and .18 billion rubles to protect the earth's interior.[33] These outlays, according to Soviet analysts, comprised 2 percent of the gross national product (**valovoi natsional'nyi produkt**),[34] and in the late 1970s 3.1 percent of all fixed assets of industry were related to environmental protection.[35]

In the GDR 1.6 billion marks were committed to environmental protection in 1973, but in recent years detailed statistics do not seem to be available. The enthusiastic attention to environmental protection in planning reports seems to have waned since the early 1970s.[36] The directives for the 1981-1985 plan period do include a section on "Water Supply and Environmental Protection," with particular attention devoted to assurance of stable drinking and agricultural water supplies. Specific indication of the magnitude of environmental investments is not provided.[37] Despite such data gaps, in many ways the economic mechanisms designed to stimulate environmental protection are more developed in the GDR than in the USSR. The **Landeskulturgesetz**, passed in 1970, provides a comprehensive framework for environmental protection policy,[38] and the Ministry for Water

Supply and Environmental Protection has, since the early 1970s, provided a coordinating point for environmental protection measures, although this ministry is mainly responsible for water resources. The upgrading in 1978 of the Main Hydrometeorological Service in the USSR to state committee level (State Committee on Hydrometeorology and the Environment – Goskomgidromet) should increase its clout and thus may accomplish the same purpose as the independent ministry in the GDR, although the Soviet Goskomgidromet is primarily assigned responsibility for air pollution control.[39]

A significant innovation in the GDR was the institution of charges for the use of land and water, in 1968 and 1971, respectively. We will leave more detailed discussion of charges for natural resource use to chapter 8. Another policy innovation in the GDR involved institution of charges for water **pollution** in 1971 and for air **pollution** in 1973. These charges are levied only when productive enterprises or associations exceed permissible emission norms.[40] (Note that these charges are distinct from the above-mentioned charges for land and water **use**.) While payments for land and water **use** are included in the planned costs of production of East German productive enterprises and associations (unless norms are exceeded), the payments for excessive **pollution** are not and should, therefore, have the effect of lowering the enterprise's profit, thus reducing incentive funds.[41] The new water law, passed in July 1982, obliges enterprises to include water-saving and water-free production processes in their long-term plans and provides for the appointment of a water commissioner (**Wasserbeauftragte**) in central state organs and enterprises. It instructs district and regional councils to develop water-protection commissions.[42] Enterprises are also responsible for reduction and safe disposal of wastes from the production process subject to the approval and oversight of local state organs and relevant ministerial bodies.[43]

East German economists point out inadequacies in many of the above policies. For example, resource use and pollution charges are too low to be effective, and present methods don't allow adequate reflection of environmental concerns in investment and capital planning.[44] Nonetheless GDR economists are less critical than their Soviet counterparts of the incentive structure for enterprises regarding environmental pollution. This may be because these East German policies do succeed to a greater degree in stimulating enterprise compliance or it may simply be due to stricter controls on public criticism in the GDR. Meanwhile many Soviet economists press for policies similar to those already in existence in the GDR. As of 1984 only minimal steps in this direction have been taken. Payment for use of water by industrial enterprises was instituted in 1982.[45] But there are no fixed charges for excessive pollution as in the GDR, and until now environmental regulations have been too poorly enforced and fines too low to be effective.[46] In response, in the late 1970s the USSR State Committee on Labor and Social Questions and the Central Council of Trade Unions approved a procedure in all branches which provides for forfeiture of bonuses of enterprise officials if plans for environmental protection measures are not carried out or if standards and rules governing natural resource use are not obeyed.[47] The new regulation loses some of its potential force because ministries which feel hampered by environmental protection regulations can often find ways to make up losses incurred by officials within their jurisdiction. Some Soviet analysts have concluded that sanctions against individual enterprise officials are, for this reason, the least effective means of enforcing environmental protection standards.

East German policies for recycling of productive wastes as secondary raw materials are also more developed. In December 1980 and May 1981 new regulations sharpened existing obligations placed on productive organizations and other state bodies regarding reuse of waste products. The decree designates "the tasks for implementing closed material cycles and for the collection [**Erfassung**], preparation, processing, and utilization of secondary raw materials as a responsibility of the first order . . .,"[48] and clarifies the obligations of the various productive and administrative organs. Reinforcing the "producer principle" (**Verursacherprinzip**) already laid down in 1975, the decree states that "the enterprise and association [**Kombinat**] in whose domain the secondary raw materials are produced is, as a general principle, responsible for the utilization [**Verwertung**] of wastes as secondary raw materials."[49] Cooperation with special associations for the collection and processing of secondary raw materials is specified in the regulation. The Ministry for Material Economy is made responsible for oversight and long-term planning. Along with a Central Commission for Secondary Raw Material Reserves, corresponding bodies at the district (**Kreis**) and regional levels (**Bezirk, Stadtbezirk**) are to be established to further territorial coordination of recycling efforts.[50] Regulations regarding reuse of waste apply not only to by-products generated in the productive process itself, but also to discarded consumption goods produced by the enterprise. For example, the chemical industry is responsible, under this principle, for the reutilization of recyclable materials in old tires, old plastic, and old oil. Already at the planning stage enterprises must take account of the disposition of waste and the enterprises or associations (**Kombinate**) must foresee development of appropriate technical processes in their scientific and technical plans.[51]

Not all branches of industry live up to the standards

established in these regulations. While collection of scrap metal has been carried out by the metallurgy industry for some time, inadequacies mark the utilization of wood product wastes, plastic refuse, and textile waste.[52] Often enterprises are spurred to action only after environmental protection organs invoke sanctions against impermissible deposition of waste.[53] To facilitate use of secondary raw materials, the Institute for Secondary Raw Material Economy (Institut für Sekundärrohstoffwirtschaft) in Berlin has, since 1976, operated an information system which provides advice to enterprises concerning use of waste products. The institute also acts as a central storage location for quantitative and qualitative data about wastes and secondary raw materials.[54] In addition, the Institute for Lightweight Construction and Economical Use of Materials (Institut für Leichtbau und Ökonomische Verwendung von Werkstoff) in Dresden has a data bank which contains information on the use of more than five thousand materials. Since 1974 this institute has operated a consultation center to assist those working in fields of technical innovation to solve problems associated with lowering of material inputs and their replacement by secondary raw materials.[55]

By 1984 about 11 percent of the GDR's requirements for economically important raw materials were covered through use of secondary sources. Seventy-three percent of raw steel, 49 percent of paper, 30 percent of copper, 73 percent of aluminum and 32 percent of glass (for bottles) were based on the use of recycled materials.[56] The reuse of scrap metal is considered particularly important since each ton is estimated to save 150 DM (valuta) of imports. Various economic incentives exist to encourage the reuse of hard-to-refine scrap metal.[57] Attention to use of waste for secondary raw materials continues to grow in the GDR, as the resource situation tightens and the cost of importing, either from the USSR or from the West, rises. At the Tenth Party Congress in 1981 and at subsequent Central Committee plenums party leader Erich Honecker has placed renewed emphasis on economical use of raw materials and fuel, and on use of secondary raw materials, important resources to achieve continued economic growth. According to GDR statistics, material and energy expenditure per unit of industrial output has declined steadily in the past several years.[58] In the period from 1981 to 1985 the utilization of secondary raw materials is to reach a level that is 128 to 130 percent of that achieved from 1976 to 1980.[59] As one GDR scholar notes, however, it ". . . is becoming ever clearer that here the issue is not preeminently a strategy of modern environmental protection, but rather simultaneously of material and energy economy."[60] Thus, in the early 1980s regulation of the utilization of secondary raw materials was removed from the framework of the **Landeskulturgesetz** and given separate legal status.[61] Meanwhile, the revised "Sixth Implementation Decree" to the **Landeskulturgesetz,** passed in 1983, deals only with "the harmless disposal of **non-utilizable** wastes" (emphasis mine), unlike the previous version which also covered recyling of wastes in the production process. Under the 1983 regulation, enterprises must demonstrate that the wastes marked for disposal are not usable as secondary raw materials.[62] Furthermore, in 1983 two GDR economists noted that "a specific coefficient [**Kennziffer**] that reflects the work of associations and enterprises in the area of environmental protection is not ... provided for," although achievements in this area may be indirectly measured if evaluation of the enterprise's recycling efforts is improved.[63]

Some recent changes in the USSR should provide a stimulus for enterprise and ministry officials to pay closer attention to material economy and productive use of wastes as secondary raw materials, but these policies still are not as well developed as in the GDR. By 1983, 22 percent of raw material for paper and cartons, 27 percent for textiles and 50 percent for steel smelting reportedly came from secondary raw materials.[64] Since 1977 norms for material expenditures in productive processes have been expanded in an effort to save expensive raw materials.[65] The metallurgy industry in particular has been reprimanded for wastefulness.[66] In 1978 bonuses went into effect for collection, storage, and shipping of scrap metal.[67] On January 25, 1980 the Council of Ministers, and the Central Committee of the CPSU passed a resolution "On Measures for Further Improvement of the Use of Secondary Raw Materials in the Economy."[68] Beginning in 1981, state, ministry, republic, and enterprise plans are to include a section on "use of secondary raw materials."[69] For the 1981-1985 period more intensive utilization of secondary raw materials is foreseen, with particular goals established for use of many productive wastes: an increase of 130 percent in use of tree bark as fuel; an increase of 140 percent in use of lignin in the metallurgy industry and as fuel; a 68.8 percent increase in utilization of ash and ash slag from electrical stations for chalking acid soils, production of cement, bricks and other materials; increased utilization of fruit and vegetable wastes, wastes from the wood-processing industry, and many others.[70] Implementation of the overall policy is to be supervised by Gossnab (the State Committee on Material-Technical Supplies), which since 1975 has run an All-Union Recyclable Materials Design and Technological Institute (Vsesoiuznyi proektiro-konstruktorskii i tekhnologicheskii institut).[71] Since 1983 formation of bonus funds depends in part on material expenditure per ruble of output. Cutbacks in material inputs supplied to enterprises are also intended to stimulate savings.[72]

Critics of past policy point out inadequacies, only some of

which are likely to be overcome by these recent Soviet policy innovations.[73] First, enterprises and ministries show little interest in these problems since they are mainly concerned with meeting planned material output quotas. Pricing policy also aggravates the problem. There is not a unified methodological document regulating estimation of prices for secondary raw materials. Inconsistencies and irrationalities result.[74] Furthermore, economists themselves seem to disagree as to whether present prices are too low or too high to stimulate recycling. Payment received by suppliers for secondary raw materials is generally based only on some expenditures such as those for storage and transport, but excludes those involved in the basic production process (the costs of whch are presumably reflected in the cost of primary output).[75] Thus, enterprises have little incentive to make their wastes available in usable form, even though buyers might find it advantageous to use them. On the other hand, other economists claim that production costs rise when scrap metal is used for steel production, due to changes required in processing. These economists argue that lower prices on recycled materials are needed, if demand is to be stimulated; they also point to lower prices for such materials on the international (capitalist) market.[76]

A second problem relates to interdepartmental coordination, which is deemed inadequate to organize the use of secondary raw materials across ministry lines. Responsibility is not clearly delegated to various organs within the ministerial and territorial structures[77] as it is in the GDR. The new Soviet regulations, in strengthening the hand of Gossnab, may bring some improvement here, but many critics see the real solution to lie in inter-branch mining and processing complexes or at least in a strong interdepartmental coordinating body in Gosplan or at the regional level.[78] Finally specialists on use of secondary raw materials are in short supply in the USSR.[79]

As mentioned previously, it is difficult to determine the role of environmentally-conscious economists in bringing about the incremental policy changes discussed above. In some cases, as with the increasing emphasis on recycling and utilization of wastes, the leadership's motivation clearly lies in the realm of material economy, not environmental protection. But this does not prevent environmental advocates from turning the situation to their advantage, as we shall see in the next chapter. The motivation for the introduction of charges on natural resource use is also only partially related to long-term considerations of resource depletion. In the GDR, implementation of charges on land and water followed only minimal public debate. The debate in the USSR on that issue was much more intense, much earlier, and with lesser effect. In fact more public discussion of natural resource pricing has developed recently in the GDR, perhaps due to the ineffectiveness of the policies adopted earlier and as a response to Soviet debate. Likewise, it is not possible to determine the importance of economists' arguments in effecting the inclusion of environmental protection in the national economic plan. It is the less radical policies, more closely linked to non-environmental goals (like reductions in the waste of valuable raw materials), which have generally been put into practice most assertively and consistently. Even here economists in both countries have expressed repeated dissatisfaction with the manner in which their concepts have been applied in practice.

While there is reason to doubt the short-term efficacy of these academic lobbyists, in the longer view economists have almost certainly helped to broaden the parameters of debate over environmental problems and gradually to introduce new perspectives which may slowly influence policy-making. In the early 1970s, as Soviet economists have noted, the very notion of the "economic effectiveness" of environmental protection measures would have seemed paradoxical to many, for environmental protection was considered to be largely a matter of **removing** elements of nature from the productive process.[80] The concept most likely still sounds odd to many, but its acceptance by Gosplan represents significant progress. A report on "Basic Directions of Work of the Economics Division of the Academy of Sciences of the USSR in Light of the Decisions of the 26th Party Congress" (1981), emphasizes the increasing demands from the party leadership for closer ties between various Academy institutes and economic-planning organs, ministries, productive enterprises and associations.[81] The document refers several times to the need for further research on environmental problems.[82] However, a resolution on activities of the Economics Division of the AN SSSR approved by the Presidium of the Academy in June 1983 seems to place somewhat less emphasis on environmental research.[83] Whether this implies a downgrading of the priority of this problem or whether it simply reflects the integration of environmental studies into existing research programs is not clear.

As economists have sought to influence environmental policy, they have become increasingly sophisticated in marshaling arguments which link their concerns to the major tasks set out by the political leadership. In the late 1960s, when a major focus of discussion among Soviet economists and jurists involved economic (monetary) evaluation of natural resources and charges for resource use, many economists linked their proposals to the basic concept of the Kosygin economic reforms (or the New Economic System in the GDR), arguing that their proposals would improve the cost-accounting character of enterprise management. In the

1970s and 1980s, with rising raw material costs and official concern with labor productivity, environmentally-conscious economists in both the USSR and (somewhat earlier) in the GDR have linked their proposals to these concerns, focusing on waste-free technology as a means of decreasing material losses and increasing productivity. Economists must phrase their arguments in terms of officially acceptable goals, both out of necessity (due to censorship), but also out of political expediency, for their arguments will be more persuasive to elites to the extent that they appeal to governing values. It is acceptable to criticize, so long as fundamentals (like the relations of production) are not directly challenged; and it is increasingly "normal" for the elite to entertain policy advice. Molding one's arguments to fit the governing political context is a sign of political sophistication and need not imply abdication of one's position if skill is used in building creatively on accepted principles. Of course there are limits to this strategy, and form does alter content, even if only subtly. Finally the opportunity to participate in even esoteric and restricted policy debates encourages the development of political sensitivities and skills which are gradually changing the terrain of Soviet politics, even if not immediately its policies.

## NOTES

1. See the exchange between Robert J. McIntyre and James R. Thornton, "On the Environmental Efficiency of Economic Systems," **Soviet Studies** 30 (April 1978): 173-92; and Charles E. Ziegler, "Soviet Environmental Policy and Soviet Central Planning: A Reply to McIntyre and Thornton," **Soviet Studies** 32 (January 1980): 124-34.

2. Such a conclusion does not invalidate Marxist analysis (which focuses on the political-economic roots of systemic problems), but only certain assumptions of Soviet Marxism-Leninism. Even here, by allowing criticism of economic policy, the official Soviet Marxist-Leninist position concedes that there are weaknesses in the existing economic structures, although not in existing relations of production.

3. N. Feitel'man, "Upravlenie prirodopol'zovaniem," **Voprosy ekonomiki** (hereafter **VEko**), no. 3 (1984), p. 63.

4. Dieter Graf notes this objection in "Zu einigen Grundfragen der ökonomischen Bewertung von Naturressourcen in entwickelten Sozialismus," **Wirtschaftswissenschaft** (hereafter **WW**) 26 (July 1978): 819.

5. See Alec Nove, **The Soviet Economic System** (London:

George Allen & Unwin, 1977), pp. 87-92.

6. For example, Erich Honecker stated the following in his report to the Tenth Party Congress of the SED in 1981: "Stable economic growth is indispensable for socialism; for the needs of humans, the requirements of our socialist society continue to develop, and only that which has been produced can be distributed." "Bericht des Zentralkomitees der Sozialistischen Einheitspartei Deutschlands an den X. Parteitag der SED," **Neues Deutschland**, April 12, 1981, p. 5.

7. N. P. Fedorenko, M. Ia. Lemeshev, and N. F. Reimers, "Sotsial'no-ekonomicheskaia effektivnost' okhrany prirody," **Priroda**, no. 10 (1980), p. 11. See also P. Ia. Baklanov, B. V. Poiarkov, and G. I. Khudiakov, "Osnovnye napravleniia sovershenstvovaniia prirodopol'zovaniia i prirodookhrannykh meropriiatii na dal'nem vostoke," **Geografiia i prirodnye resursy**, no. 3 (1981), pp. 18-19.

8. For example, Fedorenko, Lemeshev, and Reimers associate the Forestry Section of Gosplan with this viewpoint, pp. 11-12.

9. As environmental proponents in Eastern Europe point out, Marx himself did recognize the contribution of natural resources to social production.

10. For example, Thane Gustafson, **Reform in Soviet Politics** (Cambridge, ENG: Cambridge University Press, 1981), p. 51; and Ilka Norhara-Schnabel, "Zur Entwicklung der Umweltpolitik in der DDR," **Deutschland Archiv** 9 (1976): 811.

11. John Kramer, "The Politics of Conservation and Pollution in the USSR" (Ph.D. dissertation, University of Virginia, 1973), pp. 25-28.

12. On opinion groups in Soviet politics see H. Gordon Skilling, "Groups in Soviet Politics: Some Hypotheses," in H. Gordon Skilling and Franklyn Griffiths, eds., **Interest Groups in Soviet Politics** (Princeton: Princeton University Press, 1971), pp. 24-27. Skilling designates as an opinion group "those persons, within a given occupation, who have a distinguishable viewpoint of their own" (p. 25). This is distinct from an occupational group, where virtually all members of the occupation tend to share a common viewpoint. On some issues, as on the issue of pricing of natural resources, economists tend to act as an occupational group (see chapter 8). On issues of pollution control some economists form an opinion group within the profession.

13. The main articles were the following: N. Ia. Petrakov, and I. M. Khrabrov, "Nekotorye voprosy ekonomicheskoi otsenki prirodnykh resursov v svete marksistsko-leninskoi teorii zemel'noi renty," **Ekonomika i matematicheskie metody** (hereafter **EiMM**) 6 (1970): 355-62; V. S. Cherniavskii, "O sostavlenii optimal'nogo grafika vodoispol'zovaniia," 5 (1969): 590 ff; and V. I. Denisov, "O

metodakh rascheta differentsial'noi zemel'noi renty," 4 (1968): 732 ff.

14. An increase in research devoted to working out a unified methodological basis for the economic evaluation of natural resources after passage of the resolution is noted in "Obsuzhdenie problem ekonomicheskoi otsenki prirodnykh resursov," **EiMM** 11 (1975): 601.

15. Important individuals associated with TsEMI who do research on environmental problems include K. G. Gofman, M. Ia. Lemeshev, N. F. Reimers, and N. P. Fedorenko. TsEMI has cosponsored conferences on related issues and has played a particularly important role in leading research on the development of a methodology for economic evaluation of natural resources.

16. T. S. Khachaturov's scholarly interests have centered around the economics of transport and the economic effectiveness of capital investments and new technology. Since the late 1960s Khachaturov has concerned himself with economic evaluation of natural resources. In October of 1981 under his guidance **Voprosy ekonomiki** sponsored a "round table" discussion of problems of the economics of the use of nature. The round table included representatives from economic organs, research institutes, and higher educational institutions. See "Aktual'nye voprosy ekonomiki prirodopol'zovaniia," **VEko**, no. 2 (1982), pp. 81-108. For Khachaturov's articles on natural resource pricing see the bibliography. See also his recent book, **Ekonomika prirodopol'zovaniia** (Moscow: Ekonomika, 1982).

17. For example, Fedorenko, Lemeshev, and Reimers; and N. P. Fedorenko and N. F. Reimers, "Sblizhenie ekonomicheskikh i ekologicheskikh tselei v okhrane prirody," **Priroda**, no. 9 (1981), pp. 3-12.

18. The journals of these institutions are, respectively, **Wissenschaftliche Zeitschrift: Hochschule für Ökonomie Bruno Leuschner** and **Wissenschaftliche Zeitschrift der Technischen Universität Dresden.**

19. E.g., Gerhard Schmidt-Renner, Rolf Bönisch, and Harald Engelhardt.

20. Research on environmental problems at the Zentralinstitut is carried out by Harry Maier, Hans Roos, Werner Gringmuth, and Dieter Graf.

21. L. I. Abalkin, "The Economic Mechanism: Urgent Questions of Improvement," **Ekonomika i organizatsiia promyshlennogo proizvodstva**, no. 1 (January 1979), trans. (abstract) in **Current Digest of the Soviet Press** (hereafter CDSP), vol. 31, no. 19, pp. 6-7.

22. R. P. Povileiko, "Ecology and Economics (A Survey of Ideas and the Literature)," **Ekonomika i organizatsiia promyshlen-**

**nogo proizvodstva**, no. 9 (1982), trans. in **Problems of Economics** 26 (September 1983): 10.

23. See, for example, B. A. Borovskikh, **Planirovanie prirodopol'zovaniia** (Moscow: Ekonomika, 1979), chap. 3; and A. A. Gusev, "Sotsial'no-ekonomicheskaia effektivnost' prirodookhrannykh meropriiatii," in N. N. Nekrasov and E. Mateev, eds., **Sotsialisticheskoe prirodopol'zovanie: ekonomicheskie i sotsial'nye aspekty** (Moscow: Ekonomika, 1980), chap. 6.

24. "Effektivnost' zatrat na okhranu okruzhaiushchei sredy: vremennaia metodika opredeleniia ekonomicheskoi effektivnosti zatrat v meropriiatiia po okhrane okruzhaiushchei sredy," **Ekonomicheskaia gazeta**, no. 33 (August 1980), pp. 13-14. For criticism of the methodology see the comment by Varankin in the round table discussion, "Aktual'nye voprosy," pp. 83-84; and "O deiatel'nosti Otdeleniia ekonomiki Akademii Nauk SSSR," **VEko**, no. 9 (1983), p. 153.

25. O. Kolbasov, "Protect Nature," **Izvestiia**, Jan. 16, 1980, p. 2, trans. (excerpts) in **CDSP**, vol. 32, no. 2, pp. 22-23.

26. For discussion of the situation before this in the USSR see Charles E. Ziegler, "Policy Alternatives in Soviet Environmental Protection," Carl Beck Papers in Russian and East European Studies, paper 102 (Pittsburgh: University of Pittsburgh, n.d.), pp. 3-4. See also Thane Gustafson, **Reform**, pp. 116-19.

27. For the USSR see Gosplan SSSR, **Tipovaia metodika razrabotki piatiletnego plana proizvodstvennogo ob'edineniia (kombinata), predpriiatiia** (Moscow: Ekonomika, 1975), pp. 42-47, 279-86; for the GDR see "Ordnung der Planung der Volkswirtschaft der DDR 1976 bis 1980," in Akademie für Staats- und Rechtswissenschaft der DDR und Ministerium für Umweltschutz und Wasserwirtschaft, **Sozialistische Landeskultur Umweltschutz-Textausgabe ausgewählter Rechtsvorschriften mit Anmerkungen und Sachregister** (Berlin: Staatsverlag der DDR, 1978), pp. 51-64. In many productive associations (**Kombinate**) in the GDR there is a special commissioner for environmental protection (**Umweltschutzbeauftragte**), who is responsible to the director for science and technology. These commissioners are to assure implementation of environmental protection measures. See O. Ritschel, "Leitung und Planung des Umweltschutzes in VEB Bergbau- und Hüttenkombinat 'Albert Funk' Freiberg"; and H. G. Riedel and K. L. Enders, "Ökonomische und betriebswirtschaftliche Aspekte technischer Umweltschutzmassnahmen im industriellen Bereich Energie- und Chemiewirtschaft," both in **Ausgewählte Probleme der Leitung und Planung des Umweltschutzes: Vorträge des Berg- und Hüttenmännischen Tages 1975 in Freiberg**, Freiberger Forschungshefte, D102 (Leipzig: VEB Deutschen Verlag für Grundstoffindustrie, 1976), pp. 7-46.

28. Feitel'man, "Upravlenie prirodopol'zovaniem," p. 57.

29. N. Feitel'man and I. Smagarinskii, in "Regulirovanie ratsional'nogo prirodopol'zovaniia," **VEko**, no. 11 (1981), p. 37, indicate that 9.2 billion rubles were spent in the 10th Five Year Plan, whereas P. Polataev gives a figure of 9.3 billion. "Perspektivy razvitiia okhrany prirody," **Planovoe khoziaistvo** (hereafter **PKh**), no. 1 (1982), p. 40. Of course a problem with all such figures is the criterion for designating an expenditure as environmental protection, which remains unclear. This is a vexing problem precisely because nature-protecting and productive activities often overlap (e.g., returning strip-mined land to agricultural production; using wastes as secondary raw materials; and replanting of forests).

30. T. S. Khachaturov, in "Aktual'nye voprosy ekonomiki prirodopol'zovaniia" (round table), **VEko**, no. 2 (1982), p. 81; and Feitel'man, "Upravlenie prirodopol'zovaniem," p. 59.

31. P. Polataev, "Perspektivy razvitiia okhrany prirody," p. 46. In recent years the report on the fulfillment of yearly economic plans published in **Pravda** has carried a brief section on environmental protection and rational use of nature. In 1982, for example, central outlays for environmental protection were only 1.8 billion rubles according to "Broader Stride of the Five-Year Plan," **Pravda**, Jan. 23, 1983, pp. 1-2, trans. in **CDSP**, vol. 35, no. 4, p. 19.

32. See Gustafson, **Reform**, pp. 117-18.

33. P. Polataev, "Perspektivy razvitiia okhrany prirody," p. 40. Gustafson provides an excellent discussion of the leadership's special interest in problems of water quality and its relationship to Brezhnev's agricultural policies in **Reform**, especially chapter 3.

34. Feitel'man and Smagarinskii, p. 37.

35. T. Khachaturov and A. Bystrov, "In the Interests of Nature Conservation," **Ekonomicheskaia gazeta**, no. 38 (September 1980), p. 15, trans. (abstract) in **CDSP**, vol. 32, no. 38, p. 12.

36. For a discussion of expenditures see Bundesministerium für innerdeutsche Beziehungen, **DDR Handbuch** (Cologne: Verlag Wissenschaft und Politik, 1979), pp. 1095-97. The **Statistisches Jahrbuch der DDR** does contain information on the use of secondary raw materials. See, for example, Staatliche Zentralverwaltung für Statistik, **Statistisches Jahrbuch der DDR 1983** (Berlin: Staatsverlag der DDR, 1983), p. 156.

37. "Direktive des X. Parteitages der SED zum Fünfjahrplan für die Entwicklung der Volkswirtschaft der DDR in den Jahren 1981 bis 1985," **Neues Deutschland**, April 18/19, 1981, p. 6.

38. See chapter 2, pp. 43-44, for discussion of the law.

39. See "On Additional Measures to Intensify Conservation and Improve Utilization of Natural Resources," **Pravda**, Jan. 6, 1979, p. 1, trans. (condensed) in **CDSP**, vol. 31, no. 1, p. 2.

40. On the water use charge and charges for water pollution see "Zweite Durchführungsverordnung zum Wassergesetz – Abwassergeld und Wassernutzungsentgelt – vom 2. Juli 1982," **Gesetzblatt der Deutschen Demokratischen Republik** (hereafter **Gesetzblatt der DDR**), Teil I, no. 26 (July 21, 1982), pp. 485-87. This decree replaces the earlier one which introduced these charges in 1970. See "Zweite Durchführungsverordnung zum Wassergesetz," in Akademie für Staats- und Rechtswissenschaft, **Sozialistische Landeskultur**, pp. 254-60; on charges for air pollution see "Fünfte Durchführungsverordnung zum Landeskulturgesetz," in ibid., pp. 298-300. On charges for land use see **Gesetzblatt der DDR**, Teil II, no. 71 (Aug. 5, 1967), p. 487.

41. Although no reference is made to pollution charges specifically, presumably a situation ensues as described under point 4 on p. 608 in Jürgen Becher et al., **Politische Ökonomie des Kapitalismus und des Sozialismus: Lehrbuch für das marxistisch-leninistische Grundlagenstudium** (Berlin: Dietz Verlag, 1979).

42. "Wassergesetz vom 2. Juli 1982," **Gesetzblatt der DDR**, Teil I, no. 26 (July 21, 1982), pp. 469-70.

43. "Sechste Durchführungsverordnung zum Landeskulturgesetz – Schadlose Beseitigung nicht nutzbarer Abprodukte – vom 1. September 1983," **Gesetzblatt der DDR**, Teil I, no. 27 (Oct. 4, 1983), pp. 257-61.

44. On investment and capital planning see Gerhard Roth, "Betriebswirtschaftliche Probleme der Ermittlung des einmaligen Aufwandes für Umweltschutzmassnahmen," **Wissenschaftliche Zeitschrift der Karl-Marx-Universität Leipzig. Matematisch-Naturwissenschaftliche Reihe** 30 (1981): 102-103; and Wolfgang Strauss and Armin Dittmar, "Zur Berücksichtigung ökologischer Aspekte bei der Vorbereitung von Investitionen," **Wissenschaftliche Zeitschrift der Technischen Universität Dresden** 32 (1983): 136. On charges for natural resource use see chapter 8.

45. V. Afremov and T. Ekondrat'eva, "K vode – po-khoziaiski," **Ekonomicheskaia gazeta**, no. 2 (January 1982), p. 10.

46. On enforcement problems see G. Filimonov, "Protecting the Environment," **Pravda**, Aug. 22, 1978, trans. (condensed) in **CDSP**, vol. 30, no. 34, p. 18. See also Gustafson, **Reform**, pp. 112-15.

47. "To Protect Nature," **Ekonomicheskaia gazeta**, no. 42 (October 1979), p. 22, trans. in **CDSP**, vol. 31, no. 42, p. 20.

48. "Verordnung zur umfassenden Nutzung von Sekundärrohstoffen vom 11. Dezember 1980," **Gesetzblatt der DDR**, Teil I, no. 2 (Jan. 14, 1981), p. 23.

49. Ibid., p. 24. For the 1975 regulation (replaced by the regulations cited in notes 43 and 48) see "Sechste Durchführungsverordnung vom 11. September 1975 zum Landeskulturgesetz," in Akademie für Staats- und Rechtswissenschaft, **Sozialistische**

**Landeskultur**, p. 329.

50. "Verordnung zur umfassenden Nutzung von Sekundärrohstoffen vom 11. Dezember 1980," pp. 25-27. (See note 48.)

51. Ibid., p. 24. For a discussion of policies to encourage recycling see Erhard Beirau, **Wettbewerb Materialökonomie Sekundärrohstoffe** (Berlin: Verlag Tribüne, 1979), pp. 42-48; and Heinz Jurk et al., **Sekundärrohstoffe rationell erfassen und verwerten** (Berlin: Staatsverlag der DDR, 1979), pp. 42-43.

52. Ibid., pp. 17-30.

53. Horst Paucke and Günter Streibel, "Zur Verflectung von Naturprozessen und volkswirtschaftlichem Reproduktionsprozess," **WW** 28 (April 1980): 421.

54. The Institut für Sekundärrohstoffwirtschaft is a part of the Ministry of Material Economy. On the institute see "Erste Durchführungsbestimmung zur Sechsten Durchführungsverordnung zum Landeskulturgesetz – Informationssystem für Abprodukte und Sekundärrohstoffe," in Akademie für Staats- und Rechtswissenschaft, **Sozialistische Landeskultur**, pp. 330, 335-36. See also Beirau, p. 57.

55. See **Die Wirtschaft**, no. 37 (1974), p. 4; and **Neues Deutschland**, Dec. 15, 1978, p. 2. Another relevant institution is the Institute for Municipal Economy (Institut für Kommunalwirtschaft) in Dresden which deals with solid waste problems. See "Institut für Kommunalwirtschaft im Dienste des Umweltschutzes," **Urania**, no. 4 (1979), p. 47.

56. Helmut Schmidt, "Über die volkswirtschaftliche Bedeutung von Sekundärrohstoffen und Abprodukten," **Einheit**, no. 4 (1984), pp. 370, 372; and Jurk et al., pp. 12-14. For comparisons between the GDR and the US see N. Petropavlovskaia and V. Fal'tsman, "Vtorichnye resursy metalla kak faktor ekonomicheskogo rosta," **PKh**, no. 9 (1983), p. 45.

57. Jurk et al., pp. 24-25.

58. "Bericht," **Neues Deutschland**, Apr. 12, 1981, p. 6; and Schmidt, p. 370. For statistics on material economy see Staatliche Zentralverwaltung für Statistik, **Statistisches Jahrbuch 1983**, pp. 151-53.

59. "Direktive," **Neues Deutschland**, Apr. 19, 1981, p. 3.

60. Manfred Schubert, **Zur Entwicklung abproduktarmer bzw. -freier Verfahren**, Sitzungsberichte der Akademie der Wissenschaften der DDR. Mathematik – Naturwissenschaften – Technik, no. 21/N (Berlin: Akademie Verlag, 1982), p. 5.

61. Use of secondary raw materials was previously covered by the "Sechste Durchführungsverordnung vom 11. September 1975 zum Landeskulturgesetz" and is now regulated by the "Verordnung zur umfassenden Nutzung von Sekundärrohstoffen vom 11. Dezember 1980." (See notes 48 and 49.)

62. "Sechste Durchführungsverordnung zum Landeskultur-

gesetz vom 1. September 1983." (See note 43.)

63. Harry Klinger and Wolfram Leunert, "Möglichkeiten und Grenzen der ökonomisch-ökologischen Beurteilung betrieblicher Umweltschutzmassnahmen," **Wissenschaftliche Zeitschrift der Technischen Universität Dresden** 32 (1983): 182-83.

64. V. Novikov and I. Lipsits, "Sovershenstvovanie metodologii i metodiki tsenoobrazovaniia na vtorichnoe syr'e," **PKh**, no. 12 (1983), p. 48. See V. Ksintaris, "Ispol'zovanie vtorichnykh material'nykh resursov v narodnom khoziaistve," **PKh**, no. 6 (1979), p. 38.

65. G. Pokaraev and A. Zatsev note the establishment of new norms in 1977-1978 in forty branches of industry to cover 80 percent of the yearly volume of construction. The reported savings in 1978 alone were 214 tons of ferrous metals, 300 tons of cement, and 1 million cubic meters of forest materials. "Voprosy sovershenstvovaniia normirovaniia material'nykh resursov," **PKh**, no. 8 (1979), pp. 72-74.

66. See the report in **Ekonomicheskaia gazeta** on the failure of the metallurgy industry to improve the effectiveness of the use of metal and to introduce low waste technology. The report, reflecting high level party concern, calls for increased attention to the problem in the media. "V Tsentral'noi Komitete KPSS," no. 24 (June 1980), p. 4.

67. "The Second Time Around," **Pravda**, Jan. 26, 1978, p. 2, trans. (condensed) in **CDSP**, vol. 30, no. 41, pp. 14-15. This article calls for a special section of the national economic plan on recyclable resources, which was implemented three years later. Other barriers noted to the effective use of waste materials include the lack of a technical base and attitudes which view resources as inexhaustible. A lead article in **Planovoe khoziaistvo** in 1981 refers to other policies to encourage material economy. See "Ratsional'no ispol'zovat' resursy proizvodstva," no. 9 (1981), pp. 9-10.

68. A. Iurchenko, "Kompleksnoe ispol'zovanie syr'ia, pererabotka vtorichnykh resursov — vazhnyi reserv ekonomiki," **PKh**, no. 11 (1981), pp. 49-50; and B. V. Gromov et al., **Bezotkhodnoe promyshlennoe proizvodstvo — osnovnye printsipy bezotkhodnykh proizvodstv**, Itogi nauki i tekhniki: "Okhrana prirody i vosproizvodstvo prirodnykh resursov," vol. 9 (Moscow: VINTI, 1981), p. 38.

69. Iurchenko, p. 50; and V. Ksintaris, "Ispol'zovanie vtorichnykh resursov," **PKh**, no. 6 (1981), p. 82.

70. Iurchenko, pp. 51-52. Numerous articles on material economy since 1979 in **Planovoe khoziaistvo**, the publication of Gosplan, reflect the increased attention to the problem at the 26th Party Congress of the CPSU in 1981. Many of these articles include a discussion of the need to use secondary raw materials more effectively, although often no mention is made of the

environmental benefits. Recent All-Union symposia and CMEA meetings (Council for Mutual Economic Assistance) involving several of the East European countries are another indication of the high level of interest in the problem of development of low waste technological processes. Gromov et al. refer to several of these (p. 28; item 90, p. 204; and item 139, p. 207). This publication itself, put out by the All-Union Institute of Scientific and Technical Information of the State Committee for Science and Technology and the Academy of Sciences, likewise reflects growing concern with the problem. The publication includes an extensive bibliography of Soviet and Western sources on low-waste production.

71. V. Ksintaris, "Ispol'zovanie vtorichnykh material'nykh resursov v narodnom khoziaistve," **PKh**, no. 6 (June 1979), p. 41; and "The Second Time Around," **Pravda**, Jan. 26, 1978, p. 2.

72. Fyodor I. Kushnirsky, "The Limits of Soviet Economic Reform," **Problems of Communism** 33 (July-August 1984): 36-37.

73. The samizdat writer Boris Komarov (pseud.) holds a cynical view of the application of new "waste-free" technologies in the USSR. See his **Destruction of Nature in the Soviet Union** (White Plains, NY: M. E. Sharpe, 1980), pp. 91-97.

74. Novikov and Lipsits, p. 49. They do refer to a methodological instruction about the manner of establishing wholesale prices for technological industrial wastes, confirmed by the State Committee on Prices in 1975, but they consider it inadequate (pp. 49-50).

75. G. Zaikov, "Ob ekologicheskom aspekte khoziaistvennogo rascheta," **VEko**, no. 8 (1978), p. 130, does discuss some incentives to encourage utilization of wastes. On pricing problems see the discussion by V. V. Varankin and M. E. Kiabbi in "Aktual'nye voprosy" (round table), pp. 84 and 93-94. In the same round table discussion A. A. Arbatov notes that there should be two price levels for reprocessed waste materials: one for actively polluting materials so that enterprises will have a motivation to reprocess these in order to prevent environmental deterioration; and one for less harmful emissions with prices here based on the materials formed as a result of the reprocessing. In this case the prices should be adequately high to motivate production of the secondary raw materials in the cost-accounting system. (See p. 95.)

76. Petropavlovskaia and Fal'tsman, pp. 48-51; Novikov and Lipsits seem to believe that some prices are too high and some too low, p. 49.

77. V. Selivanovskii, "O planirovanii ispol'zovaniia vtorichnykh resursov," **PKh**, no. 6 (1983), p. 84.

78. See P. Polataev, "Ecology and Economic Development," **PKh**, no. 12 (1979), trans. in **Soviet Review** 22 (Spring 1981): 96.

East German economists make similar proposals for improvement, e.g., Kurt Kutzschbauch, "Ökonomische Aspekte stoff- und energiewirtschaftlich effektiverer Technologien und einer verstärkten wirtschaftlichen Nutzung von Abprodukten als Sekundärrohstoffe," **WW** 29 (June 1981): 669.

79. "The Second Time Around," **Pravda**, Jan. 26, 1978, p. 2; "Ratsional'no ispol'zovat'," p. 9; and V. Raznoshchik and V. Lobov, "Attack on Solid Waste," **Ekonomicheskaia gazeta**, no. 3 (1979), trans. (abstract) in **CDSP**, vol. 31, no. 4, p. 15.

80. Fedorenko, Lemeshev, and Reimers, p. 2.

81. "Osnovnye napravleniia raboty Otdeleniia ekonomiki Akademii Nauk SSSR po razvitiiu ekonomicheskoi nauki v svete reshenii XXVI S"ezda KPSS," **VEko**, no. 4 (1981), p. 15.

82. Ibid., pp. 12-14. See also Fedorenko's discussion of tasks facing economists. "Zadachi ekonomicheskoi nauki v svete reshenii XXVI S"ezda KPSS," **VEko**, no. 4 (1981), pp. 17-25. The attention granted to environmental problems in T. Khachaturov's opening address to the Leningrad branch of councils of the Academy of Sciences was somewhat greater, according to the report in **Voprosy ekonomiki**. See G. Seleznev, "Godichnoe sobranie Leningradskikh filialov sovetov Akademii Nauk SSSR po ekonomicheskim problemam," no. 8 (1981), pp. 158-59.

83. See "O deiatel'nosti," pp. 153-55; and D. Akhobadze, "Godichnoe sobranie Leningradskikh filialov nauchnykh sovetov AN SSSR po ekonomicheskim problemam," **VEko**, no. 8 (1983), pp. 156-57.

# 6
# Economic Growth and the Environment

The primary goal of Soviet economic policy since the First Five Year Plan (1928-1932) has been to realize a high rate of economic growth. Initially, the priority development of heavy industry was seen as the key in accomplishing this goal. Consequently the agricultural and light industrial sectors suffered, producing disproportions which still plague the Soviet economy. The priority placed on rapid industrial growth resulted from several factors: the relative backwardness of the USSR at the time of the revolution in 1917; the perceived need for a high level of military preparedness in the face of a hostile international environment; and the belief that a developed industrial society and a large working class are necessary preconditions for the building of a communist society. High growth rates were also top priority in the GDR after 1949. From the East German viewpoint growth was needed to match the "economic miracle" of the Federal Republic of Germany. Furthermore, the USSR encouraged the satellite countries of Eastern Europe to emulate Soviet development patterns, even though they might be inappropriate for the industrially advanced societies of the GDR and Czechoslovakia. As in the USSR, the Stalinist model here too placed agricultural and consumer goods production behind heavy industrial growth. As Soviet economists themselves admit, environmental protection has generally had even lower priority, some early conservation measures notwithstanding.[1] With the slowdown in the economic growth rates since the early 1960s, the importance attributed to growth in both the GDR and USSR has not diminished. One of the main motivations for the economic reforms in the GDR and USSR in the mid-1960s was to provide additional stimulus for economic growth through intensified use of available productive assets, and greater enterprise efficiency.

## NO "LIMITS TO GROWTH": RESPONSES TO THE CLUB OF ROME

A striking similarity often noted between Soviet-bloc and Western countries is this common emphasis on economic growth. Many environmentalists in the United States and Western Europe see growth fetishism as a powerful source of environmental destruction, since it leads to rampant exploitation of natural resources, creates artificial needs and a throw-away society, and breeds insensitivity to nature's own needs.[2] An influential and systematic attack on the viability of continued growth from an environmental viewpoint appeared in 1972 in the book by Donella Meadows et al., **The Limits to Growth**, published under the sponsorship of the Club of Rome.[3] Subsequent publications by the Club of Rome and J. W. Forrester's book, **World Dynamics**,[4] stimulated considerable discussion not only in the West, but also in the Soviet Union and the GDR. The overt response in these countries was an unambiguous rejection of these analyses and their prediction of world economic collapse, if existing patterns of economic development continue. GDR and Soviet analysts also gave a cool reception to some later less alarmist reports by the Club of Rome. Discussions of the Club of Rome publications, however, fueled already developing interest in questions about the relationship between optimal growth rates, protection of the environment, and depletion of valuable natural resources. Even into the 1980s responses to the controversial conclusions of these early studies continue to appear in the GDR and USSR. Examination of this debate reveals some of the parameters of public discussion in the two countries, the impact of Western literature on Soviet-bloc theory, and the broad context within which Soviet environmentalists must make their case.

The studies which have received the most critical attention in the GDR and USSR are the first two reports to the Club of Rome, **The Limits to Growth**, already mentioned above, and **Mankind at the Turning Point** by Eduard Pestel and Mihajlo Mesarovic. Both studies employed computerized mathematical models to develop prognoses of future global development in light of growing natural resource shortages, exponential population growth, and worsening pollution. The models developed in the two studies differed, as did their conclusions. The first study, carried out by an MIT research team, concluded that a state of global equilibrium is needed if ecological and economic collapse is to be averted. Such an equilibrium state would require constant population and capital plant, but the authors emphasized that such limitations need not imply stagnation, for the most fulfilling forms of human activity (e.g., social interaction, science, art, and education) could continue to develop without adverse effect.[5]

This global equilibrium or "no-growth" concept, as it is usually called in the GDR and USSR, was unambiguously rejected there. Soviet and East German theorists received the reformulation of the problem by Mesarovic and Pestel, in the second report to the Club of Rome, with more ambivalence. Mesarovic and Pestel termed unqualified arguments "for" or "against" growth "naive." Rather, they suggested the viability and necessity of "organic growth" as opposed to "undifferentiated growth." This distinction derives from an analogy to nature itself, undifferentiated growth resembling cancerous development, while organic growth "means that various groups of cells begin to differ in structure and function," as in a biological organism.[6]

> In the new conditions, exemplified by the global crisis-syndrome, the world community has been transformed into a world system, i.e., a collection of functionally interdependent parts. Each part – whether a region or a group of nations – has its own contribution to make to the organic development of mankind: resources, technology, economic potential, culture, etc. In such a system the growth of any one part depends on the growth or non-growth of others. Hence the undesirable growth of any one part threatens not only that part but the whole as well.[7]

This model also took account of regional differences, as well as of the adaptiveness of human society to meet crises, thus answering some objections raised to the Meadows' study by East German and Soviet theorists.[8] The overall reaction in the USSR and GDR was, however, ambivalent. Some critics saw the report as a justification for continued imperialist policies, making the Third World the periphery or limbs of the organic global economic system, subject to the commands of the "imperialist brain."[9] Other critics, however, even expressed mild support for the organic growth concept, noting that if the concept is to have "any reasonable meaning, then it is best understood as such planned economic growth as is realized in socialist society."[10]

Later studies for the Club of Rome have received considerably less attention, perhaps because Soviet and East German analysts find their conclusions less dramatic. But generally each subsequent study has elicited fewer objections, as Western researchers have introduced greater subtlety and differentiation into their models. Soviet researchers even participated in producing a later Club of Rome study (not specifically focusing on environmental problems), **Goals for Mankind.** The 1978 report by Dennis Gabor and Umberto Colombo, **Beyond the Age of Waste**, was also praised for placing appropriate emphasis on the **economic** and international **political** limits to future growth and on such natural limits as induced climatic changes, oxygen depletion, and an altered heat balance, rather than on shortages of natural resources.[11] Critics still maintained critical distance, however, claiming that even these experts come to utopian conclusions, due to their class-neutral approach.[12]

Several standard critiques of the early Club of Rome prognoses appear in almost all commentaries. First, most critiques interpret the studies as bourgeois apologetics in response to the developing crisis of capitalism, and at the same time as attempts by the captains of monopoly capital to come to terms with changing world conditions which require a new "strategy of survival" for capitalism.[13] In this view, the studies were triggered in the 1970s by declining growth rates and energy shortages, induced in large part by the end of cheap Third World resources which had supported the economic growth of the 1950s and 1960s. Soviet-bloc analysts argue that such futuristic projects attempt to recast the contradictions of capitalism as contradictions facing all of humanity, thus discounting the potential of socialism and masking the bankruptcy of capitalism.[14] Accordingly, these apologetic efforts take two mirror-image forms. First, the less compelling form, pseudo-optimism (exemplified by H. Kahn and his Hudson Institute) provides a hardly credible effort to whitewash the severity of the crisis.[15] Somewhat more credible are the technological pessimists like Meadows et al. and Forrester who try to channel criticism of capitalism into a harmless direction, suggesting a type of technological convergence between social systems.[16] Both tendencies have as their object "a common social and ideological goal, objectively serving the interests of the ruling class of capitalist society . . . to preserve as long as possible the status quo of the present world, to delay the inevitable fall of capitalism. . . ."[17]

Even while providing an ideological interpretation of the function and goals of these Western studies, Soviet and East German analysts are careful to exonerate the scholars involved of any intentional deceit or ill-intention. The faults of the studies are not subjectively based (i.e., in the individual failings of the researchers), but are rooted in the objective social context in which they emerge. In fact, the Soviet-bloc critics emphasize the competence and sincerity of the scholars.[18] Furthermore, many analysts admit that in spite of their propagandistic tendencies, the Club of Rome studies have made a positive contribution. They have succeeded in focusing public attention in the West on what are admittedly serious **global** problems requiring an international response.[19] Some even admit that "there is undoubtedly a realistic element in bourgeois projections of the future."[20] Specific prescriptions, such as support for long-term economic prognoses and international cooperation to combat pollution are deemed to be undeniably valid.[21]

On a substantive level, the Soviet and East German rejection of no-growth prescriptions rests on a firm assertion of the potential of scientific-technical progress under socialist conditions. The Club of Rome researchers are charged with ignoring two important factors: (1) the differences between social systems in their ability to respond to these new global challenges;[22] and (2) the unlimited potential of science and technology to develop new sources of raw materials and to respond to environmental pollution.[23] Both of these weaknesses, critics allege, turn the Club of Rome prognoses into extrapolations of past capitalist experience, leading either to unduly pessimistic or, in later studies, to utopian solutions.[24]

Finally, while most objections to the Western prognoses are directed toward the basic premises underlying the mathematical models, methodological criticisms are also raised.[25] But it is on this level that we also find the greatest enthusiasm for the studies. Many commentators feel that scientific efforts to predict future global developments are to be emulated, not rejected. In fact, such modelling "will surely in the future become one of the scientific instruments which will help communist society to develop its interaction with nature and its overall development, consciously, on a scientific basis."[26] While no comparable global models have emerged yet in the Soviet world, they seem to be in the works. And since the mid-1970s mathematical modelling of smaller eco-regions in the Soviet Union has become a new and important direction of scientific research.[27]

While the general Soviet and GDR reaction to the "limits to growth" debate has been a clear rejection of any no-growth alternative, some scholars are less critical. Two Soviet scientists stand out for their positive reception of the Club of Rome approach. Petr Kapitsa, the late, world-renowned physicist, expressed unreserved enthusiasm in 1973. Regarding the potentially explosive growth of world population, Kapitsa wrote: "In recent years the most interesting and persuasive results were those obtained by J. W. Forrester and the couple D. H. and D. L. Meadows."[28] Likewise, Petr Oldak, an economist at the Siberian Division of the Academy of Sciences of the USSR (SO AN SSSR, Novosibirsk) provided a remarkably unideological discussion of the Club of Rome studies in a 1979 popular science book. He made no reference to the apologetic function of the reports and described the studies as "none other than the first steps toward a mathematical description of the planetary bio-social system."[29] In raising some of the standard substantive objections to the studies' conclusions, Oldak saw these weaknesses as resulting from the inherent difficulty of the task, not from ideological preconceptions. Both Oldak and Kapitsa have been among the most outspoken ecological critics in the USSR on other issues as well. Other scholars are more cautious in endorsing the studies. They express their sentiments in veiled form: support for global modelling as such; praise for the competence of the researchers; or agreement with particular conclusions or prescriptions. Some provide such extensive descriptions or quotations from the studies' "objectionable" conclusions that one suspects a hidden desire to communicate them to a larger audience. Other scholars propose a more nuanced understanding of the meaning of economic growth. For example, Rychkov suggests "the possibility of a steady state for humanity in the system of nature," and a rejection of "an extensive type of development of life" or of "cancerous growth."[30] This language suggests support for the organic growth concept of Mesarovic and Pestel. Other scholars see a need to **choose** an **optimal** rate of growth, taking environmental impact into account.[31]

In the GDR even such modest support for limits to growth is unusual. On the other hand, the most radical critique of growth emanated from a GDR scholar, Wolfgang Harich. Harich's book, **Communism Without Growth**, was denied publication in the GDR in the mid-1970s.[32] The book expressed explicit agreement with the Club of Rome's assessment of environmental problems and saw the solution to be a communist society pursuing a no-growth policy. Harich sensed resonance for his views in some Soviet writings, but not in GDR publications.[33] Indeed, in the official GDR media it is difficult even to find environmental advocates who can do more than criticize "growth fetishism."[34] The optimistic commitment to continued economic growth in both the GDR and USSR is quite obviously a mandated optimism, resembling a statement of faith (or a product of censorship) rather than a result of empirical study. If Soviet scientists are allowed somewhat more leeway in responding to the Club of Rome studies, nonetheless for both countries we have discovered one of the clearest limits imposed in environmental discussion: Direct criticism of the necessity of economic growth in socialism is taboo.

It is not difficult to envisage the reason for the sacrosanct position of economic growth in Soviet and East German ideology. From the vantage point of Soviet and GDR experience, growth is closely related to the very essence of social progress; to suggest otherwise is to call into question the basis of present and past economic and political choices. In 1976, Harry Nick summarized the official position well in **Einheit** (Unity), the official theoretical journal of the SED: "Communism without growth would not be a different communism, but none at all."[35] This statement appeared just a year after the West German publication of Harich's **Communism Without Growth**. By 1983, however, the commitment to growth was becoming more differentiated, reflected in the following statement by Politburo member Günter Mittag, also published in **Einheit**:

> Our marxist-leninist economic science assumes that the economy of the German Democratic Republic is and remains growth-oriented. In socialist economics there is no alternative to the constant growth of production and of national income. With the transition from an extensive to an intensive path of growth and with the total switch to this intensive path the character of growth itself changes, however. . . . The resource-saving path of growth is then the most economically promising [**günstig**], and it is also the most promising from the standpoint of all its social effects, including the necessary steps for environmental protection.[36]

This statement, then, may be taken as a measure of the influence which environmentally-conscious economists have had in shaping the official understanding of the economy-ecology interface. Let us now examine their arguments, as they have attempted to bring subtle, but important, alterations in the understanding of that important ideological foundation of Soviet and GDR policy, the commitment to economic growth.

## THE ECONOMICS OF GROWTH

By now, the task confronting environmentalists in the USSR and GDR should be clear: to demonstrate that economic growth and environmental protection are at least compatible, if not mutually supportive. And indeed this is precisely the thrust of much Soviet and East German economic literature advocating increased attention to environmental protection. Within these constraints fairly radical positions can be taken. Some economists simply try to determine the gains and losses ensuing from environmental destruction and environmental protection; but others chart new terrain by suggesting that the problem lies in the prevailing understanding of economic growth itself. In this way they make growth itself a contested concept, and open the way for radical re-assessments of social priorities without ever questioning the unquestionable.

Official Soviet and East German definitions make "rational utilization of nature" an integral component of "nature protection."[37] From this viewpoint, there is no necessary conflict between the careful use of natural resources (one component of environmental protection) and economic growth. Certain measures primarily intended to further more intensive and complete use of limited resources are not motivated by concern for the environment, but nonetheless are consistent with efforts to prevent resource depletion or the destruction of valuable resources. From this point of view, "rational utilization of natural resources" is not only consistent with continued economic growth, but, as official statements emphasize, a prerequisite for it, particularly as now the sources of growth are understood to be intensive, not extensive (i.e., growth predicated on more efficient use of productive inputs, whether they be labor, energy, natural resources, or capital, rather than an expansion of inputs). It is probably in part because the GDR felt the crunch of resource shortages and rising import costs before the resource-rich Soviet Union that some environmental protection measures proceeded more quickly there (i.e., the passage of the **Landeskulturgesetz** in 1970, the institution of payment for resource use in the late 1960s and early 1970s, the earlier inclusion of environmental protection in national economic plans, and the assertive development of recycling programs). In the words of the Minister of Material Economy in the GDR, "strict frugality" corresponds "to the essence of socialist economic management."[38] The costs of unused waste products and inefficiently processed raw materials are too high to be ignored, from the viewpoint of productive interests, entirely apart from environmental concerns. The major problem in terms of improved utilization of natural resources involves the development of planning methods and incentives which will prevent a conflict between rational use of exhaustible natural resources and fulfillment of plan indicators at the enterprise level.[39]

The relationship between pollution control and economic growth is, however, more ambiguous. Conventional economic wisdom in the GDR and USSR, as elsewhere, holds that increases in expenditures for pollution control and nature preservation (nature preserves, protection of endangered species) will necessitate some reduction of growth rates, since such expenditures will leave less for capital investments in material production as such. The East German economist Johann Köhler is a strong advocate of this viewpoint and concludes that environmental expenditures should be delayed. They can better be afforded "when we are richer. . . . They don't make us richer in material goods, but rather draw away working forces from the area of production of material goods."[40] In recent years, however, a considerable debate has developed contesting this rather simplistic viewpoint.

Environmental proponents in both the GDR and USSR have elucidated in a complex manner the interaction between growth and environmental deterioration. Even many economists who are not strongly associated with environmental advocacy emphasize the **long run** compatibility of environmental protection and economic growth. These economists want to determine the proper balance between the two goals. For if an all-out attack is made on environmental problems now, they argue, this could hold back crucial developments in other economic sectors. In the long run, this would reduce the ability of the economy to pursue a policy of harmonious interaction with nature. On the other hand, if the balance is too one-sidedly placed on growth at the expense of nature, future generations will suffer and this will generate other costs. In the spirit of this approach, a gradual solution to environmental problems is required. The Soviet economist Loiter sees nature protection as primarily a problem of "preserving the 'productivity' of the environment for the use of future generations." He concludes that this "must not lead to the slowing of economic growth rates and thus to the postponement of the no less important goals of improving the well-being of the people."[41]

A minority of economists take a more radical approach, denying altogether the conflict between economic growth and pollution control. Their perspective implies a fundamental rethinking of economic categories, a problem we will examine more fully in the next chapter. The East German economic geographer Schmidt-Renner and the Soviet economist Oldak are outspoken proponents of this school of thought. The real choice, according to Oldak, is

> either to attain high indices in a limited time interval, which in part can be accomplished through abandonment of the expenditures for environmental protection but which in the longer run will lead to growing difficulties and a falling off of the growth rate, or to adopt somewhat lower initial indices with simultaneous provision for the needed resources for environmental protection, which will assure a high and stable rate of growth.[42]

The difference between the moderate view of Loiter and the more radical view of Oldak is one of emphasis, not type. The views of economists lie on a continuum expressing subtly differing assessments of the priority of environmental protection, which in turn lead to differing policy recommendations. For the moment we will put aside the more radical macro-economic proposals arising from the Oldak-type critique, and focus on the general efforts of economists to demonstrate the interdependence between growth and environmental protection.

The first link in the argument is to demonstrate that environmental damage itself is economically very costly. Since the mid-1970s in the USSR, and more recently in the GDR, mathematical-economists have begun generating models to evaluate the economic costs of environmental deterioration. These studies are very important, but still rudimentary.[43] Economists buttress their own efforts by pointing to foreign experience: estimated losses from environmental deterioration in countries like the U.S. and Japan provide some indication of the dimensions of the problem.[44] In 1981 Gofman and Gusev, both of TsEMI, developed a categorization of ecological expenditures: (1) measures intended to prevent damage to the environment with the help of nature-protecting measures (e.g., new technology, filtering installations); (2) measures to avert the effects of ecological disturbances on their recipients (e.g., dilution of waste water, creation of sanitary defense zones, high smoke stacks, air purifying systems, special noise-resistant windows and walls); and (3) compensation for damage done to the recipients of the ecological disturbance (e.g., increased medical services, compensation for lowered labor productivity, compensation for loss of the productivity of land, water, forest resources). Gofman and Gusev argue that foregoing type (1) expenditures may in fact produce an overall economic loss if, as a result, types (2) and (3) exceed prophylactic or type (1) costs. Unfortunately expenditures of types (2) and (3) are hard to estimate.[45]

The fundamental point, although it is not expressed in these words, is that, with state ownership of the means of production, there **can be no** externalization of costs as there can be in capitalism. As the GDR economists Paucke and Streibel explain:

> A productive enterprise can undoubtedly achieve higher effectiveness if it "saves" the necessary expenditure, for example, for purification of water emissions. But this enterprise effectiveness turns out to be only the "appearance of effectiveness" [**Scheineffektivität**] because what is involved is simply a shifting of costs to other areas of the economy.[46]

If the books of individual ministries or enterprises indicate that environmental expenditures raise costs, the fault lies not with costly environmental protection methods, but rather with the measurements of costs and benefits in the present planning and accounting system. Some costs and benefits escape the eye of the planner and accountant, thus producing dangerously low estimates of the costs of pollution and the benefits of protection. The Soviet economist Fel'zenbaum clarifies the concrete impact on enterprises. In metallurgical plants gas purification facilities reduce the output-capital ratio by 2-2.5 percent, raising the enterprise cost of commodity output by 0.5-1 percent and reducing profitability of production by 0.8-1.2 percent.[47] Thus metallurgy plants suffer poor cost-accounting indicators due to pollution control efforts. Other enterprises may benefit, of course, but the economic benefit is not then attributed in cost-accounting schedules to the environmental protection measures implemented at the metallurgy plants. The result of the accounting system is that metallurgical plants often turn off pollution control equipment at night to raise cost-accounting indicators.[48] Similar dynamics operate in other branches of the economy. In the mining industry, for example, "methods have not been developed to consider expenditures on the protection and restoration of the environment and their compensation in the enterprise cost of production and in prices."[49]

At a more general level, GDR economists emphasize that nature presently provides society with "free services." If, however, natural processes are so damaged by human intrusion that these "services" can no longer be rendered by nature, society will have to expend large amounts of money to replace them. Thus, although these natural "services" have no real economic exchange value at present, their "use value" should not be underestimated. For example, these economists calculate that society would have to come up with fifty million marks per day if the self-purification capacity of the Elbe along eighty-five kilometers above Magdeburg had to be replaced by water purification installations. Similar costs will ensue if the "free" production of wood and food by plants is reduced due to sulfur dioxide pollution and acidification of soil. If natural regulation of plant and animal populations is disrupted, expensive measures for pest control are also necessary. Increased expenditures for equal output will mean lowered labor productivity. The authors conclude that nature's "free services" must be economically evaluated and reflected as a part of national income.[50]

Several economists have developed methods to evaluate the economic effectiveness of environmental protection measures, attempting to take into account the indirect effects on production. Recently, as noted in chapter 5, a temporary methodology for this purpose was adopted by Gosplan. At this point, the methodology is intended mainly to provide a basis for central evaluation of alternative responses to environmental deterioration, rather than directly to act as a stimulus to enterprises.[51] Two Soviet economists, Fedorenko and Gofman, point out the importance of such a methodology, even if it does not immediately affect enterprise cost-accounting indicators:

> Such a method would allow the economic substantiation of an increase in the expenditures for protection of nature, that is to demonstrate that these expenditures do not reduce, but increase the effectiveness of social production. At the present time such a long-term path of protection of the environment, such as the transition to low-waste tech-

nological processes, often look economically unprofitable.[52]

Environmental investments have a short recovery time, long duration effects, and low running costs compared to most other investments.[53] Furthermore, were environmental damage evaluated accurately, this would not only justify increased overall attention, but also a shift to greater emphasis on reduction of air and noise pollution. As noted in chapter 5, about three-quarters of all centrally-funded environmental investments address water problems, which is disproportionately high relative to the economic damage caused.[54]

The type of distortions and measurement problems discussed by these economists are not unique to Eastern Europe, but plague Western economies as well. This problem is perhaps even a more difficult one for mixed capitalist economies due to the absence of an economic expression of externalities in these systems and of a centralized planning system to develop one. On the other hand, the similarities that exist in practice between the two systems could provide a starting point for a very fundamental political-economic critique of Soviet-type economies.[55] Such a critique does not, for obvious reasons, ensue in these East German and Soviet analyses. Sometimes, however, subtle comparisons between East and West are drawn. One author praises Barry Commoner for advocating production for social use and "not for private profit or 'fulfillment of the plan.'" This author raises no objection to the implicit association of profit-seeking with plan fetishism and quotes Commoner's statement verbatim.[56] This juxtaposition is probably intentional.

In addition to general arguments about the economic benefits of prophylactic environmental measures, economists and technical experts emphasize the link between productive use of wastes and secondary raw materials, on the one hand, and environmental protection, on the other. Productive use of wastes has multiple economic benefits: production costs decline; energy and material resources are saved; imports can be reduced; and expenditures on safe disposal of pollutants or compensation for their negative effects are reduced. It therefore furthers both economic growth and environmental protection, demonstrating the mutually supportive character of the two goals.[57]

The direct economic benefits of the use of productive wastes are widely recognized in the USSR and GDR, and increasingly assertive efforts to capture these benefits are being made, as noted in the previous chapter. An East German publication provides the following cost comparisons:[58]

1.0 Mark for scrap metal would cost 1.3 Marks for comparable primary raw materials

1.0 Mark for recycled paper would cost 2.4 Marks for comparable primary raw materials
1.0 Mark for wood remnant and bark would cost 2.0 Marks for the import of sawed timber
1.0 Mark for returnable glass containers would cost 1.35 Marks for new production

Likewise, V. Ksintaris, the first deputy chairman of Gossnab, provides examples of savings through use of secondary raw materials. One ton of steel produced from scrap is twenty times cheaper than one from ore; one ton of aluminum alloy produced from secondary raw materials uses twenty-three times less electrical energy, 7.4 times less conventional fuel, is 2.5 times less costly to produce, and requires one-eighth the capital investment of production from ore.[59] Furthermore, dumps presently take up over 14,000 hectares of land, which might otherwise be suitable for agricultural production.[60] Savings in the 1980s and early 1990s from reuse of major sources of secondary resources are estimated at fourteen billion rubles in the USSR.[61] Despite the national economic benefits which would ensue from such measures, economists complain that these effects are not adequately connected with the operation of economic mechanisms affecting enterprise behavior.[62] Furthermore, even existing evaluations of the use of secondary raw materials don't generally take account of environmental benefits. Were such effects taken into account, the cost-effectiveness of utilizing wastes productively, or especially of introducing new low-waste technology, would increase even further. Ksintaris estimates, for example, that production of steel from scrap would lower the emissions into the atmosphere by 6.5 times.[63] While officials or economists primarily interested in material economy sometimes take note of the environmental benefits, proper evaluation of the economic gains from low-waste technology or reuse of wastes must explicitly focus on environmental effects, taking account, for example, of the interaction of pollutants and regional differences in the economic impact of pollutants, all expressed in value terms to allow comparison between alternative strategies. At this point such a methodology is lacking, thus reducing the attractiveness of these policies, and omitting environmental consideration from comparison of alternatives.[64]

Environmentalists in the GDR and USSR attribute great theoretical importance to closed-cycle and low-waste technology. As the Soviet economist Arakelian points out, conventional environmental protection equipment is merely a palliative measure; it does not eliminate harmful effluents entirely and it often only alters the locus of the pollutants' effect.[65] Low-waste technological processes provide a more ideal, long-term solution. A group of East German economists, in an important book on environmental protection published in 1979, related closed-cycle technology to the Party's important concern with achieving increases in labor productivity.[66] Labor productivity is considered to be a key factor in maintaining economic growth rates in the USSR and GDR. These East German economists argue that in earlier times a combination of technical progress and use of the richest and most easily accessible natural resources allowed increases in worker productivity. But eventually worsening natural conditions, brought about by human activity, will choke off the continued rise of labor productivity. With more difficult access to natural reserves, less rich deposits, lower quality ore, and rising pollution, more labor must be expended in each phase of the extractive and production process, unless these worsening factors can be compensated for by technical advances.

> With increasing demands on nature and less favorable natural conditions, if the expenditure of labor for the development of resources and for the disposal of material waste increases more quickly than the expenditure for the processing of materials declines, then the danger exists that labor productivity overall will decline, and then there exists an economic compulsion to meet this danger through the better utilization of materials.[67]

This "economic compulsion" in turn requires a new technological orientation, the shift to low-waste technology. In this way the same volume of labor can be better distributed to produce a larger volume of output, since former wastes are transformed into products. The effects of worsening natural conditions on labor productivity can be counteracted. The shift to low-waste technology, these German economists argue, is therefore necessary for economic growth. Furthermore, the GDR economists Streibel and Paucke argue that low-waste technology provides the groundwork for a harmonious nature-society relationship. Until now, human productive processes have been unidirectional, with little concern for feeding wastes back into the productive cycle. According to Soviet estimates, only about 1 percent of the materials extracted from nature by humans appear in final products, leaving about 99 percent as wastes of one sort or another.[68] Natural processes, on the other hand, are largely cyclical processes of material exchange, involving an interaction of the building up of new living systems and their subsequent decomposition (a key role being played by micro-organisms here). Although nature herself does not productively use all of her wastes, the "technical imitation of nature" involved in developing closed productive processes provides the long-term basis for the harmonization of society with nature.[69] In previous time periods the human impact on the natural environment was minimal enough to allow the synchronization between linear societal production and cyclical natural processes, because nature could take up part of the recycling process herself. Now with the immense and growing impact of society on nature, this is no longer possible. The capacity of nature to break down products of human production is quickly being exceeded and indeed human influence on the environment may disrupt even nature's own adaptive mechanisms. Complete use of productive wastes is impossible, these authors argue, because the progressive inclusion of previously-generated wastes into a subsequent link in the productive cycle would require an incredibly complicated network, both technically and economically unfeasible to realize. But vast improvements can be made by introducing low-waste technology wherever possible and by making use of organic and microbiological processes.[70]

Many East German and Soviet analysts express a preference for large-scale productive complexes to realize the closed-cycle concept, although due to transport costs and gaps in the necessary large-scale technical apparatus, these may not be entirely feasible.[71] This preference is rooted in a particular understanding of the "ecologizing of production": Ecology involves interdependence ("everything is connected to everything") which implies complexity and centralization; in short, engineered closed-cycle cybernetic systems on a large scale are required to capture the complexity. This logic contrasts sharply with the ethic of the mainstream Western ecology movement. Its "small is beautiful," "back to the earth" approach to ecology is based on a different logic: Ecology means self-regulation of nature, which implies social self-regulation and local autonomy.[72] This "small is beautiful" approach in the West has its specific historical roots in the crisis of values of the 1960s which linked ecology to a general anti-authoritarian sentiment. In the USSR and GDR the preference for centralized ecological production also has political roots. The leadership undoubtedly finds local autonomy threatening. Furthermore, the prevailing ideology is based on an assumption that global production inevitably is becoming more interdependent (thus local self-regulating systems are anachronistic). The existence of GDR church writings which suggest a link between decentralization and an ecological life-style indicate that, apart from political constraints, some environmentalists in the GDR would opt for the "small is beautiful" variant. (It seems it would surely be attractive to some ethnic minorities in the USSR as well.)

The "ecologizing of production" is granted great ideological and political significance in the GDR and USSR. It forms a component in the transition to communism by providing the foundation for a harmonious relationship between society and nature. Some like Marakhov see it as heralding in a new scientific technical revolution, or like Mottek, involving "from a qualitative viewpoint, a greater technical transformation than that connected with automation," "a revolution of productive forces."[73] For these analysts, technological transformation provides a basis for social advance. The frequent implication is that capitalism will not be able to accomplish this transformation. On a concrete level, this may imply that socialist countries should not so unquestioningly import Western technology unless it meets ecological standards. Indeed, "in the future the only machines that will be sold on the internal, and especially on the export market, will be machines that are non-harmful and wastefree."[74] Thus low-waste technology offers a two-edged advantage on the international scale: immediate reduction of raw material imports, and a longer-term export market. Of course Soviet analysts do recognize that capitalist countries have produced relevant technological advances. East German economists even suggest that centralized closed-cycle complexes may be economically unfeasible because of the need to import large installations from the West.[75] Such realities blur the connection between closed-cycle technology and the advance to communism. But the more important connection for environmentalists, that between closed-cycle technology and economic growth, is easier to substantiate.

## THE LIMITS OF ECONOMIC ANALYSIS

Over ten years ago Samuel Huntington observed that at some point in the consolidation of a one-party system, "ideology ceases to be a way of achieving consensus and instead becomes a threat to consensus."[76] Marxism-Leninism provides fertile ground for dissent. Its justification is scientific and humanistic. Both science and humanism can generate myriads of alternative interpretations, as witness the contemporary neo-Marxist writings from Eastern Europe.[77] We have already seen the importance of scientists and science in the environmental debate in the USSR and GDR. Likewise the humanistic strain in Marxism-Leninism provides environmentalists with a powerful basis for critiquing the prevailing direction of economic growth.

According to official sources, the main goal of socialist policy is to do "everything for the good of the people, for the interests of the working class and all other working people."[78] Pursuit of economic growth is, then, not an end in itself, but a means of realizing this larger goal.[79] In 1981 at the Tenth Party Congress of the SED, Erich Honecker defined the main task of the Party for the 1980s as being "to raise further the material and cultural standard of living of the people on the basis of a high rate of development of socialist production, of increasing effectiveness, of scientific-technical progress, and of the growth of labor productivity."[80] Economic growth is a component of the Party's strategy in providing the means to assure "the all-around development of the socialist personality in a socialist human community."[81]

These claims are an important element of the legitimizing principles of the SED and CPSU. In setting out these humanistic purposes, the leadership not only makes a claim for its legitimacy, but also establishes a high standard for its policies. Great claims may generate great expectations or great cynicism. Cynicism is not an acceptable form of public expression on such matters. So critics of present policy may choose to treat claims as serious guides to policy-making. In so doing, they make the disparity between reality and ideology clearer. In this manner, a legitimizing ideology becomes an ideology of dissent. This does not mean that the economists discussed in these chapters are "dissidents." They surely are not in the common understanding of the word. Rather, the humanistic thrust of Marxism, both in its original articulation and in its Soviet variant, provides firm ground for dissent from a growth-maximizing attitude toward the environment. Practically and immediately the most effective strategy for environmental proponents in the USSR, in the GDR, and probably also in the West is to persuade decision-makers that environmental responsibility is economically responsible as well. But at the same time, calls by critics to live up to humanistic claims may allow injection of a more radical critique. Such appeals may be read as indictments of policy failure. Marxism-Leninism in this respect provides a rich framework for humanistic criticism of government policies and priorities. Consequently, it also provides an ambiguous source of legitimacy for the party leadership.

Some East German and Soviet economists do appeal to the ultimate humanistic goals of socialism in their discussion of the environment and economic growth. Oldak appeals to Lenin's understanding of the goal of socialism, "the complete well-being and free all-round development of all members of society." On this basis he concludes that maximizing the social product is not an adequately complete criterion for evaluating the social effects of production.[82] The final criterion should be the effect of production on well-being, not the creation of material goods. This of course requires consideration of the state of the environment.[83] The East German economic geographer Schmidt-Renner draws a similar connection between the ultimate goals of socialism and the optimal ordering of human needs: "In such an optimal ordering, the need for an intact environment is elementary, precisely because the environment is the most important of all 'use values.'"[84] The problem of environmental disturbance, he continues, is "above all a humanistic problem." Consequently quantitative economic calculation is neither possible, nor, if attempted, likely to reflect correctly the socio-cultural elements of a socialist environmental policy.[85] A movement beyond economic utility reasoning (**ökonomisches Nutzdenken**) is necessary; an ethically based approach reflecting the historical mission of the working class is called for.[86] Schmidt-Renner finally relates his proposals to declarations made at the Eighth Party Congress of the SED, that "socialism is for humanity."[87]

This type of argumentation is reflected less systematically in the writings of other environmental advocates. The East German economist Streibel argues that fulfillment of needs is measured not only by growth of national income, but by healthy working and living conditions.[88] The Soviet economist Zhamin links ecology to the Party's established formula of "securing the steady rise of the material and cultural standard of living."[89] Similar references are plentiful in the Soviet popular press, particularly in newspapers like **Literaturnaia gazeta** and **Komsomol'skaia pravda** which have a reputation **both** for outspoken environmental advocacy **and** for ideological orthodoxy.

This approach is not only noteworthy for its unorthodox appeal to ideological orthodoxy. It also involves economists arguing for the limits of economic analysis. This is a humbling confession for any professional to make about his/her own field of expertise, and it extends here beyond simply the types of appeals discussed above. Economists may refrain from such ideological flourishes, but still point out that many values are non-quantifiable, or simply not amenable to economic reasoning. Examples include the aesthetic worth of pleasant surroundings, the "pricelessness" of human health, the genetic diversity of the planet, the purification of air provided by trees and forests, or the maintenance of a stable global climate.[90] Some negative effects on the environment are simply irreversible and in such cases any economic evaluation would be infinitely high.[91] Oldak sums up the implications well: "The entry into the new era requires a change in our outlook and the assimilation of an 'ecological world view.' Great reason must correspond to the great power that has been acquired by humanity. Humanity must be extraordinarily cautious in the use of the power it has acquired."[92] In disavowing the utility of their own specialized knowledge in dealing with these questions, economists weaken their potential influence in assessing such "non-economic" and "human" values. Fortunately, as we have seen, environmentalists have allies among natural scientists, journalists, and writers who can buttress their support for a humanistic approach to economic growth.

## CONCLUSION

According to one East German scholar, the advantages of socialism cannot be fully realized at present precisely because of the need to preserve socialism through economic growth in the face of brutal and unfair capitalist competition.[93] On the other hand, the proclaimed legitimacy of socialism itself rests on its greater humanism, as some environmental advocates are quick to point out. The dilemma is an old one for Soviet leaders and is by no means limited to the area of environmental protection. Barrington Moore skillfully analyzed this dilemma of power facing the Soviet leadership more than three decades ago.[94] The conflict involves the need to deemphasize or defer the fulfillment of certain humanistic goals of socialism (e.g., fulfillment of popular consumer demands, more equal distribution of income) in the interests of the practical realities of maintaining political power. In this case, the goal of a clean and healthy environment may be sacrificed in the interest of maintaining a competitive growth rate, which the Soviet leadership apparently considers essential to the maintenance of its position. This desire to catch up with or overtake the level of economic development of the capitalist world has roots both in Soviet ideology and in international power politics. From an ideological viewpoint such an economic achievement would demonstrate the superiority of a socialist economic system over a capitalist system, an important victory in the ideological struggle, particularly in the Third World. In addition, the domestic legitimacy of the Soviet regime, and to an even greater extent of the East German regime, depends not only on environmental quality, but perhaps even more on an increased supply of high quality consumer goods for the population. As an issue of world power politics, a competitive pace of economic growth is necessary to support an expensive military in the face of the perceived threat from both the West and China. Only continued economic growth can allow success in all of these areas. Thus arguments for maintenance of environmental quality at the expense of an even marginally reduced growth rate face serious obstacles. A writer in **Komsomol'skaia pravda** has compared this deferment of attention to the needs of nature "during the 'intensive period' of construction to the 'after-us-the-deluge' type of philosophy."[95] However unsatisfying this philosophy may be, to a large extent it prevails in both East and West alike. Environmentalists in the USSR and GDR who wish to have influence must adapt. To this extent they must not openly breach the consensus unless they prefer samizdat or foreign publishers. But they **can** carefully tread beyond the limits of the ordinary to interpret the consensus in an imaginative and unorthodox manner.

## NOTES

1. M. N. Loiter, "Effektivnost' kapital'nykh vlozhenii v okhranu prirodnoi sredy," **Voprosy ekonomiki** (hereafter **VEko**), no. 1 (1976), p. 36; and "Obshchestvo i priroda," **Voprosy filosofii** (hereafter **VFil**), no. 4 (1973), pp. 53-54.

2. The similarity between social systems in regard to growth is an implicit assumption of French ecologist André Gorz in **Ecology as Politics** (Montreal: Black Rose Press, 1980). Gorz does, however, focus on the growth orientation of capitalism most explicitly. See especially chapter 1. For another critique of the growth orientation of Western society see E. F. Schumacher, **Small is Beautiful** (New York: Harper & Row, 1973). Philip R. Pryde likewise identifies the Soviet assumption that economic growth is good as one "ideological" factor which hinders priority development of environmental protection and conservation policies. **Conservation in the Soviet Union** (Cambridge, MA: Cambridge University Press, 1972), p. 165.

3. Donella H. Meadows et al., **The Limits to Growth: A Report for the Club of Rome's Project on the Predicament of Man** (New York: Universe Books, 1972). The Club of Rome is an informal international association, formed in 1968 by thirty educators, economists, scientists, industrialists, and international civil servants from ten Western countries to study the developing global system. In 1970 it initiated the "Project on the Predicament of Man," resulting in commissioning of the studies and the publication of the books discussed here. Subsequent reports, after **The Limits to Growth,** include Eduard Pestel and Mihajlo Mesarovic, **Mankind at the Turning Point: The Second Report to the Club of Rome** (New York: Dulton, 1974); Jan Tinbergen et al., **RIO: Reshaping the International Order: A Report to the Club of Rome** (New York: New American Library, 1976); Ervin Laszlo et al., **Goals for Mankind: A Report to the Club of Rome on the New Horizons of the Global Community** (New York: New American Library, 1977); Dennis Gabor and Umberto Colombo, **Beyond the Age of Waste** (New York: Pergamon Press, 1978); James W. Botkin, **No 'Limits to Learning': Bridging the Human Gap: A Report to the Club of Rome** (Oxford, Toronto: Pergamon Press, 1979); Thierry de Montbrial, **Energy, the Countdown: A Report to the Club of Rome** (New York: Pergamon Press, 1979); Orio Gianni, **Dialogue on Wealth and Welfare: An Alternative View of World Capital Formation: A Report to the Club of Rome** (New York: Pergamon Press, 1980); and Bohdan Hawrylyshyn, **Road Maps to the Future** (Oxford, Toronto: Pergamon Press, 1980).

4. Jay Wright Forrester, **World Dynamics** (Cambridge, MA: Wright-Allen Press, 1971).

5. Meadows et al., chapter V.

6. Mesarovic and Pestel, pp. 3-4.

7. Ibid., p. 5.

8. For the objections see Alfred Bönisch and Dieter Graf, "Mensch am Wendepunkt," **Deutsche Aussenpolitik** 20 (1975): 1011.

9. Ibid., pp. 1012-20.

10. Alfred Kosing, "Mensch-Gesellschaft-Umwelt," **Einheit** 31 (1976): 799. Kosing is, however, much more critical of the non-class approach taken in the model than is Soviet economist P. G. Oldak, who seems quite attracted to the "organic growth" concept. See his **Sovremennoe proizvodstvo i okruzhaiushchaia sreda** (Novosibirsk: Nauka, sibirskoe otdelenie, 1979), pp. 63-64. Other critics interpret the book as an ideological response to the realization that the no-growth concept of the Meadows' study contradicts the laws of capitalism. See Bönisch and Graf, p. 1016; Jürgen Kuczynski, **Das Gleichgewicht der Null** (Frankfurt am Main, FRG: Verlag Marxistische Blätter, 1973), p. 59; Horst Paucke and Heinz Kroske, "Weltmodelle – neue Konvergenzvarianten zu Umweltproblemen," **Deutsche Zeitschrift für Philosophie** (hereafter **DZfPh**) 23 (1975): 1506; and T. T. Timofeev, "O nekotorykh sotsial'nykh aspektakh vzaimosviazi cheloveka i okruzhaiushchei sredy," **VFil**, no. 12 (1976), p. 43.

11. See Dieter Graf, "Von 'Grenzen des Wachstums' zu den Grenzen der Verschwendung," **Wirtschaftswissenschaft** (hereafter **WW**) 25 (October 1977): 1555; and I. T. Frolov and V. V. Zagladin, "Globale Probleme und die Zukunft der Menschheit," **Einheit** 34 (1979): 716-17.

12. E.g., Frolov and Zagladin, pp. 716-18. In the same article, they also praise A. Peccei, the President of the Club of Rome, for the humanism expressed in his book **The Human Quality** (Oxford, 1977), but criticize him for its utopian conclusions. It does not seem coincidental that this article, one of the most sympathetic to the efforts of the Club of Rome found in the East German press (and published in the theoretical journal of the SED), is written by two Soviet authors.

13. See, for example, Horst Paucke and Adolf Bauer, **Umweltprobleme: Herausforderung der Menschheit** (Frankfurt am Main, FRG: Verlag Marxistische Blätter, 1980), pp. 177-78, originally published by Dietz Verlag in Berlin (East) in 1979; Harry Nick, "Mensch und Umwelt," **Einheit** 34 (1979): 710; and Harry Maier, "Ökonomisches Wachstum und gesellschaftlicher Fortschritt," **WW** 26 (January 1978): 30, 33.

14. Many critiques point out the capability of the capitalist system to respond to the crisis in the **short term**. For example, the development of an ecological industry can succeed in generating profits from environmental deterioration. For an in-depth treatment of the capitalist response to environmental problems see Alfred Leonhardt and Gerhard Speer, **Umweltreproduktion im**

**staatsmonopolistischen Kapitalismus** (Berlin: Akademie Verlag, 1977). An excellent Western analysis of the ability of capitalist economies to generate growth from the problems they produce, like environmental deterioration, is provided by Martin Jänicke, **Wie das Industriesystem von seinen Missständen profitiert** (Opladen, FRG: Westdeutscher Verlag, 1979).

15. Frolov and Zagladin, pp. 718-19, provide a critique of H. Kahn, W. Brown, and L. Martel, **The Next 200 Years** (New York, 1976).

16. Harry Maier, "Ökonomisches Wachstum," p. 30.

17. A. Shapiro, "Global'nye problemy i perspektivy mirovoi ekonomiki v burzhuaznykh kontseptsiiakh," **Mirovaia ekonomika i mezhdunarodnye otnosheniia**, no. 5 (1981), p. 113.

18. E.g., Frolov and Zagladin, "Globale Probleme," pp. 713-14; and Nick, "Mensch und Umwelt," p. 706.

19. Nick, p. 706; and A. A. Ryvkin and I. Sh. Amirov, "Kriticheskii analiz modelei global'nogo proekta 'Chelovechestvo pered vyborom,'" **Ekonomika i matematicheskie metody** (hereafter **EiMM**) 13 (1977): 456. Shapiro, as if to justify this concession, cites Lenin's comment about the ability of bourgeois professors to make valuable contributions in areas of factual, specialized research (p. 101).

20. Wolfgang Eichhorn I, "Die Zukunft der Menschheit in der weltanschaulichen Klassenauseinandersetzung unserer Zeit," **Einheit** 34 (1979): 686.

21. E. V. Gordina, "Sotsial'no-politicheskaia ogranichennost' zapadnykh modelei ekologicheskogo krizisa," **EiMM** 13 (1977): 40.

22. Virtually every article cited so far criticizes the reports for ignoring differences between social systems. See, for example, Kuczynski, **Gleichgewicht der Null**, pp. 10-11. Iu. Lipets and N. N. Chizhov describe the Meadows' model as "Malthus with computers," while Herbert Meissner characterizes no-growth as "a shame-faced expression for stagnation." Iu. Lipets and N. N. Chizhov, "Kriticheskaia sotsial'no-ekonomicheskaia otsenka nekotorykh prognozov v global'noi ekologii," in N. A. Krylova, ed., **Priroda i ekonomika: sbornik statei** (Moscow: Znanie, 1975), p. 55; and Herbert Meissner, "Das 'Wachstumsdilemma' der bürgerlichen politischen Ökonomie," **WW** 23 (March 1975): 418. See also Hermann Grosse and Manfred Puschman, **Qualität des Lebens: Ausweg oder Irreführung** (Berlin: Dietz Verlag, 1974), pp. 53-71, 125; and Dieter Klein, "Kritisches zur 'Qualität des Lebens' im Kapitalismus," **Einheit** 29 (1974): 226.

23. E.g., Oldak, **Sovremennoe proizvodstvo**, p. 58. While this is a common criticism of the Club of Rome reports, perhaps more interesting is the skepticism expressed by a few Soviet scholars about the possibilities of integrating this factor into a complex mathematical model. Because the impact of a techno-

logical advance is so dependent on its timing, it would be difficult to make long-term prognoses (e.g., one hundred years in advance) which accurately took scientific-technical innovations into account. Gordina develops this argument most fully, pp. 39-40. For this reason she calls such prognoses "unfruitful extrapolation" (p. 39).

24. Gordina, p. 40. The analysis has changed little over time. See T. S. Khachaturov, **Ekonomika prirodopol'zovaniia** (Moscow: Ekonomika, 1982), chapter 2.

25. Especially Gordina; and Ryvkin and Amirov.

26. Kosing, p. 795. See also V. Motylev, "Forecasts of the 'Club of Rome': Reality and Prophecy," **VEko**, no. 4 (1977), trans. in **Problems of Economics** 20 (October 1977): 80-81; K. V. Ananichev, **Problemy okruzhaiushchei sredy, energii, i prirodnykh resursov: mezhdunarodnyi aspekt** (Moscow: Progress, 1975), p. 157; A. K. Astaf'ev, "Sistemnyi podkhod i problema reguliatsii prirodnoi sredy," **VFil**, no. 2 (1977), pp. 74-75; Adolf Bauer and Horst Paucke, "Umweltprobleme in der Sicht des historischen Materialismus und bürgerliche Weltmodelltheorien," **DZfPh** 24 (1976): 793; and Oldak, **Sovremennoe proizvodstvo**, p. 115.

27. A report on a "school-seminar" held in September 1975 on the planning and management of natural resources includes reports on many models of regional natural resource use. See "Problemy planirovaniia i upravleniia prirodnymi resursami," **EiMM** 12 (1976); 194-99. Iu. Tupytsia discusses mathematical modelling of water resources in the Sea of Azov in "Nature Conservation as a Problem of Economic Science," **Ekonomicheskie nauki**, no. 11 (1977), trans. in **Problems of Economics** 21 (December, 1978): 20. Other examples of the use of mathematical modelling are found in the journal **EiMM**, including the articles by R. L. Raitskas and V. P. Sytkaitis, 15 (1979): 45-57; and 16 (1980): 1081-93. M. Ia. Lemeshev and K. G. Gofman call for development of long-term complex prognoses of global factors for 30 to 50 years or more ahead, in "Simpozium po ekonomicheskim problemam prirodopol'zovaniia," **EiMM** 10 (1974): 381-82. S. M. Vishnev notes that "world models" are being worked out and researched at the Computing Center (VTs — Vychislitel'nyi tsentr) of the Academy of Sciences. See S. M. Vishnev, "O modelirovanii vzaimosviazei obshchestva i prirody," **EiMM** 10 (1974): 481. N. N. Moiseev, deputy director of the Computing Center of the Academy of Sciences, also indicates that his organization has begun work on development of a mathematical language to allow formalization of a systemic approach to the interdisciplinary study of global ecological processes. Moiseev expresses unusual enthusiasm for the methods used by Jay Forrester in his study **World Dynamics**, although he also makes the usual substantive criticisms. See N. N. Moiseev, "Aktual'nye voprosy ekologicheskoi

evoliutsii i 'mirovaia dinamika' Dzheia Forrestera," **VFil**, no. 7 (1978), pp. 171-81.

28. P. L. Kapitsa, in "Chelovek i sreda" (round table), **VFil**, no. 2 (1973), p. 38 (emphasis added).

29. Oldak, **Sovremennoe proizvodstvo**, p. 52. The broader discussion is on pp. 51-58.

30. Rychkov, in "Chelovek i sreda," **VFil**, no. 4 (1973), p. 71.

31. M. Ia. Lemeshev and E. P. Ushakov, "O prirodookhrannykh zadachakh ekonomicheskogo razvitiia," **EiMM** 15 (1979): 27.

32. Wolfgang Harich, **Kommunismus ohne Wachstum? Babeuf und der 'Club of Rome'** (Reinbek bei Hamburg: Rowohlt, 1975).

33. Ibid., p. 49.

34. Günter Streibel, "Umweltschutz und Umweltgestaltung als volkswirtschaftliche Aufgabe," **WW** 23 (August 1975): 1153. He actually takes the phrase from the Soviet scholar Makliarskii.

35. Harry Nick, "Wirtschaftswachstum – seine Perspektiven im Sozialismus und seine bornierten Kritiker," **Einheit** 31 (1976): 595.

36. Günter Mittag, "Ökonomische Strategie der Partei – klares Konzept für weiteres Wachstum," **Einheit** 38 (1983): 1006. Other state and party leaders are less sensitive to these subtleties. For example, in his report on the 1981-1985 five year plan, Willi Stoph states that the goal underlying the development of the plan is "to accelerate the growth of production and of the national income compared to the results of the past five year plan." "Direktive des X. Parteitages der SED zum Fünfjahrplan für die Entwicklung der Volkswirtschaft der DDR in den Jahren 1981 bis 1985," **Neues Deutschland**, Apr. 15, 1981, p. 3.

37. See chapter 2, pp. 42-45.

38. Wolfgang Rauchfluss, "Materialökonomie – dringendes Gebot," **Einheit** 30 (1975): 167. Rauchfluss is also a member of the Council of Ministers and Central Committee of the SED in the GDR.

39. Other questions, concerning optimal rates of exploitation of mineral resources (e.g., to what extent coal should replace oil and/or gas as a fuel source, or how long nuclear fuel will be adequate without breeder reactors), may come up against conflicting estimates of the scope of reserves, but there is no principled conflict between resource depletion concerns and productive concerns regarding such questions, as long as pollution effects are ignored.

40. Johann Köhler, "Zu der Problematik der produktiven und der unproduktiven Arbeit sowie der Dienstleistungen," **WW** 22 (June 1974): 886. In the GDR this view is also expressed openly by Gerd Friedrich and Gerhard Schutz, "Effektivität, wissenschaftlich-technische Revolution und sozialistische Wirtschaftsführung,"

**Einheit** 28 (1973): 189-90. Herman Hörz also refers to environmental measures in passing as "unwirtschaftliche Projekte" (uneconomical projects). **Mensch contra Materie?** (Berlin: VEB Deutscher Verlag der Wissenschaften, 1976), p. 57. P. G. Oldak refers to many in the USSR who posit an irreconcilable conflict between economic growth and the preservation of a clean environment. See P. G. Oldak, "Voprosy ekonomicheskoi otsenki resursov biosfery," in N. P. Fedorenko, ed., **Ekonomicheskie problemy optimizatsii prirodopol'zovaniia** (Moscow: Nauka, 1973), p. 57. Likewise V. Antonov and V. Redin refer to the usual hostility toward antipollution devices based on the additional costs and labor involved. See "The Plant and the Volga," **Trud**, Sept. 16, 1972, p. 2, trans. (abstract) in **CDSP**, vol. 24, no. 37, p. 10. See also note 7 in chapter 5.

41. Loiter, "Effektivnost' kapital'nykh vlozhenii," p. 36. Along similar lines, see Siegfried Grundmann, "Mensch und Umwelt," **DZfPh** 21 (1973): 205; Otto Reinhold, "Der Charakter der gegenwärtige Krise der kapitalistischen Weltwirtschaft und die imperialistische Apologetik," **Einheit** 31 (1976): 85; Nick, "Wirtschaftswachstum," p. 423; and Kuczynski, **Gleichgewicht der Null**, p. 56.

42. P. Oldak, "Scientific-Technical Progress and the Limits of Economic Analysis," **Mirovaia ekonomika i mezhdunarodnye otnosheniia**, no. 9 (1971), trans. in **Problems of Economics** 14 (April 1972): 17. An excellent popular science article reflecting this viewpoint is N. P. Fedorenko, M. Ia. Lemeshev, and N. F. Reimers, "Sotsial'no-ekonomicheskaia effektivnost' okhrany prirody," **Priroda**, no. 10 (1980), pp. 2-13. Likewise the East German economists Horst Paucke and Günter Streibel write, "economic growth and environmental protection are integral components of a unified process, a process which has as its content rationality of the material interaction [**Stoffwechsel**] of nature and society." "Zur Verflectung von Naturprozessen und volkswirtschaftlichem Reproduktionsprozess," **WW** 28 (April 1980): 411. See also Gerhard Schmidt-Renner, "Sozialistischer Umweltschutz unter volkswirtschaftlichen Gesichtspunkten," **Petermanns Geographische Mitteilungen** 118 (1974): 11.

43. Again I refer the reader to numerous examples in the journal **Ekonomika i matematicheskie metody**. (See note 27 of this chapter.) One notable example is K. G. Gofman and A. A. Gusev, "Ekologicheskie izderzhki i kontseptsiia ekonomicheskogo optimuma kachestva okruzhaiushchei prirodnoi sredy," 17 (1981): 515-27, in which the authors try to demonstrate that calculation of the economic damage resulting from pollution must be taken into account if optimal planning decisions are to be taken. In the GDR Gesine Ende, Rolf Steffens and Jürgen Stein estimate the cost of damage to forest recreational areas from $SO_2$ emissions.

"Ökonomische Bewertung des Einflusses von $SO_2$-Immissionen auf die Erholungsleistungen der Waldlandschaft," **Wissenschaftliche Zeitschrift der Technischen Universität Dresden** 32 (1983): 219-22. In the USSR see also L. Mel'nik, "Accounting for Ecological Indicators in Solving Economic Problems," **Ekonomicheskie nauki**, no. 2 (1983), trans. in **Problems of Economics** 26 (March 1984): 20-25.

44. E.g., N. P. Fedorenko and K. G. Gofman, "Upravlenie sotsialisticheskim prirodopol'zovaniem," in N. N. Nekrasov and E. Mateev, eds., **Sotsialisticheskoe prirodopol'zovanie: Ekonomicheskie i sotsial'nye aspekty** (Moscow: Ekonomika, 1980), p. 22.

45. Gofman and Gusev, pp. 515-18.

46. Horst Paucke and Günter Streibel, "Gratisdienste der Natur — wichtiger Effektivitätsfaktor der sozialistischen Volkswirtschaft," **WW** 31 (September 1983): 1317.

47. V. Fel'zenbaum, "Cost Accounting and the Socioeconomic Effect of New Technology," **VEko**, no. 5 (1981), trans. in **Problems of Economics** 25 (June 1982): 25. In contrast, N. Feitel'man and I. Smagarinskii indicate that the fixed funds (**osnovnye fondy**) for nature-protection purposes are **not** used for calculation of the profitability of production, but that other indicators like the enterprise costs of production (**sebestoimosti**) and the return on funds (**fondootdacha**) are often influenced adversely by nature-protection expenditures. They point out that in some cases 30-40 percent of the fixed funds of the enterprise may be related to nature-protection purposes. "Regulirovanie ratsional'nogo prirodopol'zovaniia," **VEko**, no. 11 (1981), p. 43. See also N. Feitel'man, "Upravlenie prirodopol'zovaniem," **VEko**, no. 3 (1984), p. 64.

48. Fel'zenbaum, p. 27.

49. R. P. Povileiko, "Ecology and Economics (A Survey of Ideas and the Literature)," **Ekonomika i organizatsiia promyshlennogo proizvodstva**, no. 9 (1982), trans. in **Problems of Economics** 26 (September 1982): 10.

50. Paucke and Streibel, "Gratisdienste der Natur," especially pp. 1323-31.

51. See chapter 5, p. 157.

52. Fedorenko and Gofman, in Nekrasov and Mateev, **Sotsialisticheskoe prirodopol'zovanie**, p. 22.

53. N. Fedorenko and K. Gofman, "Problems of Optimization in the Planning and Control of the Environment," **VEko**, no. 10 (1972), trans. in **Soviet Review** 14 (Summer 1973): 31.

54. B. A. Borovskikh, **Planirovanie prirodopol'zovaniia** (Moscow: Ekonomika, 1979), pp. 22-23. Along similar lines, Gofman and Gusev suggest that environmental protection measures should be compared to other health expenditures in terms of cost effectiveness (p. 521). Feitel'man compares the economic

effectiveness of various types of environmental investments and finds those utilized to purify water to be the most economically effective. "Upravlenie prirodopol'zovaniem," pp. 62-63.

55. This type of criticism may grow out of a Marxist framework as well. For example, some Marxists who consider the USSR and Eastern Europe to be "bureaucratic state capitalist systems" emphasize the common drive to accumulation of capital, resulting in exploitation of the working class. These types of similarities might fit into their argument. See, for example, Chris Harman, **Bureaucracy and Revolution in Eastern Europe** (London: Pluto Press, 1974), "Introduction" and chap. 1.

56. E. K. Fedorov, **Vzaimodeistvie obshchestva i prirody** (Leningrad: Gidrometeoizdat, 1972), p. 78. It is interesting that in an English-language book put out by the Soviet Progress Publishers Fedorov and coauthor I. B. Novik are more explicit. They write: "Lamenting (and not without certain grounds) that fulfillment of the plan often prompts the heads of Soviet enterprises to treat the natural environment little better than their counterparts in the USA, Commoner rightly notes that the public and the Government of the USSR have in recent years not only fully realised the danger of an ecological crisis, but taken real and increasingly important measures to avert it." See "Ecological Aspects of Social Progress," in **Society and the Environment: A Soviet View** (Moscow: Progress Publishers, 1977), pp. 43-44. For other examples of such juxtapositions see I. Zhavoronkova, "Ob uluchshenii ispol'zovaniia mineral'no-syr'evykh resursov," **VEko**, no. 8 (1970), p. 29; and Jürgen Kuczynski, "Der unbefangene Wissenschaftler," **Das Magazin**, no. 2 (1974), p. 54. Johann Köhler criticized Kuczynski precisely on these grounds in his later rebuttal of Kuczynski's argument. See Köhler, "Problematik der produktiven Arbeit," pp. 884-85.

57. Khachaturov writes: "In the final analysis economy of material resources is an important path of ensuring protection of nature." In "Aktual'nye voprosy ekonomiki prirodopol'zovaniia" (round table), **VEko**, no. 2 (1982), p. 82. Similar statements are made by Bystrov in the same discussion, p. 87. See also B. V. Gromov et al., **Bezotkhodnoe promyshlennoe proizvodstvo**, Itogi nauki i tekhniki; "Okhrana prirody i vosproizvodstvo prirodnykh resursov," vol. 9 (Moscow: VINTI, 1981), p. 29. I. Pashko and V. Pavlov claim that reduction of material inputs in production affects the tempo of growth significantly more than improved utilization of capital. "O planirovanii pokazatelia materialoemkosti produktsii," **Planovoe khoziaistvo**, no. 8 (1979), p. 82.

58. Heinz Jurk et al., **Sekundärrohstoffe rationell erfassen und verwerten** (Berlin: Staatsverlag der DDR, 1979), p. 14.

59. V. Ksintaris, "Ispol'zovanie vtorichnykh material'nykh resursov v narodnom khoziaistve," **Planovoe khoziaistvo**, no. 6

(1979), pp. 38-39.

60. Ibid., p. 43.

61. V. Ksintaris, "Ispol'zovanie vtorichnykh resursov," **Planovoe khoziaistvo**, no. 6 (1981), p. 83.

62. V. Selivanovskii, "O planirovanii ispol'zovaniia vtorichnykh resursov," **Planovoe khoziaistvo**, no. 6 (1983), p. 85.

63. Ksintaris, "Ispol'zovanie vtorichnykh resursov" (1981), p. 83.

64. B. V. Gromov et al., pp. 32-39. For example, see the call by Harry Klinger and Wolfram Leunert for better measures of effectiveness. "Möglichkeiten und Grenzen der ökonomisch-ökologischen Beurteilung betrieblicher Umweltschutzmassnahmen," **Wissenschaftliche Zeitschrift der Technischen Universität Dresden** 32 (1983): 83-84.

65. A. Arakelian, "The Scientific-Technical Revolution and the Biosphere," **VEko**, no. 5 (1976), trans. in **Problems of Economics** 19 (March 1977): 73-77. See also Wolfgang Heinrichs, "Probleme der Vervollkommung der Materialökonomie im Prozess der verstärkten Intensivierung," **WW** 24 (July 1976): 1061, 1065-66; and Horst Paucke and Günter Streibel, "Zur Wechselbeziehung von Materialökonomie, Technologie und Umweltschutz," **WW** 25 (October 1977): 1469-70.

66. The argument which follows is summarized from Hans Roos et al., **Umweltgestaltung und Ökonomie der Naturressourcen** (Berlin: Die Wirtschaft, 1979), pp. 72-89, 147-48, 186-87. The collective of authors includes Hans Roos, Günter Streibel, Werner Gringmuth, and Kurt Kutzschbauch. In their more recent article ("Gratisdienste der Natur"), Paucke and Streibel relate environmental protection measures to labor productivity as well, pp. 1327-31.

67. Roos et al., p. 84.

68. Paucke and Streibel, "Zur Verflectung," pp. 412-13.

69. Ibid., pp. 407-9.

70. Ibid., pp. 409-11.

71. For example, ibid., pp. 417-20.

72. Just as these differing **social** principles can be derived from the principles of ecology, so differing social principles have been derived from other theories of the natural world. For example, a social Darwinist view emphasizes that society, like nature, should realize a survival of the fittest principle. Others, like Marx, have drawn contrary inferences from Darwin, namely, that societies, like natural systems, evolve to qualitatively higher stages throughout history. Thus survival of the fittest, inevitable in earlier historical periods, can be superseded in socialist and communist society.

73. V. G. Marakhov, "Nauchno-tekhnicheskaia revoliutsiia i prirodnaia sreda," **VFil**, no. 8 (1974), p. 100; and Hans Mottek, "Zu

einigen Grundfragen der Mensch-Umwelt Beziehung," **WW** 20 (January 1972): 39.

74. Arakelian, p. 82; and Mottek, "Zu einigen Grundfragen," p. 43.

75. Paucke and Streibel, "Zur Verflectung," p. 418.

76. Samuel Huntington, "Social and Institutional Dynamics of One-Party Systems," in Samuel Huntington and Clement Moore, **Authoritarian Politics in Modern Society** (New York: Basic Books, 1970), p. 28.

77. The works of Wolfgang Harich and Rudolf Bahro should provide adequate testimony to this. Both books are theoretically grounded in Marxism-Leninism. See especially Harich's **Kommunismus ohne Wachstum?** and Bahro's **The Alternative** (London: Verso, 1978).

78. **Programme of the Socialist Unity Party of Germany**, Ninth Congress of the SED of Germany (Dresden: Verlag Zeit im Bild, May 1976), p. 5.

79. See, for example, Klaus Steinitz, "Theoretische und praktische Probleme des ökonomischen Wachstums im Sozialismus," **Einheit** 22 (1967): 862. Other Soviet and East German commentators also affirm this explicitly.

80. "Bericht des Zentralkomitees der Sozialistischen Einheitspartei Deutschlands an den X. Parteitag der SED," **Neues Deutschland**, Apr. 12, 1981, p. 3.

81. Honecker as quoted by Steinitz, p. 862.

82. Oldak, **Sovremennoe proizvodstvo**, pp. 93-95. Even more interesting is Oldak's criticism of the Stalinist emphasis on "maximum satisfaction of constantly growing material and spiritual needs." Oldak argues that this formulation encourages demands on social production which exceed real possibilities in any given time period and, if not accompanied by individual development, leads to a surfeit of goods, thus weakening incentive and stimulating unhealthy use of leisure (e.g., alcoholism). Later Oldak does explain that the pre-war situation did require concentration on industrialization for national defense, in contrast to the more complex demands in the post-war period. This later qualification leaves unclear the extent to which Oldak intends a general criticism of the Stalinist economic strategy. See pp. 93-97.

83. Ibid., pp. 180-84.

84. Schmidt-Renner, "Sozialistischer Umweltschutz," p. 15.

85. Gerhard Schmidt-Renner, "Über Regelungen zum Schutz von Naturressourcen und -potenzen der gesellschaftlichen Reproduktion und Rekreation vor Störfunktionen in der DDR," **Petermanns Geographische Mitteilungen** 116 (1972): p. 110.

86. Schmidt-Renner, "Sozialistischer Umweltschutz," p. 12.

87. Ibid., p. 17.

88. Streibel, "Umweltschutz und Umweltgestaltung," p. 1153.

89. V. Zhamin, "Ecology and Economics," **VEko**, no. 7 (1975), trans. in **Problems of Economics** 20 (July 1977): 77.

90. E.g., on aesthetic values see Werner Gringmuth, "Die Mensch-Umwelt Relation bei der Schaffung der entwickelten sozialistischen Gesellschaft im Lichte der Marxschen Theorie," **WW** 21 (August 1973): 1234. On health, see Fedorenko and Gofman, "Optimization in Planning," p. 32. Particularly concerned with climate is M. I. Budyko, e.g., "Klimat budushchego," **Priroda**, no. 6 (1971), pp. 14-19.

91. E.g., Lemeshev and Ushakov, p. 28.

92. Oldak, "Scientific-Technical Progress," p. 19.

93. Kuczynski, **Gleichgewicht der Null**, p. 56.

94. Barrington Moore, **Soviet Politics: The Dilemma of Power: The Role of Ideas in Social Change** (Cambridge, MA: Harvard University Press, 1950).

95. B. Bezrukov and N. Dadabaev, "Twilight over the Aral Sea," **Komsomol'skaia pravda**, Sept. 16, 1965, trans. in **CDSP**, vol. 17, no. 38, pp. 21-22. For expression of a similar sentiment see Vladimir Soloukhin, "Creator or Spectator?" **Literaturnaia gazeta**, Feb. 25, 1965, trans. (condensed) in **CDSP**, vol. 17, no. 16, pp. 24-25.

# 7
# Macro-Economic Questions: National Accounting and Territorial Planning

Discussion of environmental policy is one of the more open arenas of debate in the USSR, less so in the GDR. Proposals for significant change appear to be gaining increasing public visibility over the years and these proposals are formulated in ever clearer terms. Debate over restructuring the planning process continues, however, to build on Marxist-Leninist concepts. This grants a seemingly esoteric character to many arguments. Most likely the reader will be struck by two things when reading the next two chapters. First, many proposals made by Soviet and East German economists are quite radical, resembling the boldest ideas coming from Western economics. At the same time, these proposals are often cloaked in a language and concepts which might normally stir only the bleary-eyed accountant or ivory-tower theorist. Of course the "cult of the expert," encouraged by bureaucratization and the expansion of science, gives many policy debates an esoteric tone in the West as well. But in the Soviet bloc controls on the media limit translation of technical debates into a form easily accessible to the uninitiated. The esoteric nature of these debates leaves controversial issues the domain of the few, shielding the public from the possibly far-reaching implications of some policy proposals. This limits potential mobilization of mass sentiment, while still allowing communication between experts and policy advice to the leadership. So, in the USSR and GDR the debate over economic growth does not produce manifestos about the "limits to growth," but rather involves academic disputes over the definition of economic concepts.

Economic obstacles to effective environmental protection in the USSR and GDR lie on two levels: (1) macro-economic problems, i.e., inadequacies in the structure of national economic planning itself and with measures of national economic performance; and (2) problems of enterprise incentives, e.g., accommodating the enterprise cost-accounting system to the needs of nature-protection. While proposals to deal with these two problem areas overlap to some extent, we will focus on proposals to remedy the first type of problem in this chapter; the next chapter examines the debate over charges for natural resource use, which is deemed by many economists to be one of the most effective ways to deal with the second type of incentive problem. Returning to the first set of problems, two macro-economic issues have gained considerable attention in the USSR: (a) problems with the system of national accounting; and (b) the need for improved territorial-branch planning. Discussion of territorial planning as a response to environmental problems is more extensive in the USSR than in the GDR, at least on a public level. The relevance of Marxist-Leninist concepts to this debate is minimal, simply because Marx and Lenin did not have cause to deal with such intricate details of socialist political economy. In contrast discussion of national accounting concepts is extensive in both countries, and the relevance of Marxist-Leninist concepts is great. These contrasting cases provide fertile ground to examine the differing impact of Marxism-Leninism on public debate in different contexts.

## NATIONAL ACCOUNTING AND PLANNING CONCEPTS

We noted in the previous chapter the call by Oldak and others for a rethinking of the meaning of economic growth itself. This call has led to proposals for changes in the definition of national income, the most generally accepted measure of economic growth in the USSR and GDR.[1] These and similar proposals involve challenges to basic propositions about the nature of productive activity, propositions which derive directly from fundamental Marxist-Leninist concepts. Consequently debate over these proposals to a great extent takes the form of debate over the correct understanding of some fundamentals of Marxism-Leninism.

### Nature Protection: Productive or Unproductive Labor?

A central focus of this debate is the distinction between productive and unproductive labor. The distinction derives from Marx's analysis in **Capital**, influenced by Adam Smith's writings.[2] Like so much of Marx's theory of capitalism, questions arise in adapting his concepts to socialist society. Economists in this case disagree over the nature of environmental protection expenditures. Do they represent the expenditure of productive or unproductive labor? The dispute sounds pedantic, but it is important because the definition of productive labor directly affects the measurement of national income, and thus of economic growth. If environmental expenditures embody productive labor, then, at least theoretically, they contribute to national income (and thus growth); if not, they at best make no contribution, and at worst divert labor from more productive (and growth-producing) endeavors. Before looking at the implications of the dispute in more detail, it is necessary to gain some background on the productive-unproductive distinction.

Marx's own explanation of the productive-unproductive distinction is ambiguous. At some points Marx defines productive labor as that which produces surplus value for the capitalist; at other points he seems to conceive of it as that which produces a **material** product.[3] This ambiguity allows Soviet analysts to designate one criterion for the distinction between productive and unproductive labor in capitalism and another in socialism, based on differences in productive relations in the two systems. According to official theory in the USSR and GDR, under capitalist conditions, productive labor is that which produces surplus value for capital accumulation and profit. Thus both a factory laborer and a Broadway actress could be productive workers in capitalism. The criterion is simply that the capitalist extracts part of the value of the labor to **produce** the expansion of capital.[4] According to official Soviet-bloc theory, such exploitative labor has been eliminated under socialist conditions. Therefore, if the distinction between the two types of labor is to be maintained, a different criterion must apply in socialism. Thus, under socialist conditions, productive labor is not distinguished by production of surplus value but by the other criterion found in Marx's writings: "Productive labor (also material productive labor) is use-value producing and useful work, marked by a material product."[5] In contrast, unproductive labor is "in general, work outside and inside material production which is not made concrete in a material product."[6] Included in unproductive labor are activities such as education, cultural work, commercial services (baths and laundries), health care, basic scientific research, and management functions. Thus our Broadway actress is excluded. Economists are careful to point out that under socialist conditions no negative value judgment attaches to unproductive labor, for it is also socially necessary in many cases. Much unproductive labor, for example education, produces indirect productive effects. Therefore "productive and unproductive labor form an inseparable unity, although productive labor is the basis of this unity. . . . Unproductive labor however useful it may be and however important it also is for the level of popular affluence, is finally, however, always dependent on productive labor and its net efficiency."[7]

Since the 1920s some Soviet economists have contested the official delimitation of productive labor. Some, like Strumilin, have consistently argued that all work performed for the socialist state, that is, all socially necessary labor, should be considered productive. According to this school of thought only the labor of priests, prostitutes, sorcerers and the like might be excluded.[8] In the 1950s and 1960s a debate raged in the USSR over the status of passenger transport. The final resolution in the USSR was to exclude it, along with personal postal services, from the sphere of productive labor. In some other East European countries, including the GDR, these are considered productive, however.[9] Presently some economists urge a reclassification of domestic services, household work, education, and/or basic scientific research as productive labor.[10]

The significance of these debates has been subject to dispute. Western students of Soviet economics are uncertain whether classification as productive labor implies higher budgetary priority.[11] On the other hand, East European economists clearly contend that the distinction between productive and unproductive labor has practical significance. This is reflected in the fact that in the GDR and USSR measures of national income are based only on the contribution of productive labor. National income is "the sum of newly created value inside a given period in material production through productive labor."[12] (Included in calculation of national income is also the turnover tax, a tax levied primarily on consumers' goods or materials predominantly for personal consumption.)[13] Great significance is in turn attributed to national income:

> Socialist society concentrates its means and forces on a high and continuous growth of national income . . . and the volume of the national income per capita is a comprehensive expression of the effectiveness of the reproduction process and growth of national income is a decisive indicator of the elevation of the national economic net efficiency.[14]

Therefore it is productive labor which provides the foundation for economic growth. Many Western economists have sharply criticized this Soviet definition of national income and of economic growth. As the British economist Alec Nove points out, however, "the Soviet definition has, in fact, one quite impressive argument in its favour: that international (and to some extent also intertemporal) comparisons of aggregates of material goods, however imperfect they may be, have a sounder basis than any aggregate of services."[15] Our purpose is not to discuss the desirability of various definitions of national income, but a clear understanding of the Soviet conception of national income is necessary if one is to grasp the potential significance of the debate over inclusion of environmental expenditures in it.

Some Soviet and GDR economists seem to believe that the classification of labor as productive does have policy implications. Johann Köhler, an East German economist at the Mining Academy in Freiberg, is the most outspoken opponent of designation of environmental protection measures as embodying productive labor. For him, as mentioned in chapter 6, the implication is that to the extent possible such measures should be deferred until later when they can be better afforded.[16] Other economists conclude the opposite: "It would be very disastrous for following generations if today these necessary investments were treated only as extra 'unproductive' costs, **therefore** to be limited to a minimum."[17] The East German economist Streibel notes that Köhler's position "is connected to a series of economic-political consequences." One such consequence is encouragement of the false perception that environmental protection and economic growth stand in antagonism to one another (presumably leading to a deemphasis on the latter).[18] These economists, although disagreeing on the nature of environmental costs, concur that the debate has important practical implications.

## Present Theory and Practice

At present the issue seems to be genuinely unresolved, both in theory and in practice. This increases the stakes in the debate. It is difficult to know how environmental expenditures are now treated in practice, although some Western analysts have assumed that they are treated as unproductive.[19] The difficulty stems in part from the general complexity of accounting practice and in part from the absence of a clear policy. According to general practice, national income accounting is influenced by the distinction between the productive and non-productive spheres (**produzierender, nichtproduzierender Bereich; proizvodstvennaia, neproizvodstvennaia sfera**) as well as by the distinction between productive and unproductive labor (**produktive, nichtproduktive Arbeit; proizvoditel'nyi, neproizvoditel'nyi trud**). The productive **sphere**, simply stated, includes those **branches** of the economy which produce material goods or services, i.e., in which a change in the physical or chemical properties of natural materials results. This includes the industrial extractive branches, construction, agriculture, forestry, and the water supply system, as well as services of a material nature, such as cargo transport, repair of household articles, communication services related to production, public food service, and some elements of trade.[20] There is not, however, a one-to-one correspondence between the productive sphere and productive labor. Some labor performed for the unproductive sphere may be productive (e..g, the manufacturing of models of machines for scientific research or the production of equipment for the health services, not to mention construction of hospitals, schools, etc.). On the other hand, some labor performed for branches in the productive sphere may be unproductive (e.g., medical services or day-care in a factory, accounting personnel).[21] The problems for national income accounting should be obvious. In the communications sector a solution has been to assign a fixed proportion to the productive sphere.[22] Generally, a simpler principle is applied: Almost all labor performed in productive branches (i.e., in the productive sphere) is counted toward calculation of national income, including the labor of enterprise management, secretarial staff, and guards. All of these personnel form a part of the so-called "collective worker" (**Gesamtarbeiter, sovokupnyi rabochii**) or productive team of the enterprise;[23] thus all are considered to contribute to material production and to national income. There are, however, some exclusions such as medical personnel and party workers.[24]

How environmental protection is classified in this scheme is unclear. Since environmental protection is not generally designated among branches of the economy, it does not clearly fall into either the productive or unproductive sphere. Environmental protection forms a part of the infrastructure of the economy which serves many branches. The water-supply system is, for example, also part of the infrastructure, but since it also forms a separate branch of the economy (with a distinct ministry and administrative structure) it can be clearly designated as part of the productive sphere for accounting purposes. In the USSR, there is no such separate ministry for environmental protection; in the GDR there is a Ministry for Water Supply and Environmental Protection, although as noted in chapter 5, it is mainly responsible for water protection. Enterprises which implement environmental protection measures may be spread throughout various other branches. Furthermore production of environmental protection devices and employment of personnel is dispersed throughout the economy, in the productive as well as the unproductive sphere. Presumably the manufacture of actual equipment would be considered productive labor. Following general practice, employees hired to fulfil nature-protecting functions within productive enterprises **might** be considered as part of the "collective worker" in branches of the productive sphere, whereas their status in unproductive branches is even less clear. Presumably resources devoted to the development of nature preserves and recreational areas would be considered unproductive. Central agencies concerned with environmental protection (like the inspectorates attached to the Ministry of Water Supply in the USSR or the

Sanitary Epidemiological Service of the Ministry of Health) would probably be considered unproductive, as are management personnel attached to all ministries in both the GDR and USSR. Unfortunately, the Soviet and East German literature does not clarify how any of these employees or expenditures are treated in terms of national income accounting, except that costs for use of scrap metal as secondary raw material are considered productive.

Even the most recent books published on the productive-unproductive problem often do not mention environmental protection.[25] One exception is a joint Soviet-GDR-Bulgarian publication, and it includes maintenance of environmental purity as part of the productive sphere.[26] As Abraham Becker has pointed out, however, writings by economists are often of little assistance in clarifying actual practice because existing policies and policy recommendations are often not clearly distinguished in these works.[27] Authoritative works on political economy (textbooks, dictionaries) are more reliable. In this case, however, they offer only minimal assistance. There are, nonetheless, a few references which would indicate that at least some expenditures on environmental protection are considered productive. For example, the East German textbook on the political economy of capitalism and socialism published in 1979 mentions protection of the environment as part of the productive sphere.[28] Exactly what part of environmental protection (i.e., does it include nature preserves, technical equipment, measures to protect rare species, etc.?) and how this is reflected in practice are unanswered questions. Less explicit is a reference in the Soviet dictionary of political economy. Here under the entry "infrastructure", "structures and objects assuring protection of the environment, rational use of water and of other resources" are included in a list of elements forming the **material** infrastructure of international cooperation.[29] This statement clarifies little, however, because the reference seems to be to equipment (the production of which would probably be considered productive in any case) and not to labor costs. These few scant references lead us to suspect that accounting policy remains poorly defined in this area, particularly for expenditures which originate below the central level, i.e., out of ministry, association, or enterprise funds. Comments by East German economists tend to corroborate this conclusion. Paucke and Kroske, for example, write that "as societal practice shows, at the present time it is not possible to place all measures that involve the environment clearly in one or the other category."[30]

**The Debate**

It is in this fluid situation that the debate over environmental protection occurs, a debate which one East German critic has described as involving "polemical exaggerations."[31] At the core of the dispute is a concern which also exercises some Western economists, namely, "what is the best way to measure the value of social production?"[32] In the Soviet bloc, as the above discussion should make clear, there is a historical and theoretical bias toward measuring its value strictly in terms of material output. A few Soviet-bloc economists, among them environmentalists like Mottek and Oldak, question this assumption. Other environmentally-conscious economists don't openly challenge this assumption but press for improved **measures** of material output, implying that the impact of production on the natural surroundings is an important component of its material effect. Into this category fall several East German economists, most notably Günter Streibel. Finally, others, like Johann Köhler, see no need to reconceptualize measures of social production at all. The dialogue between these economists about measures of social production takes the form of a disagreement over the nature of labor invested in environmental protection.

The most conservative viewpoint, that articulated by the East German economist Köhler, argues that expenditures for environmental protection produce nothing material but rather must be paid for out of what is produced:

> To do away with environmental damage or to prevent it – both amount to the same thing in principle – we must earmark a part of the produced national income. We can neither use it for means of production [**Produktionsmittel**] to expand production nor consume it individually or socially, rather we must employ a part of our labor capacity so that the environment won't be damaged or so that the damage that has occurred is remedied. The labor forces employed produce neither means of production nor wine nor bread, these are unproductive workers [**Tätige**].[33]

Expenditures for environmental protection, Köhler continues, are social consumption costs, that is, costs required in order to consume other things. This kind of social overhead is analogous to the labor required, on an individual level, to keep the house clean.[34] Köhler ultimately appeals to the labor theory of value. Since value derives only from human labor, nature has no value in itself. Therefore a change in the condition of nature, according to Köhler, can neither add nor detract anything from the value of production. By the same token neither deterioration nor improvement of the environment can subtract from or add to national income.[35]

As we shall see in the next chapter, virtually all economists in the GDR and USSR share Köhler's view of the labor theory of value, namely that nature has no value until human labor is added. What some contest is the relevance of this point to the case at hand. It is precisely the character of **labor** expended to make resources usable for production which is in dispute. Those wishing to designate such labor as productive assume that nature forms an integral part of an ecologo-economic system.[36] Thus labor required to maintain natural conditions necessary for continued production should be seen as an integral part of the productive process itself. To make this argument one need not question the basic definition of productive labor. Streibel has argued, for example, that once past damage is rectified, measures to prevent further deterioration of the environment fall into the productive sphere. Production, for Streibel, does not end with the completed product. Rather it concludes when conditions are created so that the productive process can continue with no hitches. In this sense, many environmental expenditures are comparable to equipment repair costs, which according to official theory are productive.[37]

Other economists do, at least implicitly, question the very definition of productive labor. The East German economist Mottek, for example, suggests that the value of a product should include the labor time necessary to prevent or redress environmental damage generated by production. Rectification of past damage, he concludes, is equivalent to payment of a debt. The effects of resource depletion or environmental deterioration often turn up only years or decades later.[38] These costs, for Mottek, include intangibles such as the recovery capacity (**Erholungsleistung**) of forests and changes in the living conditions of the population (e.g., polluted air or water).[39] Mottek admits that immense difficulties need to be overcome to allow such factors to be reflected in national income. Furthermore if effects on the environment are included, other questions, like the character of the service sector, should also be reexamined. Therefore he concludes

> it is understandable that the introduction of the new indices must wait until the basic questions are clarified. Should we suddenly calculate "national income" by completely new methods we would in fact have a new index which carried the same name but at least would not be completely comparable to the old. This could lead to false conclusions on a practical level.[40]

Some Soviet economists, however, advocate that the tempo of growth, national income, and other aggregate indicators of the development of the economy be determined by including the economic effectiveness of reproduction and protection of natural resources, and of economic damage resulting from ecological disturbances.[41]

If Mottek balks at the prospect of creating an entirely new comprehensive indicator for measurement of social reproduction, the Soviet economist Oldak does not. He calls this indicator "gross social wealth" (**sovokupnoe sotsial'noe bogatstvo**), almost certainly intended to offer a contrast to the official index of "gross social product" (**sovokupnyi obshchestvennyi produkt** – a measure of **material** production, including double-counting). Oldak's "gross social wealth" is an aggregate of five components: (1) accumulated material wealth; (2) flow of services; (3) accumulated knowledge; (4) condition of reproduced natural resources; and (5) condition of the health of the population. The latter two categories, according to Oldak, should be measured by change in condition over time (expressed in terms of a quantitative economic evaluation).[42] Oldak provides a radically new definition of productive labor as that which assures growth and preservation of social welfare, not material goods. Thus production of some **material** products like alcohol, tobacco, and narcotics would be excluded, as would production which harmed the environment more than the value of the material output.[43] Oldak apparently intends his measure of "gross social wealth" to replace national income or gross social product as a measure of social development. For he writes: "To evaluate the development of the biosocial system – of improvement characterizing economic, social progress and a transition to higher levels of the use of nature – is impossible with such aggregate indicators as national income and accumulated material wealth."[44] Oldak clearly intends a nearly revolutionary critique of existing economic concepts, a tendency expressed not only in regard to environmental issues but also in his demand for inclusion of social activities like education and scientific research in gross social wealth. He explicitly rejects the notion underlying the productive-unproductive distinction – the idea that material **production** serves as the basis for expansion of non-material services. On the contrary, Oldak sees knowledge as the decisive element of national wealth.[45]

If Oldak is the boldest of critics, he is not the only one who argues for the inadequacy of national income as a measure of social production. But other critics propose auxiliary concepts rather than substitutes for national income. Some, like the East German Dieter Graf, propose, for example, that national wealth is a superior measure of living conditions. National wealth reflects the long-term achievements of society, whereas national income only measures the end result of each year's production process. The long-term impact of production on the natural environment as well as problems of resource depletion are more adequately summarized in national wealth.[46] According to official definitions, national wealth "is the entirety of material and spiritual [**geistige**] goods which society has at its disposal at a definite point in time." As such it includes all material goods, all exploitable (**nutzbaren, erschliessbaren**) natural resources (e.g., forest, hydro-power, mineral resources, cultivated or cultivatible land), the skill of the workforce, scientific achievements, and art.[47] Although standard political-economic works include natural resources in national wealth, these economists insist that **effective** inclusion requires a quantitative economic evaluation of natural resources so that this information is of use to planners and managers, and so that comparisons can be made over time.[48]

## The Ecological Infrastructure

Still other proposals aim at relating environmental concerns to standard economic categories in an attempt, one suspects, to "institutionalize" attention to the condition of the environment. These ideas have generally come from Soviet economists. Fedorenko and Borovskikh, for example, have proposed creation of a separate sphere of material production charged with the task of re-creating natural resources to be used in production.[49] For purposes of determining national economic proportions, branches and productive facilities which reproduce natural wealth do not fall into either of the two major categories, extraction and manufacturing. Attention to improving extraction, these economists argue, must be accompanied by efforts to assure that what is borrowed from nature will be replaced. Otherwise grave and perhaps irreversible consequences may result. Borovskikh would define this new "nature reproduction" branch broadly: forestry, fisheries, water management, cultivation of animals and plants, land reclamation, geological prospecting, hydrotechnical construction, water and air purification, restoration of silence and open space, retrieval of valuable components from scrap, and scientific research to develop new substances would all be included.[50] While the state plan establishes targets and forecasts for withdrawal of resources from nature, so should it also set targets for their renewal.

Another approach to the same problem is the concept of the ecological infrastructure. Inclusion of environmental protection measures in the infrastructure has, unlike most other proposals discussed here, gained a degree of official acceptance.[51] The problem is that the development of the infrastructure in general, while of demonstrable importance, receives less systematic attention than one might hope. Initially some considered the infrastructure a concern primarily of bourgeois economics. But now it is officially deemed of relevance to socialist society as well.[52] The infrastructure provides the base for all production and serves all branches of the economy. It includes the transportation network; energy and water supply; postal and telecommunications systems; dwellings; institutions of learning, of trade and service; and services related to health, sports, culture, and recreation. In recent years attention to the infrastructure has increased in the USSR because its inadequacies are a source of serious dislocations in agriculture and in the development of resource-rich Siberia.

Many recent works include environmental protection measures, nature preserves and the like in the infrastructure. Zhamin, for example, divides the infrastructure into productive and social components. In the former category he includes "nature-preserving technology," while "special nature-preserving measures of general and regional significance, including the creation of national parks, preserves, and others," fall into the social infrastructure.[53] Furthermore, Zhamin suggests that the infrastructure should form a new Department III of material production.[54] Traditional Soviet and East German economics has divided material production into two Departments, based on Marx's political economic writings.[55] Nove clarifies the distinction between the two: "Department I is responsible for the production of all goods which are used in the process of production. Department II covers goods for consumption. . . . These categories are used in analysis concerning the basic 'proportions' of the economy and as a means of relating these to growth."[56] The production of the means of production (Department I) has traditionally been considered to have priority since this provides the basis for growth of consumer goods and agricultural production (Department II). As scientific-technical progress continues, the rate of growth of the two subsystems will become more equal, according to Soviet theory.[57] Designation of the infrastructure as Department III would add a new dimension to this interrelationship and to the question of proportionality. By including environmental protection in the infrastructure, Zhamin and others are attempting to find a place for it in the standard (Departments I and II) and developing (infrastructure) concepts of economic accounting and organization, and in this way to link environmental protection measures to the continued growth of the economy.

The idea of the ecological infrastructure and an earlier version of Oldak's gross social wealth (metawealth)[58] are not well received by the German economists Paucke and Kroske:

> On the contrary, through these auxiliary concepts the so-called ecological infrastructure is placed next to the production process and thereby artificially separated from it in order to establish a theoretical conception that lets the expenditures for environmental protection and environmental formation appear as a deduction from the national income.[59]

The scepticism expressed by Paucke and Kroske might well extend to a new proposal made in 1984 by the Soviet economist Feitel'man. Feitel'man seems to prefer establishing two plans for the enterprise, one a (traditional) productive plan and the other an ecological plan, rather than attempting to integrate environmental concerns directly into existing indicators of enterprise performance.[60] But perhaps Paucke and Kroske are correct in wondering whether proposals which create "special" categories for ecological effects may not simply reinforce present misconceptions about the antagonism between economic growth and environmental protection. On the other hand, integration of environmental damage and environmental improvement into enterprise and national income accounting may be a much more difficult task to accomplish politically than is establishment of "transitional" or "auxiliary" concepts like those implied in the writings of Feitel'man, Zhamin, Borovskikh, Graf, and the earlier Oldak.

**Reflections on the Debate**

Several conclusions can be drawn from examining discussion of these various proposals. First, to the participants in the discussion, classification of labor invested in environmental protection as productive or unproductive is a concrete political issue and is closely tied to the priority granted to expenditures for it. Whether this same significance applies more generally to debates regarding other types of economic activity (e.g., passenger transport) cannot be inferred from this case, although other recent research supports this conclusion.[61] My research does suggest the likelihood that more than the accountant's perfectionism is at stake. Similarly, the designation and make-up of Departments I, II and (potentially) III may have comparable practical consequences, particularly given the past theoretical significance granted to these categories in Soviet Marxism. Definitions of concepts are highly political questions in the East European context as elsewhere. And in Eastern Europe, due to the Marxist origins of these categories and concepts, definitional debates are also questions of interpretation of Marxist theory itself. The various participants in these debates build their arguments carefully drawing from Marx's own writings. The significance of the growth of **material production** as the real measure of economic growth derives from Soviet interpretations of Marxist political economy. An important issue is thus the relationship between environmental economics and the increasing or decreasing potential of material production. Our concern here is not with the ultimate correctness of the various interpretations of Marx put forward by various economists or political leaders, but to point out that a large part of the controversy between the participants centers around that issue of correct Marxist interpretation.

One cannot fail to note differences between the GDR and Soviet debates. In the GDR the primary issue has been classification of environmental protection as productive or unproductive labor. In the USSR little direct attention has been devoted to this problem. (Oldak is the exception that proves the rule.) Rather, Soviet economists have occupied themselves with developing new concepts or categories to integrate environmental protection measures into the broader planning system (e.g., ecological infrastructure, nature-reproducing sphere, Department III). The spill-over effect of the Soviet debate on the German debate is more significant than is the influence of the German debate on the Soviet. Soviet proposals are picked up and discussed by East German economists; the contrary is less common.

Why is the focus of debate different in the two countries? It is not possible to offer a definitive answer. One might suggest, however, that the overall higher level of formal integration of environmental protection into the legal and economic structure in the GDR may make the less transitional, less peripheral GDR proposals more realistic politically and economically. (For discussion of the higher level of formal integration in the GDR see chapter 5.) In the USSR, while environmental protection measures now form a component of plans at the enterprise, republic, and national levels, they appear to be further from having achieved equal status with basic productive functions than in the GDR. This may lead Soviet economists to propose auxiliary concepts. In some cases, as with Oldak's, these may be more radical than East German proposals. But to a certain extent they skirt the issue of **integrating** productive and nature-protection concerns.[62]

## BRANCH AND TERRITORIAL PLANNING

A frequently cited cause of continued pollution in the USSR is "narrow departmentalism," the tendency of economic ministries or branches to put their own interests (and those of their subordinate enterprises) before general societal interests.[63] Criticism of departmentalism is an implicit, sometimes explicit, admission that there are immediate conflicts of interest in socialist society. One Soviet critic even suggests that there is an objective contradiction between the economic interests of the "owner" (society, represented by the government) and the "user" (the enterprise) of natural resources.[64] Examples of this type of conflict abound in the Soviet press. The Volga is polluted, an editorial in **Izvestiia** charges, because of departmental reasoning which maintains that "we produce goods and all the rest is unimportant."[65] Already in 1965 the USSR Deputy Minister of Public Health and USSR Chief Sanitary Physician charged enterprises with a "barbarous attitude toward nature," due in large part to narrow departmental interests.[66] Likewise chemical industry designers are guilty; in their efforts to reduce expenditures "often savings are achieved by deliberately disregarding the requirements of environmental protection."[67]

Soviet economists do not generally attribute narrow departmentalism to the greed of individual enterprise officials. Rather, they see it as reflecting the predominance of the branch (or functional) principle in economic planning and management. Branches or ministries are the key organizational structures in the Soviet and GDR planning systems. They preside over production of particular types of products or services (petroleum, textiles, non-ferrous metals, machinery for livestock and fodder, paper and pulp). Most enterprises and associations (**Kombinate, ob'edineniia**) are answerable to a ministry and are judged primarily by fulfillment of output plans. Because of their specialized focus, ministries and their organs fail to take into adequate account the effects of their activities on other sectors of the economy or on the environment. On the other hand, territorial planning organs have relatively little influence. Therefore efforts to coordinate economic activities at the city, district (**raion**), or regional (**oblast'**) level remain captive to the specialized interests of the branches. As Oldak writes in **Izvestiia**, environmental protection "cannot be accomplished through the branch form of administration alone, since the chief aim of production departments is the exploitation of natural resources."[68] Likewise Arakelian points out that enterprises fail to introduce waste-free technology primarily because they are "subordinate to various agencies whose sphere of business interests includes only the production of certain special products," in other words, production ministries with a specialized perspective.[69]

While the dominant principle of economic organization in the USSR and GDR is functional (i.e., based on ministry structures responsible for production of particular products), ecological problems are essentially territorial in nature; they involve maintenance of viable **surroundings** for life. Although ecological imbalances can have global or national implications (e.g., disturbances to the heat balance, pollution of the ocean, production of acid rain), acute ecological stresses are generally local in nature. For this reason it is not surprising that environmental activism in the West often takes the form of local grass-roots movements focusing on particular local problems (e.g., opposition to location of nuclear power plants or of waste disposal facilities in the vicinity; reaction to local ecological catastrophes, such as the Love Canal incident; protection of local landscapes, species, or natural complexes; or resistance to large-scale construction projects). Even in the USSR, environmental activism began as a movement to preserve a particular regional ecosystem, i.e., Lake Baikal. Nonetheless, there are severe restrictions in the USSR and GDR on the ability of citizens to organize locally to lobby for environmental improvement. While state authorities encourage certain types of citizen involvement at the local level as part of "socialist democracy" (e.g., involvement in local beautification and recycling efforts), these activities do not challenge the prerogatives of productive enterprises. Autonomous grass-roots activism of the Western variety is uncommon in the Soviet bloc. To be sure, since the late 1970s, groups linked to the Evangelical Church in the GDR have engaged in local activism, placing indirect pressure on the state to respond to citizen concerns.[70] For example, beginning in May 1981 the Church Research Facility in Wittenberg has called on Christians each year to participate in "auto-less" weekends ("Mobil ohne Auto"). As part of the project local groups have organized bicycle tours to pollution-damaged areas, like the Erzgebirge. In another action in 1983 Christians gathered in Möblis, a village in Saxony adversely affected by brown coal mining, to offer support and encouragement to the residents. An ecology group in Mecklenburg has used bicycle "demonstrations" to protest the construction of a highway which would divide a recreational area on the Schwerin Lake. In June 1983 the group attempted to organize another bicycle demonstration to protest the construction of an airport near Laage, which lies in the middle of a nature protection area. This time state authorities hindered participation of citizens and finally only about two dozen participants were able to gather for the action. As these examples demonstrate, "citizen initiatives" in the GDR have used non-confrontational methods of pressure and have been limited in their ability to mobilize citizens effectively to influence regional planning. In the USSR there is even less evidence of autonomous environmental activism. Because grass-roots activism is so restricted, conflict over regional priorities is generally expressed in demands by experts for strengthening of regional planning organs to control narrow departmentalism. In the West the issue often becomes politicized in demands for local control;

in the USSR and GDR it is most often defined administratively in proposals for improved regional planning.

When environmentalists raise the question of improved territorial planning, they open an old can of worms. The coordination of functional (branch) and territorial planning has been a perennial problem in the Soviet and GDR systems. Historical experience has shown that emphasis on either the branch or territorial principle generates its own dilemmas. Yet a satisfactory method of combining the two has not been found. From 1932 to 1957 the branch principle reigned; economic ministries played the key role in the planning and administrative system, each responsible for a particular sector (e.g., coal, agricultural production) of the economy. In 1957 Khrushchev instituted a basic reform which was partly motivated by the desire to consolidate his own political power (his opponents had support in the state ministerial structure) and partly motivated by real inadequacies of the so-called branch system. This reform involved the establishment of regional economic councils (**sovnarkhozy**) as the fundamental units of economic organization, marking a shift from a functional principle of administration and planning to a territorial principle. In 1965, following the removal of Khrushchev as First Secretary of the CPSU in 1964, the new leadership again returned to the branch principle of economic organization, although the regional supply depots as well as some other aspects of the 1957 reforms were retained. This is essentially the system which still operates. GDR practice essentially has mimicked the Soviet pattern in terms of the shift between branch and territorial patterns of administration.

This rather turbulent history of change results from a fundamental problem facing any highly centralized planning system. Whatever units of economic organization and accounting are primary, these units can make decisions and allocations which maximize their own interests to the detriment of other parties. They attempt to externalize costs. The tendency is aggravated by taut planning, i.e., the establishment of often unrealistically high plan targets which breed shortages and produce a sellers' market. The initial problems with the branch principle of organization before 1957 have largely resurfaced since 1965, namely minimal sensitivity to regional coordination and subordination of national economic efficiency to the interests of the ministry. Likewise, the **sovnarkhoz** system, in effect from 1957 to 1964, led to regionalization, a degree of autarky, and a short-sighted disregard for the national impact of local decisions. While some economists now argue for increased authority for regional bodies, none advocate a revival of the **sovnarkhozy** nor an explicit reduction in the status of the existing economic ministries. Rather they urge better coordination between ministries and their subordinate organizations (enterprises, productive associations), on the one hand, and existing or new regional bodies, on the other.[71] Some suggest the establishment of coordinating bodies to ensure the subordination of specific branch interests to more general economic goals. Undoubtedly one consideration here is political, namely, a desire not to provoke opposition from powerful vested interests nor to be associated with the failures of the Khrushchev period.

Many factors besides environmental disturbances have increased the salience of the branch-territorial problem in recent years. The increasing complexity and interdependence of the economy require better coordination of communications, transport, energy networks, and other elements of the infrastructure. The infrastructure itself is in large part a territorial structure, serving diverse branches of the economy in a given region. Many of the most pressing policy dilemmas facing the Soviet and GDR leadership involve the infrastructure or have a territorial dimension. For example, rising popular expectations for improved housing, cultural facilities, and consumer goods require planning on a territorial level, to accommodate local supply to local demand.[72] The rapid development of new productive and extractive complexes in Siberia demands coordinated planning of an integrated regional structure, including housing, stores, sanitation facilities, schools, recreational centers, water supply, factories, roads, and communication lines. It simply will not do if one ministry fulfils its plan for construction of a mining complex, but no arrangements have been made to house the workers or no sanitation facilities have been installed.[73] The rapidly growing northern city of Tiumen reportedly has faced just such a situation; thousands of new workers find no shops or services anywhere near their residential areas and often housing itself is rudimentary.[74] Finally, unequal distribution of population, energy supplies, natural resource deposits, and industrial facilities grant new importance to both inter-regional and intra-regional planning and coordination. If the European part of the Soviet Union suffers from labor shortages and energy shortages, then local coordination is necessary to keep the developed industrial structure functional at optimal levels of efficiency. Likewise, in Central Asia, territorial planning is required to assure that the ample labor force is utilized effectively and that scarce water supplies are not squandered. If particular enterprises or ministries ignore the reality of local shortages (of labor and energy in European USSR, or of water in Central Asia), they damage the viability of other economic activities in the region.[75] Such problems have become so obvious that one Soviet scholar has suggested that "virtually all questions of any importance" are interbranch in nature.[76] Environmentalists who raise the question of improved territorial planning thus attach themselves to a larger chorus of voices with similar concerns.

The problem of branch-territorial coordination does not, however, lend itself to easy solution, either in the environmental area or in general. Even though local planning commissions presently draw up social and economic plans (and sometimes special plans for environmental protection), these are often no more than a summation of branch plans affecting the region.[77] Local bodies have little influence over formulation of plans of local enterprises or productive associations, who are answerable to generally more powerful branch organs.[78] Recent resolutions passed by the Council of Ministers as well as sections in the new Soviet Constitution (1977) emphasize the growing importance of regional organs in the planning process.[79] While some of this emphasis is undoubtedly polemical (to demonstrate the progressive development of democratic centralism in developed socialism), there is also a genuine concern with weaknesses in regional coordination. In the Georgian Republic, for example, territorial-branch associations have been created on an experimental basis in two districts (the city of Poti and the Abasha District) to coordinate solution of social and economic problems such as consumer goods supply, recycling of waste, and warehouse facilities.[80] Recent changes on the national level also provide regional planning bodies with an opportunity for input into enterprise planning documents before they are transmitted to higher branch authorities.[81] But without changes in the criteria for evaluation of enterprise activities it is difficult to see how this will motivate enterprises to take territorial concerns more seriously. Nor do efforts to improve material incentives designed to counteract exclusive focus on material output at the enterprise level (e.g., charges for resource use, deductions from bonuses, fines for pollution, pricing policies) assure enterprise sensitivity to regional concerns.[82] Even if such material incentives are strong enough to affect enterprise behavior (which they often have not been), they will not reflect regional concerns unless they are differentiated to take account of regional peculiarities.[83] Developing material incentives to accommodate these particularities (differential prices, fees, or penalties) simply raises the same question at a different stage, i.e., the establishment of appropriately differentiated incentives.[84] Therefore the problem is one of the structure of the planning system itself (i.e., who sets up the incentive structure, in whose interests?) and this is the manner in which many economists, particularly in the USSR, treat it.

In the GDR territorial coordination is undoubtedly less complex due to the smaller size of the country. The differing dimensions of the problem become obvious when one considers that besides the fifteen union republics, twenty economic regions are designated for purposes of territorial planning in the USSR. Each of these regions is nearly comparable to the whole of the GDR in terms of population, size, or complexity of the economic structure. The Russian Republic alone contains eleven of these regions; the Ukrainian Republic three; Kazakhstan and Belorussia each form an independent region; and finally, Central Asia, Transcaucasia, the Baltic republics, and the Moldavian Republic complete the group.[85] Despite these differences in dimension, considerable emphasis has been placed on territorial planning in the GDR, but there is substantially less discussion of the environmental component of the problem. One strategy in the GDR has been to focus central attention on selected geographical areas. The **Landeskulturgesetz** grants local councils (**Räte**) and their organs a key role in planning environmental protection measures. Recent laws on water-economy, disposal of waste, and secondary raw materials further strengthen the coordinating and supervisory functions of these bodies. Especially important are the planning commissions of the districts (**Bezirke**) and the technical organs (**Fachorgane**) for environmental protection and water economy, which are attached to the local councils.[86] In most cases, however, local decisions apparently still depend on agreement with the developmental plans of the branch enterprises for their implementation.[87] The new Water Law (July 1982) does give the council of the **Bezirk** considerable authority over protection of drinking water supplies.[88] Unfortunately, however, the observation made in 1976 by the West German analyst Ilka Norhara-Schnabel still stands: "Unfortunately one searches virtually in vain in GDR publications for information concerning, in particular, the functioning of cooperation between all the institutions dealt with in the law and the enterprises."[89] While there are some limited references to the need for improved branch-territorial coordination in the GDR, criticism is much more muted and is not generally related to environmental protection.[90]

In the Soviet Union, economists offer diverse proposals to strengthen the influence of regional organs in environmental matters. As already noted, not only ecological concerns motivate support for increased regional coordination. Although environmentalists can link their cause to other, perhaps more salient, policy concerns, the solutions they offer are sometimes unique. Some economists emphasize that existing territorial administrative units (the union republic, **raion**, **oblast'**, and **krai** divisions, and the economic regions referred to above) are inadequate for environmental planning. Administrative divisions are not coincident with ecological systems. For this reason it is suggested that it would be advisable to create new ecological regions and corresponding regional bodies, subordinate to a nationwide agency. These ecological regions might be based on climatic zones, river basins, or other natural characteristics.[91] The Soviet economist Lemeshev proposes that regional bodies should have control over a fund financed out of mandatory enterprise payments (comparable to social insurance costs) and fines. These funds could finance large-scale environmental protection measures, such as collective purification facilities, recultivation of damaged land, or planting of trees.[92] Other economists, perhaps more realistically, propose to strengthen the power of existing regional planning bodies; for example, they could be given the authority to compel enterprise support for urgent environmental protection projects.[93]

Some measures have been taken in recent years to take better account of regional interrelationships, but no underlying change in the planning system has occurred. Demands in the USSR for a central conservation agency, which could provide a coordinating point for ecological efforts, have resulted only in the creation of a state committee whose scope of authority (mainly air quality) is limited and whose powers are still not clear.[94] Numerous bodies share responsibility for environmental protection in any given region; thus location of ultimate responsibility is not always clear and coordination may be poor.[95] Calls for a national conservation law in the USSR (similar to the **Landeskulturgesetz** of the GDR) have until now been unsuccessful.[96] (Some Soviet commentators, however, proudly note attention to environmental protection in the 1977 Constitution.) Since at least 1976, however, party committees and state agencies in some cities and provinces have drawn up draft five-year and long-range plans, taking account of regional resources. The Russian Republic State Planning Committee provides methodological and practical assistance to such efforts.[97] We have already mentioned new measures to give local agencies more input into the enterprise planning process. The 1978 resolution passed by the Council of Ministers "On Additional Measures to Strengthen Environmental Protection and to Improve the Utilization of Natural Resources" requires that complex territorial schemes for environmental protection be worked out in large cities or industrial centers, union and autonomous republics, provinces (**oblast'**) and particular areas where important economic or natural complexes span more than two union republics.[98] Territorial-production complexes, advocated by many economic geographers as a means to facilitate inter-branch cooperation, have been developed in many areas of intensive economic activity, particularly in Siberia. According to critics, however, they have not realized their goal because they have no unified coordinating organ; powerful ministries (particularly the petroleum, metallurgical, chemical, and other extractive industries) continue to dominate regional development.[99] As an ad hoc response to this problem, in the early 1980s a special commission of the Council of Ministers and an inter-departmental territorial commission attached to Gosplan (seated in Tiumen) were established for the Western Siberian Territorial Production Complex.[100] Some innovations in the Baltic Republics are also promising.[101] Likewise in the GDR efforts at improving territorial planning continue. The Directives for the five-year national economic plans in the GDR include specific and somewhat detailed consideration of particular regional environmental problems.[102]

Despite these steps, it is doubtful whether the fundamental problem has been adequately dealt with, for real power of coordination and control remain with the branch agencies. While some Soviet ministries have set up special administrative units concerned with organizing conservation measures, many have not. Most ministries in the USSR have been slow in seriously analyzing the organizational and technical requirements involved in environmental protection.[103]

## CONCLUSION

The macro-economic proposals discussed in this chapter have produced only incremental changes. The bias toward material production, discussed in the first part of the chapter, is reinforced by the existing power of those ministries which provide the foundation for material production. Efforts to balance the power of these ministries with the countervailing authority of environmental agencies or regional bodies meet with multiple barriers: bureaucratically-based political opposition, technical difficulties, and ideological bias in favor of material production. Furthermore, any significant devolution of power to regional authorities in the USSR raises the specter of ethnic revivalism. When viewed from this perspective the change which has occurred is not unimpressive.

The tenets of Marxism-Leninism appear to offer little guidance to those urging greater territorial integration.[104] While Soviet environmentalists can criticize the shortsightedness of bureaucrats, in the GDR even this type of discussion is virtually non-existent. Furthermore, we do not find sensitive analyses of bureaucratic politics in the Soviet publications, an issue about which Marx and Lenin had little to say in the context of socialism. The major contribution of Marxism-Leninism to the territorial-branch debate involves the existence of central planning itself. Central economic planning and Soviet socialist ideology provide a public context for setting economic goals. Even if some goals are virtually inviolable (e.g., economic growth), these goals can be interpreted variously in terms of concepts like the social good.

Environmental proponents in the GDR and USSR need not overcome an assumption embedded in many Western societies: that the social good will result automatically from the pursuit of particular economic interests. The social good in the USSR and GDR is already understood to result from public, not private, choices. The issue can, and indeed must, be argued on these grounds. This rhetoric of the social good can, however, mask powerful organizational and personal interests. What is denied to Soviet and East German environmentalists is the opportunity to engage in a prototypical Marxist activity: the critique of private interest and public ideology.

## NOTES

1. On national income the **Wörterbuch der Ökonomie Sozialismus,** s.v. "Nationaleinkommen" (Berlin: Dietz Verlag, 1979) states: "In the socialist countries the structure, volume, rate of growth, distribution, and use of the national income reflect the development and level of the productive forces as well as the standard of living of the population" (p. 628). Likewise from **Politicheskaia ekonomiia: slovar'**, s.v. "Natsional'nyi dokhod" (Moscow: Iz. politicheskoi literatury, 1979): "The national income of socialist society is the most summary [**obobshchaiushchim**] indicator of the development of the economy and growth of the welfare of the working people . . ." (p. 237).

2. As an example see Karl Marx, **Capital** (New York: International Publishers, 1967), vol. 3, pp. 129-65, for discussion of this problem in regard to circulation, transportation, and storage costs.

3. For an exhaustive discussion of this see Helen H. M. Boss, "Productive Labour, Unproductive Labour, and the Boundary of the Economic Domain 1662-1980," vol. 1 (Ph.D. dissertation, McGill University, 1981), pp. 258-74.

4. **Ökonomisches Lexicon**, vol. 1, s.v. "Arbeit – produktive" (Berlin: Verlag Die Wirtschaft, 1969), p. 119; D. Porjasow et al., **Nationaleinkommen im Sozialismus** (Berlin: Verlag Die Wirtschaft, 1976), pp. 88-89; **Politicheskaia ekonomiia: slovar'**, s.v. "Proizvoditel'nyi i neproizvoditel'nyi trud pri sotsializme," p. 314.

5. **Ökonomisches Lexicon**, vol. 2, s.v. "Arbeit – unproduktive," p. 119.

6. Ibid., p. 119. More specifically, in socialism productive labor must produce "surplus product" (not "surplus value") and must realize the goals of socialist society. See M. V. Solodkov, **Neproizvodstvennaia sfera pri sotsializme** (Moscow: Mysl', 1978),

pp. 66 ff.; and M. V. Solodkov, ed., **Ekonomika neproizvodstvennoi sfery** (Moscow: Iz. Moskovskogo universiteta, 1980), chapter 3. The latter book is put out by the Ministry of Higher and Middle Special Education of the USSR as a textbook for students of economics specialties at universities.

7. **Ökonomisches Lexicon**, s.v. "Arbeit – produktive," pp. 119-20. See also Porjasow et al., pp. 79-80.

8. For discussion of the debate in the USSR see Porjasow et al., pp. 79-85; and Boss, vol. 2, pp. 413-43. On Strumilin in particular, see Boss, pp. 421-26. It is interesting that Strumilin includes the work of housewives as productive labor.

9. For debate within various CMEA countries, including the transportation debate, see E. S. Kudrova, who also notes differences in accounting practice between the East European countries. **Statistika natsional'nogo dokhoda evropeiskikh sotsialisticheskikh stran** (Moscow: Statistika, 1969), pp. 19-33.

10. Boss provides an excellent account of this debate, although she does not include consideration of environmental protection. See vol. 2, pp. 453-79. See also Abraham S. Becker, "National Income Accounting in the USSR," in Vladimir G. Treml and John P. Hardt, eds., **Soviet Economic Statistics** (Durham, NC: Duke University Press, 1972), pp. 115-19, for ambiguities and flux in the distinctions drawn historically, as well as on the prospects for change. For an example of recent discussion of the issue regarding spheres other than the environment, see A. Nikolaev, "Ob odnoi aktual'noi probleme," **Voprosy ekonomiki** (hereafter **VEko**), no. 11 (1971), pp. 133-36. Likewise, much of Solodkov's book (**Neproizvodstvennaia sfera pri sotsializme**) is devoted to refuting various viewpoints which he considers incorrect.

11. For example, Becker poses the question as to whether accounting concepts affect decisions on resource allocation, but offers no clear answer. See Becker, "National Income Accounting," in Treml and Hardt, pp. 116-17.

12. **Ökonomisches Lexicon**, vol. 2, s.v. "Nationaleinkommen," p. 239.

13. Alec Nove therefore suggests that a better definition of national income is "a species of 'net national material product at market prices' " (rather than at factor costs). See Alec Nove, **The Soviet Economy: An Introduction** (New York: Praeger Publishers, 1969), pp. 112-13, 284.

14. **Ökonomisches Lexicon**, s.v. "Nationaleinkommen," pp. 239-40.

15. Nove, **The Soviet Economy**, p. 285. Differences in national income accounting methods produce problems in comparing growth rates between Soviet-bloc and Western countries. Certain types of statistical conversion are required. See, for example, Stanley H. Cohn, "National Income Growth Statistics,"

in Treml and Hardt, pp. 125-27.

16. Johann Köhler, "Zu der Problematik der produktiven und der unproduktiven Arbeit sowie der Dienstleistungen," **Wirtschaftswissenschaft** (hereafter **WW**) 22 (June 1974): 881, 883-86. In this article Köhler is refuting an article by Jürgen Kuczynski, "Der unbefangene Wissenschaftler," **Das Magazin**, no. 2 (1974), p. 54.

17. Gerd Knobloch and Klaus Steinitz, "Akkumulation und Gestaltung einer effektiven Struktur der Volkswirtschaft," **WW** 19 (July 1971): 1026 (emphasis added).

18. Günter Streibel, "Umweltschutz und Unweltgestaltung als volkswirtschafliche Aufgabe," **WW** 23 (August 1975): 1141, 1149.

19. E.g., Keith Bush, "Environmental Problems in the USSR," **Problems of Communism** 21 (July-August 1972): 27. Bush offers no source for this assertion.

20. See **Politicheskaia ekonomiia: slovar'**, s.v. "Proizvodstvennaia sfera," pp. 315-16; and **Wörterbuch**, s.v. "Nichtproduzierender Bereich der Volkswirtschaft," p. 639.

21. **Politicheskaia ekonomiia: slovar'**, s.v. "Proizvoditel'nyi i neproizvoditel'nyi trud pri sotsializme," p. 315. See also Solodkov, ed., **Ekonomika neproizvodstvennoi sfery**, chapter 3, for an involved discussion of the theoretical side of this problem.

22. At least this was true in the 1950s as reported by Becker. See Boss, vol. 2, p. 450. See also Kudrova, p. 32, who refers to use of "a conventional method."

23. **Wörterbuch**, s.v. "produktive Arbeit," p. 739, describes the productive collective worker as "the entirety of all partial labor (**Teilarbeiten**) which results in a material product (or accomplishment). With the development of science and technology and the increasing socialization of the production process the designation of productive labor expands in socialism."

24. See United Nations Economic Commission for Europe, "A Note on Some Aspects of National Accounting Methodology in Eastern Europe and the Soviet Union," in R. H. Parker and G. C. Harcourt, eds., **Readings in the Concept and Measurement of Income** (Cambridge, ENG: Cambridge University Press, 1969), pp. 373-74; and on trade, in this context, see Kudrova, p. 31.

25. For example, both books by Solodkov previously cited, as well as his **Neproizvodstvennaia sfera SSSR** (Moscow: Mysl', 1981).

26. Porjasow et al., p. 99.

27. Abraham S. Becker, **Soviet National Income 1958-64** (Berkeley: University of California Press), pp. 4-5.

28. Jürgen Becher et al., **Politische Ökonomie des Kapitalismus und des Sozialismus: Lehrbuch** (Berlin: Dietz Verlag, 1979), p. 526.

29. **Politicheskaia ekonomiia: slovar'**, s.v. "Infrastruktura," p. 134.

30. Horst Paucke and Heinz Kroske, "Zum Kosten-Nutzen Problematik von Massnahmen in Umweltschutz und Umweltgestaltung," in Akademie der Wissenschaften der DDR, **Gesellschaft und Umwelt: Hans Mottek zum 65. Geburtstag**. Sitzungsbericht, no. 2/G (Berlin: Akademie Verlag, 1976), pp. 63-64. Likewise, Johann Köhler notes the need to develop a policy for practical reasons, but does not indicate what present practice is. He notes that present law in the GDR makes costs for environmental protection measures in industry part of planned production costs, but does not commit himself as to whether this is consistent or inconsistent with his own position. See Johann Köhler, "Der Charakter der Kosten für den Umweltschutz," in **Ausgewählte Probleme der Leitung und Planung des Umweltschutzes: Vorträge des Berg- und Hüttenmännischen Tages 1975 in Freiberg**, Freiberger Forschungshefte D102 (Leipzig: VEB Deutscher Verlag für Grundstoffindustrie, 1976), pp. 47-48.

31. Klaus Ahrends' review of **Umweltgestaltung und Ökonomie der Naturressourcen** in **WW** 29 (June 1981): 730. (For citation of this book see Hans Roos et al., 1979, in the bibliography.) The reviewer praises the book under review for avoiding such polemics.

32. See, for example, Edward F. Denison, "Welfare Measurement and the GNP"; and William D. Nordhaus and James Tobin, "Measures of Economic Welfare," both in Robert and Nancy S. Dorfman, eds., **Economics of the Environment: Selected Readings** (New York: W. W. Norton & Co., 1977), pp. 469-88.

33. Köhler, "Der Charackter der Kosten," p. 59.

34. Ibid., pp. 60-61.

35. Ibid., pp. 50-52.

36. Several Soviet economists use this term, suggesting the interconnectedness of ecological and economic concerns. See, for example, P. Polataev, "Ecology and Economic Development," **Planovoe khoziaistvo** (hereafter **PKh**), no. 12 (1979), trans. in **Soviet Review** 22 (Spring 1981): 105; K. A. Bagrinovskii and M. Ia. Lemeshev, "O planirovanii ekonomicheskogo razvitiia s uchetom trebovanii ekologii," **Ekonomika i matematicheskie metody** 12 (1976): 683-84; M. T. Meleshkin, A. P. Zaitsev, and Kh. Marinov, **Ekonomika i okruzhaiushchaia sreda: vzaimodeistvie i upravlenie** (Moscow: Ekonomika, 1979), p. 31; M. Ia. Lemeshev, "Problemy optimizatsii sotsialisticheskogo prirodopol'zovaniia: regional'nye aspekty," **Izvestiia Akademii Nauk SSSR, seriia geograficheskaia**, no. 2 (1979), pp. 18, 24; and P. G. Oldak, who uses a similar term, the "biosocial system", which also includes social development. P. G. Oldak, **Sovremennoe proizvodstvo i okruzhaiushchaia sreda** (Novosibirsk: Nauka, sibirskoe otdelenie, 1979), pp. 4-5. For

similar ideas in the GDR see Horst Paucke and Günter Streibel, "Gratisdienste der Natur," **WW** 31 (September 1983): 1317-32; and comments by G. Pflicke, in Ilona Herrnberger, "Wirtschaftsrechtliche Fragen des Umweltschutzes," **Staat und Recht** 32 (1983): 654.

37. Streibel, "Umweltschutz und Umweltgestaltung," pp. 1141-50. A 1979 publication by several East German economists (including Streibel) relies on the concept of "potential-creating labor," which they define as "the entirety of all social activities for conscious development (**Gestaltung**) of the natural environment . . . whose purpose lies in the maintenance and in the expansion of natural potential or in the substitution or expansion of non-renewable resources" (p. 95). At present, these authors conclude, it is not possible to say what exact relationship this "potential-creating labor" has to the traditional productive-unproductive distinction. They point out that most "potential-creating labor" is distinguished from productive labor because it neither produces a material product nor removes material from its natural context – both criteria of productive labor in their view. Nonetheless some forms of "potential-creating labor" are clearly productive (e.g., productive use of wastes) and others affect the production process much more directly than many types of unproductive labor. See Hans Roos et al., **Umweltgestaltung und Ökonomie der Naturressourcen** (Berlin: Verlag Die Wirtschaft, 1979), pp. 95-107. Several other economists not discussed in detail here also disagree with Köhler's basic position and affirm that at least some environmental expenditures should be considered as productive costs. See, for example, K. Ambree et al., **Grundfragen der sozialistischen Preistheorie** (Berlin: Verlag Die Wirtschaft, 1977), pp. 197-200; and D. Böttcher, F. Krämer, and G. Streibel, "Ökonomische Probleme des Umweltschutzes," **WW** 24 (April 1976): 594. Note the predominance of East German participants in this debate. Soviet economists like M. Ia. Lemeshev and E. P. Ushakov object to the view that environmental protection expenditures represent a deduction from national income, although they do not relate this to the productive-unproductive distinction. "O prirodookhrannykh zadachakh ekonomicheskogo razvitiia," **Ekonomika i matematicheskie metody** 15 (1979): 27. See also note 41.

38. Hans Mottek, "Zu einigen Grundfragen der Mensch-Umwelt Beziehung," **WW** 20 (January 1972): 42.

39. Hans Mottek, "Umweltschutz – ökonomisch betrachtet," in Karlheinz Lohs and Sonnhild Döring, eds., **Im Mittelpunkt der Mensch: Umweltgestaltung und Umweltschutz** (Berlin: Akademie Verlag, 1975), pp. 9-10.

40. Ibid., p. 10.

41. These economists, with the exception of Oldak, general-

ly do not relate these proposals to the productive-unproductive distinction. Because they do not elaborate in such great detail, it is difficult to determine if they are closer to the view of Streibel or of Mottek, i.e., whether they wish to include intangible costs which do not immediately affect the next stage of the productive process. See especially N. P. Fedorenko and K. G. Gofman, "Upravlenie sotsialisticheskim prirodopol'zovaniem," in N. N. Nekrasov and E. Mateev, eds., **Sotsialisticheskoe prirodopol'zovanie: ekonomicheskie i sotsial'nye aspekty** (Moscow: Ekonomika, 1980), pp. 28-29; and K. G. Gofman and A. A. Gusev, "Ekologicheskie izderzhki i kontseptsiia ekonomicheskogo optimuma kachestva okruzhaiushchei prirodnoi sredy," **Ekonomika i matematicheskie metody** 17 (1981): 527, who propose development of an indicator called "'clean' [**'chistogo'**] final product, which is obtained with a deduction for economic damage from ecological disturbance." The extra quotation marks indicate that this is a play on words, the use of "clean" here contrasting with the official indicator "pure income" (**chistyi dokhod**) of society which refers to "a part of the value of gross social product which stands apart from the social costs of production in the process of realization" (**Politicheskaia ekonomiia: slovar'**, s.v. "Chistyi dokhod obshchestva," p. 43).

42. Oldak, **Sovremennoe proizvodstvo**, pp. 179-82.

43. Ibid., pp. 182-83.

44. Ibid., p. 150.

45. Ibid., p. 182 on the critique of material production as the source of wealth; p. 172 on knowledge as the decisive element; and pp. 150-84 for the larger discussion of productive and unproductive labor.

46. Dieter Graf, "Naturressourcen und Nationalreichtum," **WW** 29 (June 1981): 671, 673, 684. Graf attributes his ideas on national wealth to Soviet sources.

47. **Wörterbuch**, s.v. "Nationalreichtum," pp. 629-30.

48. For other standard works see **Politicheskaia ekonomiia: slovar'**, s.v. "Natsional'noe bogatstvo," which includes "goods of nature which are drawn into the process of reproduction," p. 236; and Becher et al., **Politische Ökonomie**, p. 682. On economic evaluation for purposes of calculating national wealth see Graf, "Naturressourcen und Nationalreichtum," pp. 677-82; and L. I. Nesterov and E. M. Bukhval'd, "Prirodnye resursy kak element sovokupnogo obshchestvennogo bogatstva," **Izvestiia Akademii Nauk SSSR, seriia ekonomicheskaia**, no. 3 (1980), pp. 20-32. On inclusion of natural goods in national wealth see also Eva Müller, "Nichtmaterielle Güter und Naturgüter als zwei wichtige Bestandteile des Nationalreichtums," **WW** 29 (August 1981): 949-64.

49. N. P. Fedorenko, in N. P. Fedorenko, ed., **Ekonomicheskie problemy optimizatsii prirodopol'zovaniia** (Moscow: Nauka,

1973), p. 10. B. A. Borovskikh, "Urgent Problems in Planning the Reproduction of Natural Resources," **Ekonomicheskie nauki**, no. 1 (1973), trans. in **Problems of Economics** 16 (November 1973): 13-36; and B. A. Borovskikh, **Planirovanie prirodopol'zovaniia** (Moscow: Ekonomika, 1979), p. 10.

50. Borovskikh, "Urgent Problems," pp. 13-14.

51. This is reflected by its designation as part of the infrastructure in political economic dictionaries. See **Wörterbuch**, s.v. "Infrastruktur," p. 409; and **Politicheskaia ekonomiia: slovar'**, s.v. "Infrastruktura," p. 134.

52. Compare **Ökonomisches Lexikon**, vol. 1, s.v. "Infrastruktur," p. 961, published in 1969, which characterizes the term as a part of bourgeois economic science with the same listing in **Lexikon der Wirtschaft: Rechnungsführung und Statistik**, published in 1974, which defines the term generally, implying its applicability to socialist economics (Berlin: Verlag Die Wirtschaft, 1974), p. 245. The **Bol'shaia sovetskaia entsiklopediia**, 1972 ed., s.v. "Infrastruktura," indicates that the concept is relevant to socialist society. The 1953 edition had no listing.

53. V. Zhamin, "Infrastruktura pri sotsializme," **VEko**, no. 2 (1977), pp. 14-15. Oldak proposes the concept of the "ecological infrastructure" in "Scientific-Technical Progress and the Limits of Economic Analysis," **Mirovaia ekonomika i mezhdunarodnye otnosheniia**, no. 9 (1971), trans. in **Problems of Economics** 14 (April 1972): 21-22. See also his chapter "Voprosy ekonomicheskoi otsenki resursov biosfery," in Fedorenko, **Ekonomicheskie problemy**, pp. 55-56. At the first Soviet All-Union seminar on "Problems of the functioning and development of the productive infrastructure," held in April 1981, there was also discussion of the nature-protecting infrastructure. See P. Iakovlev, "Razvitie proizvodstvennoi infrastruktury," **VEko**, no. 9 (1981), p. 152.

54. See Zhamin, "Infrastruktura," pp. 15-18; and V. Zhamin, "Ecology and Economics," **VEko**, no. 7 (1975), translated in **Problems of Economics** 20 (July 1977): 80. B. M. Rabinovich proposes to accomplish the same goal by adding a new subdivision for reproduction of natural resources to Department I. See his comments in the round table discussion, "Aktual'nye voprosy ekonomiki prirodopol'zovaniia," **VEko**, no. 2 (1982), p. 97.

55. Marx, **Capital**, vol. 2, pp. 395 ff.

56. Nove, **The Soviet Economy**, p. 294.

57. See N. A. Tsagolov, ed., **Kurs politicheskoi ekonomii**, vol. 2 "Sotsializm" (Moscow: Iz. Ekonomicheskoi literatury, 1963), pp. 451-62; and **Lexikon der Wirtschaft: Industrie**, s.v. "Abteilungen I und II der gesellschaftlichen Produktion" (Berlin: Verlag Die Wirtschaft, 1970), pp. 25-26.

58. Oldak, "Limits of Economic Analysis," pp. 19-20.

59. Paucke and Kroske, "Zum Kosten-Nutzen Problematik,"

p. 63.

60. N. Feitel'man, "Upravlenie prirodopol'zovaniem," **VEko**, no. 3 (1984), pp. 64-66.

61. See Boss, vol. 2, p. 455. Boss also provides an interesting discussion of the impact of what she calls the "materialist fallacy" on Soviet industrialization policy in the 1930s, vol. 2, pp. 372-404.

62. Even Oldak's "aggregate social wealth" leaves material output and condition of nature as separate subdivisions, whereas Streibel's conception aims to include effects of nature in the measurement of material output itself.

63. See B. Bykhovskii and N. Gladkov, "The Rights of Nature," **Izvestiia**, July 5, 1967, trans. (condensed) in **Current Digest of the Soviet Press (CDSP)**, vol. 19, no. 27, pp. 26-27.

64. M. Ia. Lemeshev, "Problemy optimizatsii sotsialisticheskogo prirodopol'zovaniia: regional'nye aspekty," **Izvestiia Akademii Nauk SSSR, seriia geograficheskaia**, no. 2 (1979), p. 26. N. G. Zhavoronkova also refers to the "sometimes contradictory interests of departments, enterprises, institutions, and citizens," in "Planirovanie gorodskimi sovetami okhrany okruzhaiushchei sredy," **Vestnik Moskovskogo Universiteta**, seriia 11 (Pravo) no. 5 (1981), p. 68. Likewise Oleg Volkov writes in **Pravda** that "protected natural territories are created and exist not in some abstract environment, but in an atmosphere of clashes between conflicting interests, changes, and population growth. . . ." In "The Child has Outgrown his Clothes," **Pravda**, Feb. 22, 1977, trans. (condensed) in **CDSP**, vol. 29, no. 8, p. 7.

65. "The Volga Must Be Clean" (editorial), **Izvestiia**, May 15, 1968, trans. (condensed) in **CDSP**, vol. 20, no. 20, p. 20.

66. Iu. Danilov, "Let Us Protect the Water, Air, and Soil from Pollution," **Pravda**, June 21, 1965, trans. in **CDSP**, vol. 17, no. 24, pp. 13-14.

67. M. Mullagaliamov, "The City's Air," **Trud**, Dec. 2, 1977, trans. (condensed) in **CDSP**, vol. 29, no. 49, p. 8. Likewise, a Moscow official attributes difficulties in relocating older, polluting production facilities to narrow departmentalism of ministries. See A. Blokhnin, "Limitations on Purity," **Izvestiia**, Apr. 26, 1973, trans. in **CDSP**, vol. 25, no. 17, p. 24.

68. P. Oldak, "Environmental Protection is a Comprehensive Task," **Izvestiia**, Feb. 4, 1976, p. 3, trans. (condensed) in **CDSP**, vol. 28, no. 5, p. 5. For a similar viewpoint, see the article by jurists G. Aksenenok, M. Kozyr, N. Krasnov, and O. Kolbasov, "Land and the Law," **Pravda**, Aug. 1, 1968, trans. in **CDSP**, vol. 20, no. 32, pp. 11-12; and G. Aksenenok and N. Syrodoev, "Law Protects Nature," **Pravda**, June 19, 1973, trans. (condensed) in **CDSP**, vol. 25, no. 25, pp. 19-20.

69. A. Arakelian, "The Scientific-Technical Revolution and

the Biosphere," **VEko**, no. 5 (1976), trans. in **Problems of Economics** 19 (March 1977): 78.

70. For more on environmental activism in the GDR's Evangelical Church, see chapter 3, pp. 87-91. For an excellent discussion of specific activities discussed here see Wolfgang Büscher and Peter Wensierski, **Null Bock auf DDR** (Reinbek bei Hamburg: Spiegel Verlag, 1984), pp. 37-42, 53-54, 64-73.

71. See, for example, G. Popov, "Improving Administration," **Pravda**, July 6, 1977, trans. (condensed) in **CDSP**, vol. 29, no. 27, p. 12.

72. N. Zenchenko, "Kompleksnost' ekonomicheskogo i sotsial'nogo razvitiia regionov," **PKh**, no. 4 (1982), pp. 28-29, 33-34.

73. Critics also point out that the natural balance in the Far East and in Siberia is much more sensitive due to the extreme climatic conditions. Therefore the mechanistic transfer of development patterns from European USSR to these areas can produce negative effects on the environment. Special regional plans are needed. See P. Ia. Baklanov, B. V. Poiarkov, and G. I. Khudiakov, "Osnovnye napravleniia sovershenstvovaniia prirodopol'zovaniia i prirodookhrannykh meropriiatii na Dal'nem Vostoke," **Geografiia i prirodnye resursy**, no. 3 (1981), pp. 13-15; P. Polataev, "Ecology and Economic Development," **PKh**, no. 12 (1979), trans. in **Soviet Review** 22 (Spring 1981): 97-98; and F. Martynov, "Metodologicheskie problemy regional'nogo prirodopol'zovaniia (na primere Kuzbasskogo TPK)," **VEko**, no. 11 (1979), pp. 55-59. Boris Komarov (pseud.), in his samizdat publication, takes a less than sanguine view of the sensitivity of planners to special ecological requirements of Siberian development. **The Destruction of Nature in the Soviet Union** (White Plains, NY: M. E. Sharpe, Inc., 1980), pp. 14-16 and chapter 9.

74. S. Vtorushin, "Streets Become Dead Ends," **Pravda**, Oct. 19, 1979, trans. (condensed) in **CDSP**, vol. 31, no. 42, pp. 5-6. Vtorushin urges the provincial soviet executive committee and provincial party committee to support the city soviet and executive committee in their efforts to pressure the ministries to respond. In another article Vtorushin accuses organs of the Ministry of Power and Electrification of insensitivity to geographic and climatic features of the Tiumen region. See his "The North's Electric Rivers," **Pravda**, Oct. 6, 1979, trans. (condensed) in **CDSP**, vol. 31, no. 40, pp. 7-8.

75. Soviet economist N. N. Nekrasov discusses the interconnection between interregional and intra-regional planning in regard to environmental protection in his **Territorialökonomie: Theorie–Probleme–Aufgaben** (Berlin: Verlag Die Wirtschaft, 1979), pp. 141-45. Lemeshev also addresses this question in "Problemy optimizatsii prirodopol'zovaniia," pp. 20-21.

76. G. Popov, "How Reliable are the Interfaces?" **Pravda**,

July 27, 1976, trans. (condensed) in **CDSP**, vol. 28, no. 30, p. 29; and G. Popov, "Improving Administration," p. 12.

77. See N. Zenchenko, "Rol' mestnykh organov v territorial'nom planirovanii," **PKh**, no. 2 (1979), pp. 63-64.

78. N. G. Zhavoronkova provides a good discussion of the problems facing city planning commissions in "Planirovanie gorodskimi sovetami," pp. 70-71. The weakness of city soviet and party officials in Novokuznetsk is a good example. Novokuznetsk has among the highest levels of air and water pollution in the USSR, due largely to the metallurgy industry. Environmental protection plans drawn up by city officials are low on the priority list of enterprises and ministries. See "The Face of a Veteran," **Ekonomika i organizatsiia promyshlennogo proizvodstva**, no. 2 (1978), trans. (abstract) in **CDSP**, vol. 30, no. 25, pp. 4-6; and G. Birskii, "Decisions Must Be Carried Out," **Ekonomicheskaia gazeta**, no. 20 (May 1980), trans. (abstract) in **CDSP**, vol. 32, no. 23, pp. 11-12.

79. See Article 147 of the 1977 Soviet Constitution which, among other things, charges local Soviets of People's Deputies with supervising and coordinating activities of enterprises, organizations, and institutions in their area as regards land use, nature conservation, employment of manpower, production of consumer goods, and social, cultural and communal services. For explanation of these efforts to an East German audience see K. F. Sheremet, "Die neue Verfassung der UdSSR und die Entwicklung des Territorialprinzips der staatlichen Leitung," **Staat und Recht** 29 (1980): 836-44. (Shemeret is deputy director of the Institute for State and Law of the Academy of Sciences of the USSR.)

80. G. Lebanidze, "Poti Experiment," **Pravda**, Feb. 18, 1983, p. 2, trans. (condensed) in **CDSP**, vol. 35, no. 7, p. 9.

81. In 1981 the Council of Ministers, CPSU Central Committee, and Presidium of the Supreme Soviet passed a resolution "On Further Enhancing the Role of the Soviets of People's Deputies in Economic Construction." See "In the CPSU Central Committee, the Presidium of the USSR Supreme Soviet, and the USSR Council of Ministers," **Izvestiia**, March 29, 1981, p. 1, trans. in **CDSP**, vol. 33, no. 13, p. 11. The resolution also provides that lists of construction projects for environmental protection facilities must be submitted to local authorities so that their completion can be coordinated with construction of production facilities. Furthermore, local soviets are to receive increased funding from deductions from payments for water taken from water-management systems amounting to 50% of these payments (see chapter 8, p. 245). For discussion of these measures see Zenchenko, "Kompleksnost' razvitiia regionov," pp. 27-30; and V. Pavlenko, "Novyi etap v sovershenstvovanii territorial'nogo planirovaniia," **PKh**, no. 8 (1980), pp. 9-10, on efforts to improve coordination between Union Republic ministries and local planning

organs.

82. For discussion of the role of material incentives in this context see Rolf Bönisch, Gerhard Mohs, and Werner Ostwald, eds., **Territorialplanung** (Berlin: Verlag Die Wirtschaft, 1976).

83. Feitel'man ("Upravlenie prirodopol'zovaniem," pp. 60-61) points out that maximum permissible concentrations of pollutants in particular regions will be exceeded if standards for individual enterprises are not based on a regional plan.

84. M. N. Loiter, **Prirodnye resursy i effektivnost' kapital'nykh vlozhenii** (Moscow: Nauka, 1974), p. 193; and V. Leksin, "National-Economic Effectiveness and Branch Interests," **Ekonomika i organizatsiia promyshlennogo proizvodstva**, no. 4 (July-August 1977), trans. (abstract) in **CDSP**, vol. 29, no. 36, p. 2.

85. N. Kopylov, "A New Economic District," **Ekonomicheskaia gazeta**, no. 11 (March 1983), p. 14, trans. (excerpts) in **CDSP**, vol. 35, no. 11, p. 22. See also Nekrasov, **Territorialökonomie**, pp. 29, 131.

86. Stephan Supranowitz et al., **Landeskulturgesetz: Kommentar zum Gesetz über die planmässige Gestaltung der sozialistischen Landeskultur in der DDR vom 14. Mai 1970** (Berlin: Staatsverlag, 1973), pp. 64-75, 88-89; Konrad Scherf, "Landschaftsplanung aus der Sicht der Territorialplanung und Territorialforschung," in Hans Richter, ed., **Beiträge zur planmässigen Gestaltung der Landschaft** (Gotha/Leipzig: VEB Hermann Haack, 1978), p. 15; and Ellenor Oehler, "Zu staatlichen und rechtlichen Problemen der Landschaftsplanung," in ibid., pp. 21-22. See also "Dritte Durchführungsverordnung zum Wassergesetz – Schutzgebiete und Vorbehaltsgebiete – vom 2. Juli 1982," **Gesetzblatt der DDR**, Teil I, no. 26, p. 487.

87. See Ilka Norhara-Schnabel, "Zur Entwicklung der Umweltpolitik in der DDR," **Deutschland Archiv** 9 (1976): 813-14.

88. "Dritte Durchführungsverordnung zum Wassergesetz" (see note 86), pp. 487-89.

89. Norhara-Schnabel, p. 817. For a reference to the problem in the environmental area see comments by H. Bartz and B. Schneeman in Herrnberger, pp. 684-85.

90. Unlike Soviet analysts who admit conflicts of interest in socialist society (see note 64), cooperation and commonality of interest are emphasized in East German articles on territorial/branch coordination. Typical of this generally more uncritical tone is the article by Dieter Hösel and Gerhard Schulze, "Zur Verantwortung der örtlichen Räte," **Staat und Recht** 29 (1980): 866-76. In the GDR there is more concern with "territorial rationalization," i.e., to improve efficiency through cooperation in use of local resources. See Günter Gerlach and Kurt Schubert, "Die weitere Verwirklichung der territorialen Rationalisierung," **Staat und Recht** 27 (1978): 1083-94.

91. Oldak, **Sovremennoe proizvodstvo**, pp. 143-44; and M. Ia. Lemeshev and N. S. Kozerskaia, review of **Aktual'nye problemy okhrany okruzhaiushchei sredy: ekonomicheskie aspekty**, by N. G. Chumachenko et al., in **Ekonomika i matematicheskie metody** 17 (1981): 1227-28. The reviewers criticize the book for inadequate attention to the territorial aspect of environmental problems. The book under review does, however, provide a useful summary of local organs with environmental protection functions. See N. G. Chumachenko and L. A. Belashova, eds., pp. 244-56.

92. M. Ia. Lemeshev, "Problemy optimizatsii sotsialisticheskogo prirodopol'zovaniia," p. 27; and M. Ia. Lemeshev, in "Aktual'nye voprosy" (round table), p. 100. For a similar view see Feitel'man, "Upravlenie prirodopol'zovaniem," pp. 57-58.

93. N. G. Zhavoronkova, "Planirovanie gorodskimi sovetami," pp. 70-71; and S. G. Zhukov and Kh. V. Luik, in "Aktual'nye voprosy" (round table), pp. 98, 102. N. Feitel'man proposes payments by enterprises to finance environmental protection measures as well. "Socioeconomic Problems of Ecological Equilibrium in Western Siberia," **VEko**, no. 10 (1978), trans. in **Problems of Economics** 21 (April 1979): 18.

94. See chapter 2, p. 43.

95. M. Dolishnii, V. Kravtsiv, and S. Kuzyk, "Vzaimosviaz' ekonomicheskikh i ekologicheskikh problem v regione," **PKh**, no. 12 (1983), p. 87.

96. See O. S. Kolbasov, "Osnovnye napravleniia pravotvorchestva v oblasti okhrany okruzhaiushchei sredy," **Sovetskoe gosudarstvo i pravo** (hereafter **SGiP**), no. 3 (1980), pp. 72-75; and his comments in the round table discussion, "Pravovaia okhrana okruzhaiushchei sredy," **SGiP**, no. 9 (1979), p. 109. Calls for a nationwide conservation law are widespread, but among its proponents there are also disagreements regarding such issues as its relationship to already existent Union Republic legislation. See V. V. Petrov, "Ob"ekt i predmet pravovoi okhrany prirody v SSSR," **SGiP**, no. 4 (1976), p. 64; O. S. Kolbasov, "Pravovaia okhrana prirody v SSSR," **SGiP**, no. 9 (1967), p. 47; and N. A. Syrodoev, "Leninskie idei ob okhrane prirody i prirodookhranitel'noe zakonodatel'stvo," **SGiP**, no. 4 (1974), pp. 71, 74. G. N. Polianskaia specifically cites the GDR law as an example for the USSR to follow. "Poniatie okhrany prirody i prirodookhranitel'noe zakonodatel'stvo," **SGiP**, no. 9 (1975), pp. 73-74.

97. V. Tolmachev, "The Important Role of Integrated Planning," **Ekonomicheskaia gazeta**, no. 31 (July 1976), trans. (abstract) in **CDSP**, vol. 28, no. 35, p. 16.

98. "On Additional Measures to Intensify Conservation and Improve Utilization of Natural Resources," **Pravda**, Jan. 6, 1979, trans. (condensed) in **CDSP**, vol. 31, no. 1, pp. 1-2; and Fedorenko

and Gofman, "Upravlenie sotsialisticheskim prirodopol'zovaniem," p. 15.

99. Strong support for the development of territorial-production complexes (TPC) comes from leading economic geographer Iu. G. Saushkin. See, for example, his comments in E. G. Panfilov, ed., **Ekologiia i politika: Problemy zashchity prirodnoi sredy** (Prague: Mir i sotsializm, 1972), p. 57. Criticism of coordination in territorial-production complexes is found, for example, in O. Latifi, "Vakhsh Energy," **Pravda**, Aug. 4, 1979, trans. (condensed) in **CDSP**, vol. 31, no. 31, pp. 13-14 (referring to the South Tadzhik TPC) and S. Vtorushin and V. Sevastianov, "Tiumen – The Second Stage," **Pravda**, May 17, 1979, trans. (condensed) in **CDSP**, vol. 31, no. 20, p. 8.

100. Feitel'man, "Upravlenie prirodopol'zovaniem," p. 58.

101. M. L. Ramans, "In the Framework of a Single Complex," **Ekonomicheskaia gazeta**, no. 49 (December 1977), trans. (abstract) in **CDSP**, vol. 29, no. 5, p. 20.

102. E.g., **Directives for the Five-Year Plan for the GDR's National Economy 1976-1980** (Dresden: Verlag Zeit im Bild, 1976), pp. 154, 161, 195.

103. A. Tsygankov, "Don't Procrastinate: On the Branch Administration of Conservation Measures," **Ekonomicheskaia gazeta**, no. 22 (May 1977), trans. (abstract) in **CDSP**, vol. 29, no. 23, p. 13.

104. Bönisch et al. in **Territorialplanung** place a fair amount of emphasis on the awareness of Marx and particularly of Lenin of the problems of branch-territorial interaction. See pp. 20, 25-26, 32-33, 109. Undoubtedly on a general level this is valid, but the use of quotes from Marx and Engels appears here to be mainly polemical and to provide no concrete guidance in solving problems of branch-territorial coordination.

# 8
# Natural Resource Pricing and the Labor Theory of Value*

Probably the most frequent Western criticism of Soviet environmental policy is that Soviet enterprises are granted free use of natural resources.[1] Despite past policy fluctuations and some recent innovations, Soviet enterprises have generally been able to use land, water, mineral resources, and, of course, air, without incurring any cost; enterprises therefore have often squandered these resources and paid little heed to their conservation. Western critics often suggest that market mechanisms (i.e., prices responsive to supply and demand) are better tools for regulating resource use than are central plans and that the best way to gain compliance is to appeal to economic self-interest. Furthermore, according to some critics, ideological factors have played an important role in inhibiting the introduction of natural resource pricing in the USSR.[2] For example, the Marxist labor theory of value allegedly has led the Soviet leadership to view natural resources as "free gifts of nature" because they are not the product of human labor; therefore these "free gifts" should be provided to enterprises at no charge.[3] (According to the labor theory of value the exchange value of a commodity derives from the human labor invested in its production.) In short, this Western critique of Soviet environmental policy may appear to be an indictment both of central planning and of Marxist theory itself.[4]

Recent policy innovations in the USSR and the experience of other East European countries suggest, however, that the labor theory of value does not present an insuperable barrier to the institution of charges for natural resource use in the Soviet bloc.

---

*This chapter originally appeared as an article by Joan DeBardeleben, "Marxism-Leninism and Economic Policy: Natural Resource Pricing in the USSR and GDR," **Soviet Studies** 35 (January 1983): 36-52, and is reprinted by permission of Longman Group Ltd. (UK), publisher.

As other scholars have noted, already in the 1960s many Soviet scholars themselves openly attributed natural resource waste in part to the absence of charges for natural resource use.[5] Not only did some economists embrace material stimuli as instruments of central policy, but they denied any contradiction between natural resource pricing and the labor theory of value. Beginning in the mid-1960s appeals for natural resource pricing have appeared frequently in the popular as well as the scholarly press. Soon economists began arguing over the **method** of establishing charges rather than over the **principle**. Soviet legal scholars **did**, however, raise strong theoretical objections to any monetary evaluation of natural resources, claiming it would be both ineffective and inconsistent with socialist political economy. Furthermore, the principle of free use of natural resources was affirmed in Soviet framework laws governing use of land, water, and mineral resources passed in the late 1960s and 1970s. Nonetheless, charges for land and water use were introduced in the GDR in 1968 and 1971, respectively, and in 1982 Soviet authorities also announced introduction of water use charges for industrial organizations. All of these circumstances should suggest that the impact of Marxist theory on Soviet and East German policy-making is more complex than previously assumed. Constraints imposed on policy-makers by Marxist-Leninist precepts are not determinant, at least in this area. On the other hand, Soviet and East German scholars have devoted considerable attention to the theoretical justifiability of such policies in socialism. Ideological factors, therefore, seem to have some weight.

In this chapter we will examine the debate over pricing of natural resources in the GDR and USSR with the purpose of clarifying the impact of Marxist-Leninist theory on policy-making. Do policy-makers simply ignore Marxist theory if pragmatic considerations intervene?[6] Or does Marxist theory constrain their actions, leading leaders to pursue irrational and ineffective courses of action? These questions are difficult to answer for several reasons. First, the interaction of "ideological" and "pragmatic" concerns is very complex. Soviet experience amply demonstrates that even the most authoritative formulations are subject to modification in light of immediate challenges facing regime and party.[7] The problem is further complicated by the sanctioned flexibility introduced into official ideology by the Leninist distinction between a minimum and maximum program. The distinction between short-term and long-term goals is often at the heart of Western discussion of pragmatic versus ideological factors in Soviet politics. In any particular situation both types of concerns affect the definition of problems, the setting of goals, and the evaluation of alternatives. Further, the language in which short-term responses are conceptualized may implicitly introduce constraints and values which are derived from the broader concerns of Marxism-Leninism. Ideological and pragmatic concerns need not be distinct, but may overlap and intertwine with one another, often in subtle ways. The significant question usually involves the level and manner in which political constraints, shared assumptions, and a common conceptual and normative language shape alternatives and arguments. The shared assumptions and language derive in part from the socially inherited understanding of Marxism-Leninism and in part from the more immediate experiences of specific decision-makers in differing political contexts. In labelling alternatives and arguments as either ideological or pragmatic, one may overlook the role of values in setting short-term goals; values are rarely, if ever, neatly separated from practical response to problematic situations.

As noted, charges for natural resource use have generally been seen by Western analysts as a component of Soviet environmental policy. In the Soviet context this issue, however, embodies much broader concerns about the need for more effective utilization of material inputs and increased productivity. As discussed in chapter 2, the mainstream Soviet and East German approach defines "protection of nature" as one element in the larger complex of concern with "rational utilization of nature."

It was precisely this concern with economic rationality which served as a background to the economic reforms of the 1960s in the GDR and the USSR. To encourage intensive rather than extensive growth, economic levers were introduced to stimulate optimization of resource use (primarily labor and capital) at the enterprise level. Advocacy of natural resource pricing gained momentum within the context of economic reform. Proposals for an economic evaluation (**ekonomicheskaia otsenka**) of natural resources were often advanced in the late 1960s as a means of perfecting the basic idea of enterprise economic accountability which was central to the reform. Although many of the reform measures introduced in the 1960s have been weakened, those economists concerned with conservation in the USSR and GDR continue to make their case by demonstrating that an effective policy of nature protection is necessary to maintain growth and to assure effective utilization of material inputs and investments, that is, to realize more traditional economic goals.[8] In the late 1970s intensified concern with issues of rational resource use has rekindled interest in proposals for pricing of natural resources in both the GDR and USSR. Discussion will most likely continue following the introduction of charges for water use in 1982 in the USSR.[9] The reaction to the issue mirrors not only the views of a minority of interested individuals, but reflects a wider range of implicit and explicit assumptions and attitudes.

## BACKGROUND: SOVIET AND GDR POLICIES

Before proceeding to an analysis of the debate over the pricing issue in the two countries, it is important to establish the framework in terms of actual public policy. This is necessary not only to make sense of the various arguments, but also to clarify the relationship between ideological and policy concerns. Simple observation of policy differences between the two countries will not in itself tell us anything about the role of Marxism-Leninism in policy formulation, for such differences may be explained in a variety of ways. But analysis of the types of arguments offered for differing policies, the apparent persuasiveness of these arguments, and the terms in which these policy differences are explained may tell us much about the role of Marxism-Leninism.

As noted above, important initiatives in the direction of natural resource pricing occurred in 1968 in the GDR and in 1982 in the USSR. However, previous to this some limited measures which might be construed as serving the same function as charges for natural resource use have been in effect in both countries. The turnover tax on many mineral resources, timber, and electricity; stumpage charges in the timber industry; charges made by Machine Tractor Stations (until 1956) to agricultural enterprises roughly in accordance with the productivity of land; differential procurement prices paid by the state for agricultural products; and experiments with differential rent payments in the Baltic Republics might accomplish goals similar to pricing of resources. These measures do not, however, represent a policy of charges for resource use. With the exception of the Baltic experiment, they appear to reflect attempts to stimulate effective use of resources without reverting to direct charges for resource use.[10]

The Principles of Land Legislation (1968), of Water Legislation (1970), and of Legislation of the USSR and the Union Republics on the Earth's Interior (1975) all contain similar language affirming "use free of charge."[11] This policy was presented in 1968 to the Soviet public in **Pravda** and **Izvestiia** with the explanation that: "The committees deemed it impossible to agree to the proposals on establishing payment for land use. Use of land free of charge is one of the greatest achievements of the Great October Socialist Revolution."[12] Despite this affirmation of principle certain exceptions have been allowed with the approval of the Council of Ministers. Some minor policy changes have also occurred. Until 1974 in the USSR land was evaluated when it was removed from agricultural production and in some cases compensation for losses to agricultural users was required. The system was not consistent and methods, where applied at all, differed between various union republics. In 1974 the Council of Ministers passed a resolution requiring compensation to land users for losses to agricultural production associated with expropriation by non-agricultural land users on the basis of the cost of preparing new land of equal size for agricultural production. Apparently this resolution has not been completely successful in accomplishing its purposes, and in any case does not constitute introduction of a "price" on land.[13]

A charge for water used by agricultural producers was in effect from 1949 to 1955 in the USSR, but the charge was not even sufficient to compensate for government outlays needed to construct the irrigation systems and was abolished.[14] Subsequent policy financed all capital expenditures for construction and maintenance of large irrigation and melioration systems, as well as large water canals, out of the state budget. Enterprise production expenditures only included direct costs of operation of water-use installations belonging to the enterprise itself. Special water use, that is use affecting the condition of the water, was in some cases subject to a fee established by the Council of Ministers. Recent discussion of the problem, however, indicates that such charges were applied only selectively.[15] As of January 1, 1982 all industrial enterprises must pay for water taken from water-resource systems. Rates are based on costs of material outlays, depreciation, and wages for provision of water and protection of water resources. They are to be differentiated by region and are considered planned outlays in enterprise expenditures. If enterprises use more water than allowed by established norms, charges are 400 percent higher. Revenues from enterprise payments for water use are to be divided between the state budget and local agencies, which should help implement the measures.[16] Already, however, economists are criticizing the new charges as "negligible and formal," too low to stimulate effective use of natural resources.[17]

Economic levers have been applied somewhat more systematically in some mineral extractive industries. Until July 1967 no charge was levied for extracted minerals and there was no penalization for loss of valuable resources as waste products.[18] Since 1967 payment for use of some mineral resources is required. Economist Shkatov expresses the common criticisms of the 1967 measure: The payments do not proceed from objective differences in natural conditions, but are based on overall profit level; they only apply in enterprises with high profitability; they don't apply to coal; and they do not reflect losses during the extractive process. Furthermore, costs of geological prospecting are excluded.[19] And indeed, as noted, "The Principles of Legislation of the USSR and the Union Republics on the Earth's Interior" of 1975 affirmed in Article 6 that "the earth's interior is assigned for use free of charge except in instances established by the USSR Council of Ministers."[20]

Meanwhile in the GDR more significant changes were taking place. Until 1964 administrative measures were applied to attempt to reduce removal of valuable agricultural land from agricultural production. The "Land Use Decree," effective February 1, 1965, introduced payment to owners and agricultural producers for removal of land for non-agricultural uses, similar to the charges instituted in 1974 in the USSR. These limited measures were criticized by economists and legal experts as ineffective, and a more comprehensive system of economic levers was called for. In 1968 a charge on the use of land removed from the agricultural and forest land fund was implemented, and in 1971 a similar charge for the use of water.[21] The "Land Use Charge" is differentiated according to the quality of the land and the use to which the land is put. For enterprises on economic cost-accounting the charge must be paid to the state budget out of the land user's own resources or from investment credits. Such enterprises make a single payment, which reflects itself in all future years as a part of the calculation of the enterprise's productive funds, leading to a permanent increase in the productive funds tax (**Produktionsfondabgabe**) which is paid each year. Some land users directly involved with intensification of agricultural production make the payment to a special fund to be used for soil improvement measures.[22] This charge functions very much as a price on land. The "Water Use Charge" introduced in 1971 corresponds to the costs of water provision, the maintenance of waterways, and water supply installations. Agricultural enterprises are excluded. If the authorized amount of water use is exceeded by an enterprise, an extra penalty is charged. In addition, enterprises causing pollution of air or water are required to pay the damaged party compensation and are charged for emission of pollutants beyond the permissible standard. Questions have been raised subsequently by economists as to the adequacy of the charges.[23] But the principle of charges for natural resource use has been accepted in the GDR.

The fact that prices in the GDR and USSR do not reflect variations in supply and demand or in marginal costs aggravates the difficulty of attempts to define charges for natural resources which will encourage rational use. Furthermore the return to a more centralized system of planning and price-setting in 1971 in the GDR was especially damaging to the effectiveness of the Land and Water Use Charges. The flexibility granted to enterprises which was to be the basic condition of this economic lever was greatly reduced. These constraints imposed by the larger economic system on the effective operation of resource charges are themselves the product of considerations worthy of analysis in their own right.[24] Our analysis of but one specific aspect of that economic system (i.e., resource pricing) does not exhaust the possible larger role which Marxism-Leninism plays in policy-making in that system, but rather points to a method of understanding the inter-relationship of so-called ideological and pragmatic concerns.

The short-term ("pragmatic") goal of the proposed policies for economic evaluation of natural resources is clear: namely, to reduce waste of valuable and increasingly scarce resources, and to assure optimal utilization by regulating demand or providing a measure of opportunity costs of various options for resource use. The purpose here is to relate discussion of measures to accomplish these goals to the understanding which the participants have of the basic nature of the productive forces and relations of production, ownership, and property in a socialist state. These, of course, are the basic categories which underlie the Marxist analysis of capitalist society and of the longer-run move toward a socialist and communist society. Indeed, the Soviet and East German discussion since 1965 has paid attention not only to the immediate practical effects of various proposals, but also to these theoretical issues.

I suggested in chapter 1 that the categories which permeate scholarly debate are most likely perceived by scholars to be close to the values which motivate policy-making (page 9). If this is so, the continued attention to theoretical issues in scholarly discussion of natural resource pricing would indicate that Marxist concepts such as the labor theory of value are perceived to have some impact on policy formation in this area. Nonetheless, non-ideological factors may also play an important role. For example, a major factor which led the GDR to adopt a policy of charges for natural resource use before the USSR did so was probably the greater severity of resource shortages in the GDR, necessitating a quicker policy response to immediate dilemmas facing the regime. Such short-term ("pragmatic") considerations are, however, intertwined with basic systemic (long-term) commitments in most policy choices. The weight granted to theoretical (or long-term) considerations will vary depending on the salience of shorter-term concerns about resource misuse. This implies both that non-ideological influences help shape the interpretation of basic theoretical assumptions, but also that theoretical/ideological concepts may mediate the impact of non-ideological factors on policy-making. In this particular context, an examination of the debate over resource charges suggests that differing Soviet and GDR experiences, reflected in different theoretical interpretations, have provided somewhat distinct normative contexts for discussion of the ideological justifiability of payment for natural resource use in socialist society. Thus, Marxist-Leninist concepts provide a framework of analysis within which shorter-term considerations are placed. Just as liberalism provides a normative context of discussion (i.e., shared assumptions and values) which differs from France to Great Britain to the USA, so also with the categories of Marxism-Leninism in Eastern Europe. To be sure, the analogy is limited, as the constraints on deviation from Marxist-Leninist norms in Eastern Europe differ both in degree and form from the constraints on deviation from liberalism in the Western world. Censorship through a state-controlled media plays a more important role in the former, while public opinion plays a larger role in the latter.

We will now try to test this hypothesis against the experience of the GDR and the USSR in the area of economic evaluation of natural resources.

## THE SOVIET DEBATE

In the USSR discussion of economic evaluation of natural resources has proceeded on two levels. The first level concerns theoretical questions relating to the economic character of natural resources and the overall justifiability of placing a monetary evaluation or price on them. The second set of issues involves the development of a unified methodology for ascertaining the appropriate evaluation of natural resources. The response to this second issue of course may depend upon an answer to the first set of questions. This second set of issues is treated briefly at the end of this chapter.

In regard to the first set of issues, certain assumptions appear to be universally accepted in the published discussion, the most important being the following:

a) Natural resources untouched by human activity have no economic (exchange) value.

b) The Soviet government is the exclusive owner of all natural resources, and absolute land rent has been abolished.

c) Some sort of economic evaluation of natural resources is needed to assure rational use. The meaning of the term "economic evaluation" (**ekonomicheskaia otsenka**) is however sufficiently broad to allow much room for disagreement. Fedorenko points out that use of economic evaluation may be limited to the central planning process and is not synonymous with payment for resource use. It is over the monetary evaluation (**denezhnaia otsenka**) that dispute centers.[25]

d) Commodity relations are appropriate and necessary in socialism.[26]

These assumptions represent the limits of acceptable debate in the USSR (and GDR). To understand the nature of the disagreement over the application of these positions we must gain a bit of historical perspective.

The policy of free use of natural resources, according to both economists and legal experts, arose as a result of the decree "O Zemle" ("On Land"), enacted shortly after the Bolshevik victory. This decree and the later decree "On Socialization of Land" abolished private ownership and provided the basis for the exclusive government ownership of land. With this decree land fell out of commodity exchange and became considered a "free gift of nature," payment for which was considered inappropriate. As early as the mid-1920s theoretical discussion of the economic treatment of natural resources began, centered around their position in the accounting of national wealth. In the 1950s the discussion revived, spurred by Kantorovich's argument for the application of differential rent to agricultural land.[27]

By the late 1960s most economists came to argue in favor of monetary evaluation of natural resources, emphasizing that while untouched resources have no value, once labor is expended in their development or reproduction, they become carriers of value. Thus they should not be considered "free gifts" any longer. Soviet lawyers, on the other hand, generally continued to oppose monetary evaluation of natural resources, basing their view on what they considered the primary factor, the exclusive government ownership of land and natural resources. As exclusive government property, they reasoned, natural resources cannot be treated as commodities, cannot be bought and sold, therefore do not have exchange value, and cannot be evaluated in monetary terms. Thus the issue for economists became not the justifiability of a monetary evaluation, but rather the method of its realization. For legal scholars, the significant issue remains stimulation of more rational resource use through the planning mechanism, in which non-monetary methods (e.g., on the basis of a detailed and comprehensive cadastre) may play a role. The issue is thus defined differently for the two groups.[28]

While some economists agree that natural resources cannot be treated as commodities, this point does not hold the significance for them that it does for most legal analysts. Likewise, for these economists, the exclusive government ownership of natural resources presents no obstacle to their monetary evaluation. Bronshtein writes: "A payment for land in conditions of socialism has nothing in common with the purchase and sale of land in a capitalist society or its concentration in the hands of some class of exploiters."[29] Rather payment for land use will help assure implementation of economic accountability principles in the planned economy, he continues. Loiter suggests that natural resources should be treated like other productive assets. These assets, he notes, are also owned by the government, but have a monetary evaluation **(stoimostnaia otsenka)** used in estimating the productive funds of the enterprise. Without such an evaluation, he argues, "the carrying out of the planned economy is unthinkable."[30]

Other economists make an almost Rousseauian argument regarding the significance of national ownership. Rakitskii, for example, makes no mention of the labor theory of value, but relies on the role of the socialist state:

> Payment into the budget for resources presented to the enterprises is a concrete form of the formation of centralized funds of means of the government budget. Consequently payment for resources is one form of realization by the government of its function as carrier of the economic interest of society as a whole.[31]

This line of argument appears to be capable of justifying any action of the government.

Soviet legal scholar Kolbasov is perhaps the most outspoken proponent of the opposite view. As a pioneer in teaching "the legal protection of nature" at Tomsk University, he can by no means be considered indifferent to the problem of environmental protection. Kolbasov recognizes the inadequacies of the present system, but generally emphasizes the need for organizational changes and for better coordination of nature protection functions. These, of course, are the types of solutions in which the economists can place little faith. In addition, however, Kolbasov does concede that some form of material stimulus and economic evaluation is necessary to further rational use of resources. However, the definition of economic evaluation offered by Kolbasov is "evaluation of their meaning as a condition of material production" which can take a variety of forms. His objection is principally to a monetary evaluation (**stoimostnaia otsenka**), as this implies the transforming of natural objects into commodities. Further, Kolbasov concludes that such evaluation would be ineffective, in that enterprises would be oriented to increasing profitability in their use of natural resources in the current moment, but without concern for long-run consequences.[32]

Kolbasov draws a rigid line between natural objects and commodity-material values, and similarly between law governing natural objects and that governing civil relations. Uses of water regulated by civil law, such as between the providers of drinking water from natural reservoirs and subscribers to the services, may be subject to monetary evaluation. In regard to the use of natural reserves, however, Kolbasov designates the following general principles: (a) "the principle of the exclusive right of government ownership of natural objects, which includes water resources"; (b) development according to plan; (c) the principle of complex use;

and (d) "inherent in the law of water use in the USSR are the principles of free and indefinite use of water resources."[33]

While Kolbasov accepts application of the principles of commodity exchange to water removed from its natural medium, other legal scholars take an even more extreme position. According to Zhakipov, even when removed from the natural complexes in which it is found, water does not lose the characteristic of a natural object, since it returns to nature in the same form for reuse. Thus all water is excluded from commodity relations, according to Zhakipov.[34] Arguing along similar lines, Krasnov in 1967 pointed to the GDR as a model, which ironically implemented payment for land use just a few months later.[35]

If the definitions and distinctions drawn by these legal scholars appear to fetishize the differences between natural objects and commodities and between land and civil law, they appear to be also somewhat tautological. Kolbasov argues that what the economists in reality mean by value is exchange value, which can apply only to commodities. Since natural resources are not commodities (as exclusive government property) they cannot have an exchange value. Thus not all objects which are products of labor have value; consequently the economists are wrong. The difficulty seems to be that once it is admitted that commodity relations are in fact necessary in socialism, distinctions like those which Kolbasov and his legal colleagues try to draw become very problematic. Thus economists Karnaukhova, Fedorenko, and Gofman theoretically justify monetary evaluation of land on the basis that the USSR has retained commodity-money relations, which Marx specifically connected with the development of differential rent.[36] This of course is Krasnov's argument turned to the opposite conclusion. This illustrates well the common framework of shared concepts in which the dispute takes place. The interpretation of these concepts and their significance is at least in part a product of the training and background of the particular analysts. The American legal expert Zigurds Zile has observed the generally legalistic style of Soviet juridical thought; he explains it in part by codified law tradition, the relative underdevelopment of the study of political and administrative behavior in the USSR, and the Stalinist legacy of intolerance toward empirical studies in the political field.[37] Soviet economists, on the other hand, have had greater leeway in conducting empirical analyses of Soviet economic deficiencies, at least since the 1960s. Thus they may apply concepts less formalistically than jurists.

One suspects that the influence of legal scholars in formulation of the principles of legislation for land, water, and the earth's interior may in part explain the adherence to the "free use" principle in these documents.[38] Economists, on the other hand, are likely to have had a greater impact on the formulation of the 1979 economic reform which includes acceptance of payment for water use. Introduction of this idea in the context of the economic reform does not appear to have led to changes in the "free use" article of the Principles of Water Legislation. Thus the process of incremental change in this policy area may produce seeming contradictions between stated principles (free use) and policy (movement toward natural resource charges). The contradiction likely reflects the differential influence of the two scholarly circles and their institutes in particular phases of policy-making.

## THE DEBATE IN THE GDR

Public discussion of economic evaluation of natural resources in the GDR has involved a narrower range of opinions and fewer participants than in the USSR. Presumably much of this discussion has occurred on a less visible level. But since 1964 both economists and legal scholars have called for monetary evaluation of natural resources. Since 1975, after adoption of the land and water use charges (in 1968 and 1971), East German economists have been even more outspoken than earlier, suggesting improvements in the system of natural resource charges.

In 1965 the economists Streibel and Bachman questioned whether it was appropriate to grant free use of land and water to productive enterprises.[39] Streibel articulated his position more fully in 1966 in the leading economic journal. Like his Soviet counterparts, he expressed skepticism over the efficacy of increased administrative controls and argued for prices "which cover the costs and the further reproduction of the installations." While affirming that water has value when labor is expended in making it usable, Streibel went on to argue that a charge is justified even for the use of "valueless" water (e.g., as in the case of underground water supplies).[40] Again in 1983 Streibel, along with other GDR economists, affirmed that natural resources in themselves have no economic (exchange) value, but nonetheless they should be given an economic evaluation, and charges for their use are justifiable.[41] The justification offered by Streibel in 1966 seems to underlie these more recent views as well. Even if natural resources have no economic (exchange) value they do have use value and use of "free resources" saves social expenditures which would otherwise be necessary. Appealing to Marx, Streibel concludes that "the socialist state, as owner of land and thereby also of water, can set a charge through its organs on 'valueless' water, pieces of land, etc. . . . as an expression of the rent which derives from its monopoly."[42]

While GDR economists share the view of Soviet economists, GDR legal scholars do not oppose charges for resources as do Soviet jurists. This appears to be related to the history of land ownership in the two countries. In the Soviet bloc only Mongolia and the USSR have effected complete nationalization of land. The remaining East European countries retain a mixed system of land ownership.[43] This is largely a result of the fact that capitalism was further developed in these countries and the system of private ownership was consequently more highly developed when the communist parties came to power. Although exact figures are not available, Hoffmann and Schinke estimate that 30 percent of the land area in the GDR is nationally owned, the remainder privately or cooperatively owned.[44] This difference in land ownership systems is of some significance for the debate which we are considering. Some of the viewpoints which hold sway among Soviet legal scholars do not prevail in the GDR, as a result, specifically, of the lack of nationalized land in the GDR.

Rainer Arlt, a legal scholar, considers differences between the GDR and USSR in relationship to nationalization of land. Basing his argument on Lenin's analysis, Arlt contends that nationalization of land is not synonymous with socialization of agricultural production. While nationalization was necessary in the Soviet Union because of the remaining half-feudal relations at the time of the Bolshevik revolution, in the GDR the path to full socialization of agricultural production had to take another form, Arlt suggests. Initially private ownership rights were maintained, but combined with the full right of land use by the agricultural collectives.

> If Marx and Engels made the determination that capitalism in agriculture does not depend on the forms of private ownership and of land use, so one can draw a similar conclusion for socialist agriculture. The socialization of agricultural production develops for the present independently of the form of land ownership.[45]

Rather, Arlt continues, "the use relationships determine directly to a very essential degree the socialist relations of production in collectively run agriculture."[46] Thus, according to Arlt, aspects of the GDR land system such as compensation to private owners for use of land, the inclusion of privately owned land in commodity relations expressed in a price, and the remnants of ground rent are all necessary and justifiable aspects of GDR land relations for the time being. However limited the practical meaning of such ownership rights, the existence of formal legal categories totally absent in the Soviet Union creates a significantly different context for the consideration of applying a monetary evaluation to land or to other natural resources.

Arlt argues that in terms of political economy, land has no value. But nonetheless he warns against the view that land is not wealth and not an asset **(Vermögen)** in the economy.[47] Thus whereas in the Soviet Union legal scholars seem caught in a process of fetishization of forms, GDR legal scholars focus on the content behind those forms. The result is perhaps a fundamentally different type of attitude toward proposals for economic evaluation of natural resources, which in some manner may appear to resemble older **forms** of land and resource use.

The explanation which Edgar Schramm provides for the land use charge confirms the importance of the considerations mentioned above. Specifically he cites the fact that until its implementation, a transfer in land utilization between nationally owned entities was carried out without payment. When privately owned land was laid claim to, payment was made on the basis of the 1944 ceiling prices. "These ceiling prices are differentiated in accordance with the principle of capitalist price formation related very sharply to the location of land."[48] Thus, Schramm concludes, the investor was motivated to make use of valuable agricultural and forest land for nonagricultural purposes. Even though the new land use charge applies to all types of land ownership, it probably did not appear to East German scholars to involve introduction of a qualitatively new principle of land relations and indeed it was consistent with other aspects of the economic reform in the GDR.

## METHODS OF ECONOMIC EVALUATION

Among economists in the Soviet Union, the main issue of dispute over economic evaluation by 1967 came to be the development of a satisfactory methodology. In the late 1970s, East German economists also began debating the merits of various methodologies, presumably due to dissatisfaction with the previous reforms. Marx and Engels do not provide much guidance on such questions regarding micro-economic policy in socialism. Nonetheless most economists build their arguments on concepts derived from Marx's writings, such as the labor theory of value and differential rent. Three general approaches to the economic evaluation of natural resources appear to be predominant in the USSR.[49] The first is based on calculation of the labor expenditure necessary to develop the resource. Strumilin claims to derive this method directly from the labor theory of value and conceives of the evaluation as a price, although he does not consider land a commodity.[50] The second approach involves application of Marx's concept of Differential Rent I. According to this method land with higher productivity due to natural soil or climatic conditions would be evaluated accordingly higher. Proponents of differential rent claim that this method derives from the labor theory of value, in that the productivity of labor depends on the quality of the land. Thus, without consideration of Differential Rent I, application of equal amounts of labor would produce differential income for agricultural enterprises with the same overall level of efficiency in organization and production.[51] While the first method of evaluation (Strumilin's) would most likely lead to a higher evaluation for poorer quality land (since more labor would be expended to render it productive), the second method (differential rent) would have the opposite outcome. A major objection to Strumilin's method is precisely this factor, that it would not be effective in practice to stimulate rational use of resources. It might be somewhat more effective if charges were based on the value of labor needed to replace the natural resources used, under present conditions.

The third proposed method of evaluation is based on measures such as gross output or net profit. This method does not as clearly derive from the labor theory of value and is criticized on that basis. Factors other than differences in natural resource quality, such as superior enterprise organization, could account for differences in output or profit.[52]

The concept of differential rent seems to have the most proponents.[53] It is defended on two grounds. First, conditions of more equal socialist competition would be created for agricultural enterprises working with differing natural conditions. Secondly, rational utilization of resources would be stimulated as richer land would presumably be used more effectively if its use involved payment of a higher charge. None of the methods reflect relative scarcity of resources in a systematic manner, although some proponents of each attempt to adjust the level of charges to the general condition of resource shortage.[54] Since the purpose of resource pricing is directly related to the problem of regulating demand for scarce resources, the marginal nature of this concern seems odd. This relative insensitivity to the scarcity issue may well be rooted in the economists' attempts to defend their proposals in the framework of traditional Marxist political-economic concepts, such as the labor theory of value, which do not intrinsically reflect scarcity values.

The purpose here is not to provide a detailed analysis of these and other less commonly discussed methods of evaluation in the Soviet Union. The interesting point is the difficulty of drawing the distinction between so-called pragmatic and ideological concerns, even in an area where the Marxist classics offer little specific guidance. The information we have suggests that the criteria for evaluating methods involve an overall value framework in which concepts like the labor theory of value, Marx's understanding of differential rent, and socialist competition all play a role. These concepts are not separate from evaluation of the effectiveness of the particular method in terms of an accepted short-run goal, but serve as a viewpoint from within which that effectiveness is assessed.

The East German debate over methods, which has blossomed in the late 1970s, has a somewhat different character. East German economists also focus on three methodologies: (1) socially necessary costs to reproduce natural resources (a variant of the Strumilin proposal); (2) some form of differential rent; and (3) a non-price form (sometimes called a monopoly price form) based on the **use-value** of the natural resources.[55] The latter method provides the foundation for the existing land and water use charges adopted in 1968 and 1971. As Schramm explained when the land use charge was adopted in 1968, it is based on "non-price" forms (**ausserpreisliche Formen**) which reflect the differing use-value qualities and differing economic roles of land. The principles underlying it are twofold: (1) the evaluation must be high enough that it will serve as an effective incentive for rational land use, necessitating differentiation by land quality; and (2) the evaluation cannot be applied uniformly to all branches of the economy, but must take account of special conditions and the nature of land use.[56] Schramm apparently hesitates to call this monetary evaluation a "price" because he understands the term "price" to refer to Marx's concept of a capitalist land price or to a monetary expression of exchange value. The land use charge represents neither. However, it resembles a price in terms of function and differs from a price as we know it primarily due to its method of derivation, which is conditioned by the absence of extensive market relations in the larger economic system.

While most GDR economists concerned with natural resource use argue that existing charges are too low to be effective, some support the principle underlying existing policy, namely, basing the charges on the use-value of natural resources. Charges based on the economic exchange value (**Wert**) of natural resources (derived from the labor expended in their development, i.e., the labor theory of value) would be too low to be effective stimuli to enterprises. These economists therefore conclude that charges for resource use need not derive directly from the exchange value of natural resources or from differential rent but can also reflect that part of their use-value which is freely granted by nature (their gratis effect).[57] If use-value is the criterion for setting charges, then they can be adjusted to reflect the differing uses to which one and the same natural resource can be put.[58] This method attempts to take account of scarcity. It is deemed consistent with socialism not because it derives directly from the labor theory of value or from other specific Marxist economic concepts, but because it facilitates the realization of social goals. Thus Schramm describes it as a form which fully corresponds to the character of the economic system of socialism in the GDR and which improves it.[59]

This viewpoint has gained some supporters in the USSR as well, mainly because Soviet economists share the concern that other methods of evaluating natural resources, if applied as charges for their use, would be too low to stimulate enterprises to effective use.[60] Nonetheless, in general, Soviet economists stick much more closely to the labor theory of value in developing proposals. We again see that East German scholars interpret Marxist-Leninist categories more flexibly than their Soviet counterparts.

## CONCLUSION

Despite the fact that Marxism-Leninism serves as the legitimizing value framework in both the GDR and USSR, it is applied and interpreted differently in the area of natural resource pricing in the two countries. It does not appear that these divergent interpretations were simply developed ex post facto to justify policies adopted for other reasons; rather, Marxist-Leninist concepts have been part and parcel of the debate over policy alternatives. This suggests that policy differences do not derive exclusively from the greater intensity of the resource problems facing the GDR. Rather Marxism-Leninism and the nature of socialism are interpreted more flexibly in the GDR at least in part as a result of differing national experiences of the two countries. These experiences have produced differing perceptions as to what policies are compatible with socialism and Marxism-Leninism. Specifically, the GDR maintained a wider variety of forms of land ownership after 1949. This may make it easier for East German legal analysts to see natural resource pricing as compatible with the evolving East German socialist system. Also in proposing methods for establishing resource charges, East German economists are less doctrinaire than their Soviet counterparts. This pattern may seem to contradict what we have found to be the case generally in comparing discussion of environmental policy in the two countries, namely that East German discussion generally is more ideological and sticks unimaginatively to Marxist-Leninist fundamentals. The explanation for this apparent contradiction may rest in the level of control exercised over different elements of environmental debate. When constraints are lifted on East German economists, they may interpret Marxist theory more flexibly, for they have absorbed basic Marxist-

Leninist doctrine less fully. In this case, constraints were eased in 1967 when Schramm, presumably articulating official policy, explained the land use charge as deriving from use relationships, not from the labor theory of value. The fact that East German discussion blossomed after this article, not before, seems significant. Everything indicates that nothing prevents Soviet economists from wholeheartedly endorsing Schramm's position, or Soviet jurists from endorsing natural resource charges. (Indeed a few do.) Most **choose** not to. It is important to remember that even the older generation of economists and jurists in the USSR have been almost fully educated under Soviet power, many in the Stalinist period, unlike some of their East German counterparts. They have almost certainly absorbed a more doctrinaire Marxist-Leninist position than their East German comrades.

Finally, it seems clear that ideological (or theoretical) considerations interact with pragmatic considerations. What led East German policy-makers to interpret Marxism-Leninism flexibly, as consistent with natural resource charges based on use-value, we cannot know. The tight resource situation was certainly a major factor, however. Nonetheless in both countries the categories of discussion and legitimation remain Marxist-Leninist in form, even in concrete consideration of specific policies. These categories and the larger socio-economic framework in which they are applied do place certain kinds of constraints on available alternatives and on their effectiveness. But the constraints are not inflexible or unchanging, any more than the constraints of the liberal heritage are inflexible and unchanging in Western liberal economic systems. They rather form part of a normative framework of argument and evaluation.

## NOTES

1. Philip R. Pryde, **Conservation in the Soviet Union** (Cambridge, ENG: Cambridge University Press, 1972), pp. 42-44. See also Marshall Goldman, **The Spoils of Progress** (Cambridge, MA: MIT Press, 1972), pp. 47-48, 112-15; Josef Füllenbach, **Umweltschutz zwischen Ost und West** (Bonn: Europa Union Verlag, 1977), pp. 39-40, 88-91; Boris Komarov (pseud.), **The Destruction of Nature in the Soviet Union** (White Plains, NY: M. E. Sharpe, 1980), p. 40; and Manfred Hoffmann, **Instrumente zur Lenkung der landwirtschaftlichen Bodennutzung in der DDR** (Berlin [West]: in Kommission bei Duncker und Humblot, 1977), pp. 42-44. For a general discussion of resource pricing in the USSR, see John M. Kramer, "Prices and the Conservation of

Natural Resources in the Soviet Union," **Soviet Studies** 24 (January 1973): 365-68.

2. Pryde, **Conservation**, pp. 42-43; Goldman, **Spoils of Progress**, pp. 47, 112; Komarov, p. 40; and Christiane Busch-Lüty, "Zur Umweltproblematik in sozialistischen Systemen," **Aus Politik und Zeitgeschichte**, no. 27 (1981), pp. 21-22.

3. Komarov, p. 40; Pryde, **Conservation**, p. 42; and Busch-Lüty, pp. 21-22.

4. Philip Pryde does note, however, that "it is possible that this problem could be resolved without resorting to the pricing of land, a solution that would be more acceptable to Marxist political theorists. It could be done, for example, by a very thorough set of resource management regulations, enforced (rigidly) by a much more severe schedule of fines and penalties than exist at present. More comprehensive land use planning would be required, as well." See Pryde, **Conservation**, p. 43.

5. Ibid., p. 43; and Goldman, **Spoils of Progress**, pp. 51, 114-16. Goldman notes past instances of application of some fees for water use in the USSR, but indicates that "while ideology did not then prevent such a fee from being introduced, it shackled its effectiveness to such an extent that the whole concept was discredited" (p. 112).

6. For example, Samuel Huntington and others have suggested that in the USSR and Eastern Europe ideology is losing its relevance and is increasingly being replaced by a more pragmatic approach. See Samuel Huntington, "Social and Institutional Dynamics of One-Party Systems," in Samuel Huntington and Clement Moore, eds., **Authoritarian Politics in Modern Society** (New York: Basic Books, 1970), p. 27-28.

7. For an excellent analysis of the impact of Marxist-Leninist ideology on Soviet policy-making from a historical perspective see Barrington Moore, **Soviet Politics: The Dilemma of Power** (Cambridge, MA: Harvard University Press, 1950).

8. See, for example, Günter Streibel, "Die ökonomische Stimulierung der rationellen Nutzung natürlicher Gebietsressourcen," **Wirtschaftswissenschaft** (hereafter **WW**) 14 (December 1966): 1969. Also N. P. Fedorenko and K. G. Gofman, "Vvedenie," in N. P. Fedorenko, ed., **Ekonomicheskie problemy optimizatsii prirodopol'zovaniia** (Moscow: Nauka, 1973), p. 6; and L. Il'ev, "Nature Conservation and Cost Accounting," **Ekonomicheskie nauki**, no. 11 (1977), trans. in **Problems of Economics** 21 (April 1979): 3-9.

9. On new Soviet charges for water use see V. Afremov and T. Ekondrat'eva, "K vody – po-khoziaiski," **Ekonomicheskaia gazeta**, no. 2 (January 1982), p. 10. For recent Soviet discussion of resource pricing see L. Ia. Osipovich, "Khozraschet i ispol'zovanie resursov," **Planovoe khoziaistvo** (hereafter **PKh**), no. 9

(1979), p. 86; S. Kadetov, "Kompleksnoe ispol'zovanie mestorozhdenii poleznykh iskopaemykh," **Voprosy ekonomiki** (hereafter **VEko**), no. 3 (1980), pp. 34-38; N. Feitel'man, "Ekonomicheskaia otsenka prirodnykh resursov," **VEko**, no. 10 (1980), pp. 63-73; K. G. Gofman, **Ekonomicheskaia otsenka prirodnykh resursov v usloviiakh sotsialisticheskoi ekonomiki** (Moscow: Nauka, 1977); G. Zaikov, "Ob ekologicheskom aspekte khoziaistvennogo rascheta," **VEko**, no. 8 (1978), pp. 129-31; and T. S. Khachaturov, **Ekonomika prirodopol'zovaniia** (Moscow: Ekonomika, 1982), chap. 6. Recent GDR discussion includes Dieter Graf, "Zu einigen Grundfragen der ökonomischen Bewertung von Naturressourcen im entwickelten Sozialismus," **WW** 26 (July 1978): 811-23; Horst Paucke and Günter Streibel, "Zur ökonomischen Bewertung von Naturressourcen," **WW** 29 (September 1981): 1063-80, and Klaus Ahrends, "Zur Rentenbewertung von Naturressourcen," **WW** 31 (September 1983): 1133-52.

10. A lucid Soviet description of such policies is found in Iu. V. Iakovets, "Intensification of Production and the Evaluation of Resources," **Ekonomika i organizatsiia promyshlennogo proizvodstva**, no. 9 (1979), trans. in **Problems of Economics** 22 (March 1980): 62-65, 67. Iakovets claims that stumpage fees compensate roughly half of the outlays in forestry. For criticisms of these policies see J. Wilczynski, "Toward Rationality in Land Economics under Central Planning," **The Economic Journal** 79 (September 1969): 542; M. N. Loiter, **Prirodynye resursy i effektivnost' kapital'nykh vlozhenii** (Moscow: Nauka, 1974), pp. 49-50; E. Karnaukhova, "Ekonomicheskaia otsenka zemel' v sel'skom khoziaistve," **VEko**, no. 8 (1968), p. 90; and M. Bronshtein, "K voprosu o 'tsene' zemli," **VEko**, no. 5 (1968), pp. 111-12.

11. See "Fundamental Principles of Land Legislation of the USSR and Union Republics," Article 8; "Fundamental Principles of Water Law of the USSR and the Union Republics," Article 16; and "Principles of Legislation of the USSR and Union Republics on Minerals," Article 9, in William E. Butler, **The Soviet Legal System: Legislation and Documentation** (Dobbs Ferry, NY: Oceana Publications, 1978), pp. 469, 488, and 523, respectively.

12. F. A. Surganov (Chairman, Agriculture Committee, Council of the Union of the Supreme Soviet), "On Draft Principles of Land Legislation," **Pravda**, Dec. 14, 1968, trans. in **CDSP**, vol. 21, no. 1, p. 12.

13. "Protect the Land, the People's Wealth," **Izvestiia**, Aug. 23, 1974, trans. in **CDSP**, vol. 26, no. 34, p. 21. For criticisms see P. A. Makhin, N. F. Sedlovskii, and V. D. Kireev, "Strip Mining and Recultivation," **Ekonomika i organizatsiia promyshlennogo proizvodstva**, no. 1 (January-February 1977), pp. 160-62, trans. (abstract) in **CDSP**, vol. 29, no. 8, p. 16.

14. Loiter, **Prirodnye resursy**, pp. 189-97.

15. See ibid., pp. 189-97; and G. Kirianov, "Thoughts About Water," **Pravda**, Dec. 12, 1972, trans. (excerpts) in **CDSP**, vol. 24, no. 50, pp. 13-14. Afremov and Ekondrat'eva provide examples of instances in which water use has been subject to charges: water taken from the Moscow, Volga-Uvod and Irtysk-Karaganda canals and other artificial bodies of water; and water used by metallurgical plants in the Ukraine.

16. Afremov and Ekondrat'eva; and N. Bystritskaia and V. Mikhura, "O plate za vodu promyshlennymi predpriiatiiami," **PKh**, no. 1 (1983), pp. 123-24.

17. R. P. Povileiko, "Ecology and Economics," **Ekonomika i organizatsiia promyshlennogo proizvodstva**, no. 9 (1982), trans. in **Problems of Economics** 26 (September 1983), p. 11; and V. Sidorova, "Ratsional'no ispol'zovat' prirodnye resursy," **PKh**, no. 1 (1983), pp. 120-21.

18. Kramer, "Prices and Conservation," p. 372; and N. Feitel'man, "Ob ekonomicheskoi otsenke mineral'nykh resursov," **VEko**, no. 1 (1968), p. 110.

19. V. Shkatov, "Tseny na prirodnye bogatstva i sovershenstvovanie planovogo tsenoobrazovaniia," **VEko**, no. 9 (1968), pp. 67-77. See also A. Bozhedomov, "Fiksirovannye rentnye platezhi v neftianoi promyshlennosti," **VEko**, no. 1 (1980), pp. 128-32, for proposals for improvement.

20. Butler, p. 523.

21. On charges for use of land see Hoffmann, **Instrumente zur Lenkung**, pp. 695-96. See also "Verordnung über Bodenutzungsgebühr vom 15. Juni 1967," **Gesetzblatt der DDR**, Teil II, no. 71 (August 5, 1967); and "Zweite Durchführungsverordnung zum Wassergesetz," **Gesetzblatt der DDR**, Teil II, no. 3 (January 13, 1971). The water charges were retained in the new Water Law. See "Zweite Durchführungsverordnung zum Wassergesetz vom 2. Juli, 1982," **Gesetzblatt der DDR**, Teil I, no. 26 (July 21, 1982), pp. 485-86.

22. Edgar Schramm, "Die Bodennutzungsgebühr," **WW** 17 (May 1969): 695-96. See also Hans Franzy, **Landeskultur und Umweltschutz in der DDR** (Hamburg: Gesamtdeutsches Institut, Bundesanstalt für gesamtdeutsche Aufgaben, 1971), pp. 62-67, for discussion of this measure.

23. Franzy, pp. 95-97. Also Gottfried Schnurrbusch, "Landwirtschaft und Landeskultur," in Karlheinz Lohs and Sonnhild Döring, eds., **Im Mittelpunkt der Mensch** (Berlin: Akademie Verlag, 1975), p. 308; Graf, "Zu Grundfragen der ökonomischen Bewertung," pp. 819, 820-21; and Paucke and Streibel, "Zur ökonomischen Bewertung," p. 1078.

24. For examination of such problems in developing a methodology for natural resource evaluation see Judith A. Thornton, "Soviet Methodology for the Valuation of Natural

Resources," **Journal of Comparative Economics** 2 (1978): 321-33.

25. Fedorenko, **Ekonomicheskie problemy**, p. 21. See also A. A. Mints, **Ekonomicheskaia otsenka estestvennykh resursov** (Moscow: Mysl', 1972), p. 48. Graf likewise cautions against identifying economic evaluation of natural resources with payment for their use ("Zu Grundfragen der ökonomischen Bewertung," p. 819). Often the term **stoimostnaia otsenka** is used in addition. It seems that this term is used almost synonymously with **denezhnaia otsenka**, but at points its use will be noted.

26. See Walfried Schliesser, "Probleme der Warenproduktion und des Wertgesetzes im Sozialismus," **WW** 15 (June 1967): 901, 907, for a discussion of the commodity form. Also Fedorenko and Gofman, "Vvedenie," p. 6.

27. Wilczynski explains reasons for the renewed interest, p. 544. See also L. V. Kantorovich, **The Best Use of Economic Resources** (Cambridge, MA: Harvard University Press, 1965), pp. 97-101. For the significance of resource depletion in stimulating the debate see Iu. Sukhotin, "Ob otsenkakh prirodnykh resursov," **VEko**, no. 12 (1967), p. 37. For a history of the treatment of natural resources see Mints, **Ekonomicheskaia otsenka**, pp. 50-55.

28. Some legal scholars do favor monetary evaluation of natural resources, but they appear to be a definite minority. See for example N. N. Bedenin, "Okhrana prirody v SSSR," **Sovetskoe gosudarstvo i pravo** (hereafter **SGiP**), no. 10 (October 1969), pp. 151-53.

29. Bronshtein, p. 103.

30. Loiter, **Prirodnye resursy**, p. 53.

31. B. Rakitskii, "Ekonomicheskie funktsii platy za resursy," **VEko**, no. 12 (1966), pp. 21-23. M. Vilenskii makes a similar argument in "Zemle – denezhnuiu otsenku," **VEko**, no. 3 (1968), p. 106.

32. O. S. Kolbasov, "Pravovaia okhrana prirody v SSSR," **SGiP**, no. 9 (1967), pp. 46-49; O. S. Kolbasov, "Okhrana prirody," **SGiP**, no. 2 (1972), pp. 12-20; and O. S. Kolbasov, "Iskliuchitel'naia sobstvennost' gosudarstva na ob"ekty prirody," in Viktor M. Chkhikvadze, ed., **Gosudarstvo, pravo, ekonomika**, Akademiia Nauk SSSR, Institut gosudarstva i prava (Moscow: Iuridicheskaia literatura, 1970), p. 227.

33. O. S. Kolbasov, "Legislation on Water Use in the USSR," translated in Irving Fox, ed., **Water Law in the Soviet Union** (Madison, WI: University of Wisconsin Press, 1971), pp. 104-8. The Soviet edition was published in 1966.

34. N. Zhakipov, "Nekotorye voprosy teorii sovetskogo vodnogo zakonodatel'stva," **SGiP**, no. 1 (January 1965), pp. 127-28.

35. N. I. Krasnov, "Pravovoe obespechenie ratsional'nogo ispol'zovaniia zemli," **SGiP**, no. 3 (March 1967), pp. 50, 53. M. I. Kozyr' takes a similar position. "Gosudarstvennoe upravlenie

zemel'nym fondom v SSSR," **SGiP**, no. 8 (August 1968), p. 51.

36. E. Karnaukhova, "Ekonomicheskaia otsenka," p. 90; and Fedorenko and Gofman, "Vvedenie," p. 6.

37. See Zigurds L. Zile, "Kolbasov's **Legislation on Water Use in the USSR** from the Perspective of Recent Trends in Soviet Law," in Irving Fox, ed., **Water Law**, p. 76, for a discussion of patterns of thought of Soviet jurists.

38. Kolbasov explains that the draft law on air quality was drawn up at the Academy of Sciences' Institute of State and Law, then amended in response to comments from ministries, state committees, departments, and Union Republic Councils of Ministers. Likely a similar procedure was followed for the framework laws on land, water, and minerals. This suggests a key role for legal analysts in formulation of these laws. See O. Kolbasov, "Protect Nature," **Izvestiia**, Jan. 16, 1980, trans. (excerpts) in **CDSP**, vol. 32, no. 2, pp. 22-23.

39. Günter Streibel and Georg Bachman, "Wasser und Bodenrationelle Nutzung," **Die Wirtschaft**, Oct. 21, 1965, p. 16.

40. Streibel, "Ökonomische Stimulierung," p. 1979.

41. Ahrends, "Zur Rentenbewertung," pp. 1334-35; and Horst Paucke and Günter Streibel, "Gratisdienste der Natur," **WW** 31 (September 1983): 1321, 1331.

42. Streibel, "Ökonomische Stimulierung," p. 1979. Likewise Dieter Graf supports his argument with ample reference to Marx. These examples illustrate attempts to stretch familiar interpretation of concepts by appealing to Marxist classics ("Zu einigen Grundfragen," pp. 811-23).

43. Loiter, **Prirodnye resursy**, p. 51.

44. Manfred Hoffmann and Eberhard Schinke, **Bodenrecht und Bodennutzung in der Landwirtschaft der DDR**, Bericht des Bundesinstituts für ostwissenschaftliche und internationale Studien, no. 57 (Cologne, 1973), pp. 3-21, include a discussion of the land-owning system in the GDR. See also Franzy, pp. 55-57, and Günther Rohde, ed., **Bodenrecht: Lehrbuch** (Berlin: Staatsverlag der DDR, 1976), pp. 353-58.

45. Rainer Arlt, "Die Lehre Lenins über die Nationalisierung des Bodens und sein Genossenschaftsplan in ihrer Bedeutung für die Entwicklung des Bodenrechts der DDR," **Staat und Recht** 16 (October 1967): 1582.

46. Ibid., p. 1585. More recently see Ellenor Oehler, "Aktuelle Fragen der Lehren von Karl Marx zur Gestaltung der Umweltbeziehungen," **Staat und Recht** 32 (1983): 309-12.

47. As cited by Ellenor Oehler in a discussion of Arlt's doctoral defense. "Die genossenschaftliche Bodennutzung und ihre rechtliche Gestaltung im neuen ökonomischen System der Planung und Leitung der Volkswirtschaft in der Landwirtschaft," **Staat und Recht** 14 (February 1965): 266.

48. Schramm, p. 704.

49. It should be noted that Soviet economists themselves categorize the options differently. For example, Fedorenko and Gofman see the two options as (1) based on expenditures for development, i.e., Strumilin's view, and (2) differential rent ("Vvedenie," p. 6), while Bronshtein refers to three proposed indices: (1) value of gross output; (2) net income; and (3) differential rent ("K voprosu o 'tsene' zemli," pp. 102-7).

50. S. Strumilin, "O tsene 'darovykh blag' prirody," **VEko**, no. 8 (August 1967), pp. 60-72.

51. F. A. Loiter, "Ekonomicheskie mery po ratsional'nomu ispol'zovaniiu vodnogo resursov," **VEko**, no. 12 (1967), pp. 75-86; L. Braginskii, "O vzaimodeistvii ekonomicheskikh rychagov v upravlenii kolkhozami," **VEko**, no. 3 (1971), pp. 78-80; and Gofman, **Ekonomicheskaia otsenka**. There are, however, disagreements among proponents of this method. See, e.g., I. Mikhasiuk, "The Economic Evaluation of Land and the Improvement of Rental Relationships in the USSR," **Ekonomicheskie nauki**, no. 7 (1978), pp. 40-47, trans. in **Problems of Economics** 21 (February 1979): 65-78.

52. See views expressed by S. D. Cheremushkin and S. V. Lyashenko, in E. Petrova, "Obsuzhdenie problem ekonomicheskoi otsenki zemli," **VEko**, no. 6 (1967), pp. 146-47; and Vilenskii, pp. 106-7.

53. This is confirmed by Graf in his critique of Soviet proposals. See Dieter Graf, "Zur ökonomischen Bewertung von Naturressourcen und Naturpotentialien," in Hans Richter, ed., **Beiträge zur planmässigen Gestaltung der Landschaft** (Gotha/Leipzig: Hermann Haack, 1978), p. 36.

54. For example, proponents of Differential Rent I disagree among themselves about the base standard of differential evaluation. Some, like Loiter, advocate using the worst land in use as the base standard (giving it an evaluation of zero), while Shkatov argues for using land embodying average developmental expenditures as the base standard. (This would involve a system of polarized prices in which the best land would receive a positive evaluation, obligating the user to pay a charge for its use, while the worst land in use would receive a negative evaluation, providing the user with a subsidy.) Shkatov objects to Loiter's approach because he thinks it would lead to unacceptable price increases. Bronshtein, on the other hand, raises the opposite objection to both proposals. Because good land is scarce and usable land is finite, the prices that would result from both Loiter's and Shkatov's proposals would result in too low an evaluation of the worst land in use. Therefore Bronshtein rejects Differential Rent I as an adequate basis for evaluation of land. See Loiter, "Ekonomicheskie mery," p. 54; Shkatov, "Tseny na

prirodnye bogatstva," p. 71; and Bronshtein, "K voprosu o 'tsene' zemli," pp. 108-9.

55. Again, some GDR economists categorize the options differently. Ahrends, for example, sees two options: (1) based on differential rent (but not distinguishing Differential Rent I and II, thus apparently including differing results deriving from differences in enterprise efficiency as well as in natural conditions); and (2) based on expenditure for development of the resource. Ahrends supports differential rent. "Zur Rentenbewertung," pp. 1335-36.

56. Schramm, p. 706.

57. Graf, "Zu einigen Grundfragen," pp. 819-22; Eva Müller, "Nichtmaterielle Güter und Naturgüter als zwei wichtige Bestandteile des Nationalreichtums," **WW** 29 (August 1981): 960-64; Paucke and Streibel, "Zur ökonomischen Bewertung," p. 1072; and idem, "Gratisdienste der Natur," all introduce elements of these arguments. Ahrends distinguishes economic evaluation of natural resources (which he wants to base on differential rent) from other economic levers like the land use charge. While he does not clarify the methods for establishing the latter, he implies that they should in some way reflect the use-value of the resource. "Zur Rentenbewertung," pp. 1337-38.

58. Müller, pp. 960-64.

59. Schramm, p. 706.

60. B. A. Borovskikh, **Planirovanie prirodopol'zovaniia: Voprosy metodologii** (Moscow: Ekonomika, 1979), pp. 32-35.

# 9
# Conclusions

Marxists, and no less so Soviet Marxists, have suffered great historical disappointments. Alongside the apparent victories of world socialism in the post-war revolutions of Eastern Europe and Asia stand repeated historical setbacks. In fact, the history of the Soviet Marxist party has been a history of confrontation with recurrent crises, often unforeseen and unexplained by Marxist theory. The failure of the European revolution in the period during and following the Bolshevik Revolution was but the first of these. Subsequent failures – often unacknowledged – belie hope that the Marxist vision will be realized in the Soviet socialist world. Public ownership of the means of production has secured neither prosperity nor class liberation. "Socialist culture" has not produced the new socialist person; greed, corruption, and petty individualism flourish. "Class unity" has not obliterated national chauvinism or the desire for national autonomy. The proletarian revolution seems further from realization in the West than it was after World War I, and the greatest outpourings of working class rebellion occur in the "socialist" world itself. International war threatens not to set imperialist nations against one another, but to render all the earth barren. The most fundamental precondition for building a communist world order, the elimination of material scarcity, seems as far from realization as ever, with millions starving and ill-housed in the world, and with international competition increasingly centered around access to ever more scarce and expensive raw materials. If the crisis of capitalism should, in theory, offer some consolation to Soviet Marxists, its actual consequences are ambiguous. The internationalization of production and commerce, foreseen by Marx, subjects the Soviet bloc as well to the effects of the capitalist economic cycle, and capitalist concern with raw material sources intensifies the danger of nuclear confrontation. Resource depletion and environmental deterioration are but one more dilemma to add to the list of problems neither foreseen by Marxist theoreticians nor easily solved by "socialist" society.

These historical disappointments have left Marxist theory in crisis. Solutions to these dilemmas evade Marxist theorists as they do liberal theorists. Even under the best of circumstances Marxists today have not found answers to troubling questions of bureaucratization, class consciousness, nuclear war, and ecological destruction. In the Soviet bloc prospects are even dimmer. Despite an army of scientists and intellectuals supposedly engaged to apply Marxist theory to the solution of pressing theoretical and practical problems, the results, as we have seen, show few advances in Marxist theory and have generated few fresh responses to major social dilemmas. In fact, in the environmental area, Soviet-bloc arguments are surprisingly similar to those found in the West. Appropriate measures of GNP or national income, incentives to obstruct polluters, development of a new ecological ethic, effective territorial planning, and synthesis of scientific research all preoccupy Soviet and Western thinkers in remarkably similar form. The same dialogues appear to be taking place, albeit in somewhat different language.

It is perhaps not surprising that the Soviet environmental debate has produced so few theoretical advances. For Soviet and East German theorists are forced to apply a castrated variant of Marxist theory. They are deeply constrained by ideological assumptions. The most debilitating of these is the claim that the USSR and GDR are socialist societies in the Marxist sense and the Rousseauian-like inferences that the potential of socialism is coincidental with state/Party interests. In our study we have seen repeatedly the shackles which this assumption places on Soviet and East German social scientists. As we observed in chapter 6, economists can criticize "departmentalism" but they may not expose the structural or socio-economic causes which lead enterprise and ministry officials to externalize environmental costs. In chapter 7 we noted that while Soviet economists explore sources of inefficiency resulting from the primacy of branch economic organization they dare not consider the social and political effects of centralized, as opposed to local, control and planning. While Soviet and East German analysts speak of the need for waste-free technology, they may not consider its relationship to political control and local self-determination. There is no political-economic analysis of these core problems in the USSR and GDR. These are problems of scientific administration, not politics. Thus, even where public ownership of the means of production permits public policy experiments impossible in the West, these may not be tried, or if tried, are subject to only the most incremental criticism and improvement. While it is beyond the scope of this study to seek an authentic Marxist interpretation of environmental problems in the Soviet bloc, it is beyond doubt that genuine Marxist analysis cannot stop short of posing basic political-economic questions. While debate in the Soviet bloc is imbued with Marxist categories, it is a peculiarly uncritical and limited application of Marxist theory.

What, then, is the significance of Marxist categories in the Soviet-bloc debate? More often than not, dispute over Marxist concepts is a way of debating policy priorities and options. Such categories form the framework for dialogue and disagreement, and as such they introduce a subtle but important bias into the debate. As we observed in chapter 8, because the labor theory of value provides the foundation for debate over natural resource pricing, considerations of scarcity are implicitly downgraded. This does not mean they are eliminated entirely, for other Marxist concepts, such as differential rent, may reintroduce these considerations. But the conceptual framework focuses attention away from considerations of supply and demand to considerations of productive costs. Likewise, debates over the priority of economic expenditures occur through the prism of Marxist categories of productive labor and national income, as we observed in chapter 7. This context introduces a materialist-productionist bias, if only because non-material values (e.g., health, aesthetic, and recreational concerns; protection of untouched nature and rare species) are more difficult, if not impossible, to measure quantitatively. Other strains in Marxism-Leninism, namely its humanistic foundations, allow environmentalists to appeal to non-material values as well, but their arguments must confront a materialist definition of progress. As these examples illustrate, Marxism-Leninism helps to shape the "discourse region" within which the Soviet and GDR debates occur.[1] As such it sets the normative terrain for the dialogue, influencing what constitutes a legitimate argument and providing the tools available to participants as they seek to persuade others.

It is often difficult to determine whether arguments over Marxist categories are derived from genuine disputes over Marxist theory or are legitimizing cloaks for other motivations. In some cases it may make no difference. For if everyone participates in the drama, the script becomes the stage of life. "The difference between illusion and reality becomes just as obliterated as that between truth and falsehood, if illusions guide behavior that shapes and changes reality."[2] When the public (Marxist-Leninist) language pervades public expression, it is difficult to imagine that an entirely distinct private language and conceptual structure can be fully developed. Nonetheless, there is very likely some bifurcation in the consciousness of Soviet scholars and leaders. The way a point is defended may not express the real motivations. Furthermore, there are different types of uses of Marxist arguments. For example, when geographers argue over the nature of historical determinism in the dispute over unified geography (chapter 4) one suspects that the appeal to Marxist categories is legitimatory. For the arguments are highly polemical and the nature of historical determinism is of great significance in the evolution of Stalinist and post-Stalinist Marxist doctrine. On the other hand, when economists appeal to productiveness theory to justify higher or lower environmental expenditures one suspects that the application of Marxist categories is instrumental, to achieve certain policy ends. As discussed in chapter 7, some economists have implied this. Finally, when economists discuss the application of the labor theory of value to natural resource pricing (chapter 8) they apply Marxist concepts in an almost routine, matter-of-fact fashion. In this case the application of Marxist categories to highly technical issues in a non-polemical manner suggests that the categories are embedded in the cognitive structure. From these examples we can see that Marxist-Leninist concepts perform a variety of functions. But all of these functions suggest that Soviet and East German analysts believe them to remain at the core of political legitimization. As long as this belief persists, Marxist-Leninist categories will continue to pervade policy debate in the USSR and GDR.

Equally interesting are those cases in which Marxist categories do not seem to intrude at all, or only superficially. Prime examples are East German discussion of methodologies for natural resource pricing, particularly in the 1960s (chapter 8), discussion of branch-territorial coordination problems (chapter 7), and discussion of the organization of environmental research (chapter 4). In the case of territorial planning this pattern may simply reflect the fact that Marx, Engels, and Lenin had little to say on the problem and that any lessons from Stalin or Khrushchev would be of dubious ideological legitimacy. In other cases, such as East German discussion of natural resource pricing, Marxist definitions were never institutionalized in government policies for reasons discussed in chapter 8 (and East German economists were given the cue in 1969 that they need not be). The contrast with the intrusion of Marxist categories in East German discussion of national income accounting is instructive here, for in this case national income was defined in terms of the Marxist categories of productive and unproductive labor before the environmental debate began. The case of the organization of environmental research is perhaps the most interesting of the three cases, for as experience in the Soviet geography discipline demonstrates, Marxist-Leninist categories probably could intrude, but neither Soviet scientists nor the political leadership have attempted to fit the debate to that mold. The Soviet leaders either trust their scientists to apply Marxist methods, as Gouldner suggests,[3] or they hesitate to risk the consequences of ill-fated political intrusion in scientific research. For whatever reason, Soviet science has achieved a significant degree of autonomy from direct political intrusion.

As suggested in chapters 1 and 2, in both the USSR and GDR environmental problems present both legitimacy challenges and concrete policy dilemmas to the political leadership. Thus this arena of public debate represents in microcosm a larger political problem facing the leaders, namely the need to balance control and flexibility, to prevent political disruption and discontent, while assuring necessary policy deliberation. How do Soviet and East German authorities mediate these conflicting demands?

In the GDR concern with legitimation produces a public debate which is more consistently and explicitly ideological than in the USSR. As noted in chapter 2, this is particularly noticeable in the popular media. Less criticism of governmental policy surfaces, the superiority of socialism is repeatedly reiterated, and the inadequacies of capitalist policies are more strongly asserted in the GDR. Nonetheless there are clear limits to popular presentation of environmental problems in the USSR as well. Top leaders (i.e., Politburo members) may not be criticized (although heads of ministries may, in some cases), economic growth is a sacrosanct value, and the superiority of socialism over capitalism is not questioned. Nonetheless the range of permissible disagreement and substantive criticism of present policies in the popular media is considerably broader in the USSR than in the GDR.

Our examination of the scholarly debate has revealed the same pattern. In the GDR scholars generally are less imaginative in their application of Marxism-Leninism and are more concerned to demonstrate ideological orthodoxy, often imitating Soviet arguments. Theoretical discussion in the GDR represents a fairly sophisticated explication of emerging doctrine, but does not generally delve into subtleties and ambiguities. On the other hand, East German scholars have made significant contributions to practical policy debate regarding, for example, development of technical processes to improve recycling and economic policies to provide incentives for rational natural resource use. The most innovative East German theoretical discussion has built on these foundations, analyzing the theoretical basis and significance of waste-free technology. While much of the debate in the GDR evidently occurs outside public view, the results, such as the **Landeskulturgesetz**, its implementation decrees, and introduction of charges for natural resource use and for pollution, demonstrate substantial concrete attention devoted to environmental issues by legal and economic experts. The intense impact of environmental deterioration and particularly of resource shortages in the GDR appears to have facilitated serious consideration of policy remedies. Discussion of such remedies is less deeply imbued with Marxist-Leninist categories of substantive content than is much Soviet policy debate. While East German treatment of environmental problems is fairly innovative on a practical policy level and may indeed influence future Soviet policy choices, the theoretical treatment is generally absent or derivative from Soviet sources. Silence may be the best guard against heresy, but imposed silence may be dysfunctional in the long run. While GDR scientists may have made advances in treating certain technical problems, larger questions about the ecological viability of the present economic strategy are not addressed. The longer-term costs, in terms of resource depletion and damage to health, forests, water, and land, may place even greater strain on the system's political and economic stability than would tolerance of more open ecological debate. In the USSR, scholars are given more leeway in grappling with the broad-ranging implications of environmental deterioration. Some Soviet scientists feel that ecological crisis is impending, both in East and West, but their warnings seem to have little impact on policy-makers. Furthermore, most Soviet scholars do not dissociate themselves from the dominant Western tradition favoring technological expansion and anthropocentric values.

In the face of pressing policy problems the Soviet and East German leaders continue to cling to an optimistic doctrine of final socialist victory. The optimism wanes only in official admissions about the danger of nuclear annihilation. Regarding every other issue, Soviet-bloc theoreticians and leaders continue to assert the onward march of progress and to blame setbacks either on external imperialist threats or temporary aberrations. Below the level of official doctrine, however, there is ambivalence, frustration, and conflict in Soviet-bloc attitudes, reflected clearly in the writings we examined in this study. Official optimism is the umbrella over unofficial uncertainty. The veil of official ideology only dimly masks the doubts of scientists, intellectuals, national minorities, and the working population, not only in the environmental area but in others as well. Intellectuals and scientists, being both visible and thoughtful, express most fully the ambivalence between mandated optimism and genuine uncertainty. Soviet ideologues assert that the Club of Rome reports reflect an effort of the capitalist establishment to come to terms with new challenges to the world order; the writings of scientists and intellectuals we have examined in this study reflect similar efforts on the part of the Soviet leadership. What we have observed is a balancing act on both sides: the leadership seeking insights by allowing some range of creativity without admitting uncertainty; and scientists seeking to use that margin of freedom to persuade other scientists, the leadership, and, in some cases, the population of the need for change. To be sure that margin of freedom can be shrunk, expanded, or eliminated by the leadership, but the longer it exists the more accustomed the intellectuals and population become to the politics of persuasion and compromise, even within carefully defined parameters.

Official and unofficial analyses of the ecological crisis in the Soviet bloc take on a dichotomous character. On the one level is over-arching optimism, based on a firm belief in technology and the victory of world socialism. On this level, capitalism is the villain. On another level is the recognition that in socialist society solution of these problems, like all others, will require arduous and consistent attention, involving difficult trade-offs and choices, many of which cannot be made expeditiously. Problems of ecological deterioration are admitted to be as difficult and treacherous as are problems of agricultural productivity, production efficiency, consumer goods provision, nationality conflict, and technical innovation. Nonetheless, it cannot be admitted directly and openly that they are essentially the same environmental problems that plague Western countries. This dichotomous view of environmental problems serves to isolate the Soviet bloc from the outside world, for it maintains that truth for the Soviet bloc is different from truth elsewhere because social structures differ. This insulation from the outside world is a key underpinning of Soviet and especially East German legitimacy, for it shields the population from aspirations and critiques which the regime feels it can ill afford. The great irony is that it is precisely the economic and technological processes which Marx did analyze that make this isolation so hard to maintain. The costs of autarky have become too high. And pollution, like radio waves, honors no borders. In admitting that **global** problems exist, Soviet leaders suggest the possibility of global dialogue, and this implicitly breaks down the dichotomous view of the world and undermines the isolation of Soviet society. Soviet-bloc scientists also skirt this isolation by capitalizing on official support for international cooperation and by alluding to common sources of environmental problems in East and West. Soviet and East German leaders again must perform a balancing act, and, as we observed in chapter 4, Soviet scientists attempt to use their range of liberty to expand international contacts and learn about international perspectives.

These contradictions produce a dynamic for change in the Soviet bloc. The Soviet and East German leaders only hesitantly allow their intellectuals to analyze that dialectic in fear that the theories which emerge might serve not simply to interpret the world, but indeed to change it.

## NOTES

1. This notion was stimulated by Hanna Pitkin's discussion of "language regions" in **Wittgenstein and Justice** (Berkeley: University of California Press, 1972), especially chapter 7.

2. Herbert Marcuse, **Soviet Marxism** (New York: Columbia University Press, 1958), p. 88.

3. Alvin Gouldner, **The Coming Crisis in Western Sociology** (New York: Basic Books, 1970), p. 470.

# List of Abbreviations

## ABBREVIATIONS OF JOURNAL TITLES USED IN THE NOTES AND BIBLIOGRAPHY

| | |
|---|---|
| **CDSP** | Current Digest of the Soviet Press |
| **DZfPh** | Deutsche Zeitschrift für Philosophie |
| **EiMM** | Ekonomika i matematicheskie metody |
| **Geographische Berichte** | Geographische Berichte der Geographischen Gesellschaft der Deutschen Demokratischen Republik |
| **PKh** | Planovoe khoziaistvo |
| **SGiP** | Sovetskoe gosudarstvo i pravo |
| **SGRT** | Soviet Geography: Review and Translation |
| **VEko** | Voprosy ekonomiki |
| **VFil** | Voprosy filosofii |
| **WW** | Wirtschaftswissenschaft |

## OTHER ABBREVIATIONS

| | |
|---|---|
| **AdW DDR** | Academy of Sciences of the German Democratic Republic (Akademie der Wissenschaften der DDR) |
| **AN SSSR** | Academy of Sciences of the Union of Soviet Socialist Republics (Akademiia nauk SSSR) |
| **COMECON (CMEA)** | Council for Mutual Economic Assistance |
| **CPSU** | Communist Party of the Soviet Union |
| **ECE** | Economic Commission for Europe (United Nations) |
| **FRG** | Federal Republic of Germany |
| **GDR** | German Democratic Republic |
| **GNKT** | State Committee of the USSR on Science and Technology (Gosudarstvennyi komitet SSSR po |

| | |
|---|---|
| | nauke i tekhnike) |
| **Goskom-gidromet** | State Committee of the USSR on Hydrometeorology and the Environment (Gosudarstvennyi komitet SSSR po gidrometeorologii i kontroliu prirodnoi sredy) |
| **Gosplan** | State Planning Commission (USSR) (Gosudarstvennaia planovaia komissiia) |
| **Gossnab** | State Committee on Material-Technical Supplies (USSR Council of Ministers) (Gosudarstvennyi komitet po material'no-tekhnicheskomy snabzheniiu) |
| **Iz.** | Publishing house (izdatel'stvo) |
| **IIASA** | International Institute for Applied Systems Analysis (Laxenburg, Austria) |
| **SCOPE** | Scientific Committee for Problems of the Environment (International Council of Scientific Unions) |
| **SED** | Socialist Unity Party (GDR) (Sozialistische Einheitspartei Deutschlands) |
| **SO AN USSR** | Siberian Division of the Academy of Sciences of the USSR (Sibirskoe otdelenie, Akademiia Nauk SSSR) |
| **STR** | Scientific-technical revolution |
| **TsEMI** | Central Mathematical-Economics Institute (USSR) (Tsentral'nyi ekonomiko-matematicheskii institut) |
| **TPC (TPK)** | Territorial-production complex (Territorial'no-proizvodstvennyi kompleks) |
| **VINTI** | All-Union Institute for Scientific and Technical Information (USSR) (Vsesoiuznyi institut nauchnoi i tekhnicheskoi informatsii) |
| **VNIISI** | All-Union Scientific-Research Institute for Systems Research (USSR) (Vsesoiuznyi nauchno-issledovatel'skii institut sistemnykh issledovanii) |

# Bibliography

This bibliography by no means includes all sources consulted for this study, but primarily those cited in the notes. Articles from **Current Digest of the Soviet Press** (CDSP), from popular magazines, and from newspapers are generally not included here unless they are cited repeatedly in the notes. Titles of newspapers and magazines consulted are listed separately at the end of the bibliography. For abbreviations of journals see the list on page 275.


Abramov, L. S. "Problema vzaimodeistviia prirody i obshchestva i mesto geograficheskikh nauk v ee razrabotke." **VFil**, no. 2 (1965), pp. 143-48.

Academy of Sciences of the USSR and Academy of Sciences of Czechoslovakia. **Man, Science, and Technology: A Marxist Analysis of the Scientific-Technological Revolution.** Prague: Academia Prague, 1973.

Afremov, V., and Ekondrat'eva, T. "K vode – po-khoziaiski." **Ekonomicheskaia gazeta**, no. 2 (January 1982), p. 10.

Agafonov, N. T., and Lavrov, S. B. Review of **Ekonomicheskaia geografiia: teorii, metody**, by Iu. G. Saushkin. **Vestnik Moskovskogo Universiteta**, no. 2 (1974). Translated in **SGRT** 15 (November 1974): 582-87.

Ahrends, Klaus. Review of **Umweltgestaltung und Ökonomie der Naturressourcen**, by Hans Roos et al. **Wirtschaftswissenschaft** 29 (June 1981): 727-34.

_______. "Zur Rentenbewertung von Naturressourcen." **WW** 31 (September 1983): 1333-52.

Akademie der Wissenschaften der DDR. **Gesellschaft und Umwelt: Hans Mottek zum 65. Geburtstag.** Sitzungsbericht, no. 2/G. Berlin: Akademie Verlag, 1976.

Akademie für Staats- und Rechtswissenschaft der DDR und Ministerium für Umweltschutz und Wasserwirtschaft. **Sozialistische Landeskultur Umweltschutz-Textausgabe ausgewählter Rechtsvorschriften mit Anmerkungen und Sachregister.** Berlin: Staatsverlag der DDR, 1978.

Akademiia Nauk Turkmenskoi SSR. **Okhrana prirody Turkmenii i nashi zadachi.** Ashkhabad: Ylym, 1972.

Akhobadze, D. "Godichnoe sobranie Leningradskikh filialov nauchnykh sovetov AN SSSR po ekonomicheskim problemam." **VEko**, no. 8 (1983), pp. 157-58.

Aksenenok, G. A. "Obespechenie ratsional'nogo ispol'zovaniia zemli." **SGiP**, no. 10 (1968), pp. 12-19.

______. "Poniatie sovetskogo zemel'nogo prava i ego sistemy." **SGiP**, no. 3 (1969), pp. 28-36.

______. "Priroda i chelovek v usloviiakh nauchno-tekhnicheskogo progressa." **VFil**, no. 10 (1973), pp. 54-63.

"Aktual'nye voprosy ekonomiki prirodopol'zovaniia." Round table. **VEko**, no. 2 (1982), pp. 81-108.

Alekseev, V. "Protection of the Natural Environment (Economic Aspect)." **Ekonomicheskie nauki**, no. 1 (1973). Translated in **Problems of Economics** 16 (November 1973): 24-35.

Alpat'ev, A. M., and Riabchikov, A. M. "Rol' geografii v reshenii problemy sokhraneniia i uluchsheniia prirodnoi sredy." **Vestnik Moskovskogo Universiteta, seriia 5, geografiia**, no. 3 (1981), pp. 11-17.

Ambree, K. et al. **Grundfragen der sozialistischen Preistheorie.** Berlin: Verlag Die Wirtschaft, 1977.

Ananichev, K. V. **Problemy okruzhaiushchei sredy, energii, i prirodnykh resursov: mezhdunarodnyi aspekt.** Moscow: Progress, 1975.

Anokhin, Iu. A., and Izrael', Iu. A. "Sistemnyi analiz i imitatsionnoe matematicheskoe modelirovanie kak metodologicheskaia osnova opredeleniia dopustimykh nagruzok antropogennykh zagriaznenii okruzhaiushchei sredy – regional'nyi podkhod." In **Comprehensive Analysis of the Environment – Proceedings of the Soviet-American Symposium**, pp. 68-78. Tbilsi, March 25-29, 1975. Leningrad: Hydrometeoizdat, 1975.

Anuchin, V. A. **Osnovy prirodopol'zovaniia. Teoreticheskii aspekt.** Moscow: Mysl', 1978.

______. "A Sad Tale about Geography." **Literaturnaia gazeta**, February 18, 1965. Translated in **SGRT** 6 (September 1965): 27-31.

______. "Sootnoshenie obshchestva i prirody v geograficheskoi srede i filosofskie problemy geografii." **VFil**, no. 4 (1975), pp. 80-91.

______. "Straddling the Boundaries between Sciences." **Priroda**, no. 4 (1972). Translated in **SGRT** 13 (September 1972): 432-40.

______. **Theoretical Problems of Geography.** Introduction by David Hoosen. Edited by Ronald J. Fuchs and George J. Demko. Columbus, OH: Ohio State University, 1977.

Apter, David E. **The Politics of Modernization.** Chicago: The University of Chicago Press, 1965.

Arakelian, A. "The Scientific-Technical Revolution and the Biosphere." **VEko**, no. 5 (1976). Translated in **Problems of Economics** 19 (March 1977): 68-85.

Arendt, Hannah. **The Origins of Totalitarianism.** New York: Harcourt, Brace, Jovanovich, 1951.

Arlt, Rainer. "Die Lehre Lenins über die Nationalisierung des Bodens und sein Genossenschaftsplan in ihrer Bedeutung für die Entwicklung des Bodenrechts der DDR." **Staat und Recht** 16 (1967): 1577-88.

Armstrong, John A. **The European Administrative Elite.** Princeton: Princeton University Press, 1973.

Artanovskii, S. N. "Obshchestvenno-kul'turnye aspekty ekologii cheloveka." **VFil**, no. 3 (1975), pp. 124-30.

Arutiunov, S. A. "Kul'turologicheskie issledovaniia i global'naia ekologiia." **Vestnik Akademii Nauk SSSR**, no. 12 (1980), pp. 92-98.

Astaf'ev, A. K. "Sistemnyi podkhod i problema reguliatsii prirodnoi sredy." **VFil**, no. 2 (1977), pp. 71-81.

**Ausgewählte Probleme der Leitung und Planung des Umweltschutzes: Vorträge des Berg- und Hüttenmännischen Tages 1975 in Freiberg.** Freiberger Forschungshefte D102, Wirtschaftswissenschaften, sozialistische Betriebswirtschaft. Leipzig: VEB Deutscher Verlag für Grundstoffindustrie, 1976.

Bagrinovskii, K. A., and Lemeshev, M. Ia. "O planirovanii ekonomicheskogo razvitiia s uchetom trebovanii ekologii." **EiMM** 12 (1976): 681-91.

Bahro, Rudolf. **The Alternative in Eastern Europe.** London: Verso, 1978.

Baklanov, P. Ia.; Poiarkov, B. V.; and Khudiakov, G. I. "Osnovnye napravleniia sovershenstvovaniia prirodopol'zovaniia i prirodookhrannykh meropriiatii na Dal'nem Vostoke." **Geografiia i prirodnye resursy**, no. 3 (1981), pp. 13-19.

Barm, Werner. "Umweltschutz in der DDR." **Deutsche Studien** 10, no. 38 (June 1972): 194-203.

Barth, Ludwig, and Sowade, Anneliese. "Probleme der sozialistischen Landeskultur im Geographieunterricht." **Zeitschrift für den Erdkundeunterricht**, part 2 (1973), pp. 63-77.

Bartz, Heinz, and Voss, Sigrid. "Einflussnahme der wirtschaftsrechtlichen Regelung auf eine rationelle Rohstoff- und Materialwirtschaft zur intensiveren Naturressourcennutzung." **Staat und Recht** 32 (June 1983): 443-50.

Bauer, Adolf. "Probleme der Naturaneignung." **DZfPh** 30 (1982): 1323-33.

Bauer, Adolf, and Paucke, Horst. "Einheit und Kampf zwischen Natur und Gesellschaft." **DZfPh** 27 (1979): 593-602.

_______. "Umweltprobleme in der Sicht des historischen Materialismus und bürgerliche Weltmodelltheorien." **DZfPh** 24 (1976): 783-98.

Bauer, Ludwig. "Aufgaben, Begriffsbestimmungen und Publikationsorgane auf dem Gebeit der Landeskultur und des Naturschutzes." **Geographische Berichte** 12, no. 42 (1967): 58-60.

_______. "Zu einigen Problemen und Aufgaben der Landeskultur und Umweltgestaltung." **Geographische Berichte** 17, no. 64-65 (1972): 174-80.

Becher, Jürgen et al. **Politische Ökonomie des Kapitalismus und des Sozialismus: Lehrbuch für das marxistische-leninistische Grundlagenstudium.** Berlin: Dietz Verlag, 1979.

Becker, Abraham S. "National Income Accounting in the USSR." In **Soviet Economic Statistics**, pp. 69-119.. Edited by Vladimir G. Treml and John P. Hardt. Durham, NC: Duke University Press, 1972.

_______. **Soviet National Income 1958-64.** Berkeley: University of California Press, 1969.

Bedenin, N. N. "Okhrana prirody v SSSR." **SGiP**, no. 10 (1969), pp. 151-53.

Beirau, Erhard. **Wettbewerb Materialökonomie Sekundärrohstoff.** Berlin: Verlag Tribüne, 1979.

Berger, Peter L., and Luckman, Thomas. **The Social Construction of Reality.** Garden City, NY: Doubleday & Co., 1966.

Bergkhol'tsas, I. I. "Mestnye Sovety i okhrana prirody." **SGiP**, no. 6 (1973), pp. 78-82.

"Bericht des Zentralkomitees der Sozialistischen Einheitspartei Deutschlands an den X. Parteitag der SED." Delivered by Erich Honecker. **Neues Deutschland**, April 12, 1981, pp. 3-14.

Bialer, Seweryn. **Stalin's Successors: Leadership, Stability, and Change in the Soviet Union.** Cambridge, ENG: Cambridge University Press, 1980.

Biermann, Wolf. **Für meine Genossen.** Berlin (West): Verlag K. Wagenbach, 1972.

_______. **Mit Marx- und Engelszungen.** Berlin (West): Verlag K. Wagenbach, 1968.

Billwitz, K. "Grundfragen und technische Massnahmen zur Reinhaltung der Biosphäre." **Geographische Berichte** 18, no. 68 (1973): 218-19.

Bittighöfer, Bernd; Edeling, Herbert; and Kulow, Hans. "Theoretische und politisch-ideologische Fragen der Beziehung von Mensch und Umwelt." **DZfPh** 20 (1972): 60-79.

Blagosklonov, K. N.; Inozemtsev, A. A.; and Tikhomirov, V. K. **Okhrana prirody**. Moscow: Vyshaia Shkola, 1967.

Bobylev, S. "Puti uluchsheniia ispol'zovaniia zemel'nykh resursov." **VEko**, no. 8 (1980), pp. 49-59.

Bogachev, V. "O gornoi rente i otsenke mestorozhdenii syr'ia i topliva." **VEko**, no. 9 (1974), pp. 25-38.

**Bol'shaia sovetskaia entsiklopediia**, 1972 ed.

Bönisch, Alfred, and Graf, Dieter. "Mensch am Wendepunkt." **Deutsche Aussenpolitik** 20 (1975): 1007-26.

Bönisch, Rolf; Mohs, Gerhard; and Ostwald, Werner, eds., **Territorialplanung**. Berlin: Verlag Die Wirtschaft, 1976.

Borovskikh, B. A. **Planirovanie prirodopol'zovaniia: voprosy metodologii**. Moscow: Ekonomika, 1979.

_______. "Urgent Problems in Planning the Reproduction of Natural Resources." **Ekonomicheskie nauki**, no. 1 (1973). Translated in **Problems of Economics** 16 (November 1973): 3-23.

Boss, Helen H. M. "Productive Labour, Unproductive Labour, and the Boundary of the Economic Domain 1662-1980." Ph.D. dissertation, McGill University, 1981.

Böttcher, D.; Krämer, F.; and Streibel, G. "Ökonomische Probleme des Umweltschutzes." **WW** 24 (April 1976): 592-96.

Box, Hans Jürgen; Haalck, Jorgen; and Joachim, Siegfried. "Maritimer Umweltschutz (Küstenbereich DDR) in rechtlicher Sicht." **Wissenschaftliche Zeitschrift der Universität Rostock: Gesellschafts- und sprachwissenschaftliche Reihe** 22, no. 4 (1973): 371 ff.

Bozhedomov, A. "Fiksirovannye rentnye platezhi v neftianoi promyshlennosti." **VEko**, no. 1 (1980), pp. 128-32.

Braginskii, L. "O vzaimodeistvii ekonomicheskikh rychagov v upravlenii kolkhozami." **VEko**, no. 3 (1971), pp. 75-82.

Bronshtein, M. "K voprosu o 'tsene' zemli." **VEko**, no. 5 (1968), pp. 102-12.

Bruns, Wilhelm. "Deutsch-deutscher Umweltschutz. Notwendig und möglich!" **DDR-Report** 17, no. 2 (1984): 70-74.

Brzezinski, Zbigniew K. "Soviet Politics: From the Future to the Past." In **The Dynamics of Soviet Politics**, pp. 337-51. Edited by Paul Cocks, Robert W. Daniels, and Nancy Whittier Heer. Cambridge, MA: Harvard University Press, 1976.

_______. **The Soviet Bloc: Unity and Conflict**. Cambridge, MA: Harvard University Press, 1967.

Buck, Hannsjörg F., and Spindler, Bernd. "Luftbelastung in der DDR durch Schadstoffemissionen." **Deutschland Archiv** 15 (1982): 943-58.

Bundesministerium für innerdeutsche Beziehungen. **DDR-Handbuch**. Cologne: Verlag Wissenschaft und Politik, 1979.

Burrichter, Clemens; Förtsch, Eckart; and Müller, Hans-Joachim. "Die wissenschaftlich-technische Revolution – Kriterien und Konsequenzen." **Deutschland Archiv** 9 (1976): 516-29.

Busch, Karl-Franz. "Probleme der sozialistischen Umweltgestaltung." **Wissenschaftliche Zeitschrift der Technischen Universität Dresden** 32 (1983): 99-100.

Büscher, Wolfgang, and Wensierski, Peter. **Null Bock auf DDR: Aussteigerjugend im anderen Deutschland.** Reinbek bei Hamburg: Spiegel Verlag, 1984.

Busch-Lüty, Christiane. "Zur Umweltproblematik in sozialistischen Systemen: Ideologie und Realität." **Aus Politik und Zeitgeschichte,** no. 27 (1981), pp. 18-46.

Bush, Keith. "Environmental Problems in the USSR." **Problems of Communism** 21 (July-August 1972): 21-31.

_______. "The Soviet Response to Environmental Disruption." In **Environmental Deterioration in the Soviet Union and Eastern Europe,** pp. 8-36. Edited by Ivan Volgyes. New York: Praeger, 1974.

Butler, William E. **The Soviet Legal System: Legislation and Documentation.** Dobbs Ferry, NY: Oceana Publications, 1978.

Bykhovskii, A. V. "Ekologiia kak kompleksnaia nauchnaia distsiplina." **VFil,** no. 8 (1979), pp. 153-55.

Bykhovskii, A. V., and Los', V. A. Review of **Razmyshleniia naturalista. Prostranstvo i vremia v zhivoi i nezhivoi prirode** and of **Razmyshleniia naturalista. Nauchnaia mysl' kak planetnoe iavlenie,** by V. I. Vernadskii. **VFil,** no. 1 (1978), pp. 173-76.

Bystritskaia, N., and Mikhura, V. "O plate za vodu promyshlennymi predpriiatiiami." **PKh,** no. 1 (1983), pp. 123-25.

Campbell, Robert W. **Soviet Energy Technologies: Planning, Policy, Research and Development.** Bloomington: Indiana University Press, 1980.

Chappell, John E., Jr. "The Ecological Dimension: Russian and American Views." **Annals of the Association of American Geographers** 65 (March 1975): 144-62.

_______. "Marxism and Geography." **Problems of Communism** 14 (November-December 1965): 12-22.

**Chelovek i ego sotsial'naia i prirodnaia sreda.** Materialy respublikanskoi nauchno-tekhnicheskoi konferentsii. Alma-Ata: Nauka, 1975.

"Chelovek i sreda ego obitaniia." Round table. **VFil,** no. 1 (1973), pp. 48-60; no. 2 (1973), pp. 37-52; no. 3 (1973), pp. 51-73; and no. 4 (1973), pp. 57-79.

Cherniavskii, V. S. "O sostavlenii optimal'nogo grafika vodoispol'zovaniia." **EiMM** 5 (1969): 590 ff.

Chilinski, Karlheinz. "Konferenzbericht: Ökonomische Aspekte des Umweltschutzes." **Wissenschaftliche Zeitschrift: Hochschule für Ökonomie Bruno Leuschner Berlin** 19 (1974): 89-93.

Chkhikvadze, Viktor M., ed. **Gosudarstvo, pravo, ekonomika.** Akademiia Nauk SSSR, Institut gosudarstva i prava. Moscow: Iuridicheskaia literatura, 1970.

Chumachenko, N. G., and Belashova, L. A., eds. **Aktual'nye problemy okhrany okruzhaiushchei sredy: Ekonomicheskie aspekty.** Kiev: Naukova Dumka, 1979.

Cibulka, Hanns. "Swantow." **Neue Deutsche Literatur** 29 (April 1981): 23-52.

Clawson, Robert W., and Kolarik, William. "Soviet Resource Management: Political Aspects of Water Pollution Control." In **Soviet Resource Management and the Environment**, pp. 105-22. Edited by W. A. Douglas Jackson. Columbus, OH: American Association for the Advancement of Slavic Studies, 1978.

Cohn, Stanley H. "National Income Growth Statistics." In **Soviet Economic Statistics**, pp. 120-47. Edited by Vladimir G. Treml and John P. Hardt. Durham, SC: Duke University Press, 1972.

Connolly, William. **The Terms of Political Discourse.** Lexington, MA: D.C. Heath, 1974.

Cooper, Julian. "The Scientific-Technical Revolution in Soviet Theory." In **Technology and Communist Culture: The Socio-Cultural Impact of Technology under Socialism**, pp. 146-75. Edited by Frederic J. Fleron, Jr. New York: Praeger Publishers, 1977.

Croan, Melvin. **East Germany: The Soviet Connection.** The Washington Papers, vol. 9. Beverly Hills: Sage Publications, 1976.

_______. "Some Constraints on Change in Eastern Europe." **Slavic Review** 33 (June 1974): 241-45.

Dallin, Alexander, and Breslauer, George W. "Political Terror in the Post-Mobilization Stage." In **Change in Communist Systems**, pp. 196-214. Edited by Chalmers Johnson. Stanford, CA: Stanford University Press, 1970.

DeBardeleben, Joan. "Esoteric Policy Debate: Nuclear Safety Issues in the USSR and GDR." **British Journal of Political Science**, forthcoming.

Demko, George. "Trends in Soviet Geography." **Survey**, no. 55 (April 1965), pp. 163-70.

de Montbrial, Thierry. **Energy, the Countdown: A Report to the Club of Rome.** New York: Pergamon Press, 1979.

Denison, Edward F. "Welfare Measurement and the GNP." In **Economics of the Environment: Selected Readings**, 2nd edition, pp. 469-78. Edited by Robert and Nancy S.

Dorfman. New York: W. W. Norton & Co., 1977.
Denisov, V. I. "O metodakh rascheta differentsial'noi zemel'noi renty." **EiMM** 4 (1968): 732 ff.
**Directive of the Eighth Congress of the SED of Germany for the Development of the National Economy of the GDR from 1971 to 1975**. Dresden: Verlag Zeit im Bild, 1971.
**Directives for the Five-Year Plan for the GDR's National Economy 1976-1980**. Dresden: Verlag Zeit im Bild, May 1976.
"Direktive des X. Parteitages der SED zum Fünfjahrplan für die Entwicklung der Volkswirtschaft der DDR in den Jahren 1981 bis 1985." Presented by Willi Stoph. **Neues Deutschland**, April 15, 1981, pp. 3-6.
"Direktive des X. Parteitages der SED zum Fünfjahrplan für die Entwicklung der Volkswirtschaft der DDR in den Jahren 1981 bis 1985." **Neues Deutschland**, April 18/19, 1981, pp. 3-8.
Dolishnii, M.; Kravtsiv, V.; and Kuzyk, S. "Vsaimosviaz' ekonomicheskikh i ekologicheskikh problem v regione." **PKh**, no. 12 (1983), pp. 84-88.
Doskach, A. G.; Riabchikov, A. M.; Gvozdetskii, N. A.; Vendrov, S. L.; and Pokshishevskii, V. V. "Rejoinders." **Izvestiia Vsesoiuznogo Geograficheskogo Obshchestva**, no. 4 (1975). Translated in **SGRT** 17 (December 1976): 705-8.
Doskach, A. G.; Trusov, Iu. P.; and Faddeev, E. T. "Problema vzaimodeistviia prirody i obshchestva i sovremennaia geografiia." **VFil**, no. 4 (1965), pp. 104-15.
Duffy, Gloria. **Soviet Nuclear Energy: Domestic and International Politics**. Prepared for the U.S. Department of Energy. Santa Monica: Rand Corp., 1979.
"Effektivnost' zatrat na okhranu okruzhaiushchei sredy: vremennaia metodika opredeleniia ekonomicheskoi effektivnosti zatrat v meropriiatiia po okhrane okruzhaiushchei sredy." **Ekonomicheskaia gazeta**, no. 33 (August 1980), pp. 13-14.
Efremov, Iu. K. "Preobrazovanie prirody kak sostavnaia chast' prirodopol'zovaniia." **Voprosy geografii** 108 ("Prirodopol'zovanie: geograficheskii aspekt"), pp. 14-26. Edited by S. A. Kovalev. Moscow: Mysl', 1978.
Efremov, L. "Planet's Health." **Izvestiia**, June 5, 1974. Translated in **CDSP** 26, no. 23 (1974), p. 30.
Eichhorn, Wolfgang I. "Die Zukunft der Menschheit in der weltanschaulichen Klassenauseinandersetzung unserer Zeit." **Einheit** 34 (1979): 685-94.
"Ekologiia, nauka, obshchestvo." **VFil**, no. 9 (1974), pp. 98-101.
"Ekonomicheskaia otsenka prirodnykh resursov — obzor statei postupivshikh v redaktsiiu." **VEko**, no. 1 (1969), pp. 75-110.
Eliseev, I., and Kondratenko, A. "Nashe krovnoe delo." **Kommunist**, no. 5 (1968), pp. 112-17.

Eliseeva, L. D. "Institutu geografii Sibiri i Dal'nego Vostoka 20 let (1959-1979)." **Geografiia i prirodnye resursy**, no. 2 (1980), pp. 184-85.

Ende, Gesine; Steffens, Rolf; and Stein, Jürgen. "Ökonomische Bewertung des Einflusses von $SO_2$-Immissionen auf die Erholungsleistungen der Waldlandschaft." **Wissenschaftliche Zeitschrift der Technischen Universität Dresden** 32 (1983): 219-22.

Engelhardt, Harald. "Zur Berücksichtigung des Aufwands für Umweltschutz in der Preisbildung." **Wissenschaftliche Zeitschrift: Hochschule für Ökonomie Bruno Leuschner Berlin** 19 (1974): 93-97.

Engels, Frederick. **Dialectics of Nature**, 2d revised ed. Moscow: Progress Publishers, 1954.

Enloe, Cynthia H. **The Politics of Pollution in a Comparative Perspective**. New York: David McKay Company, 1975.

Evans, Alfred B., Jr. "Developed Socialism in Soviet Ideology." **Soviet Studies** 29 (July 1977): 409-28.

Falcke, Heino. "Gerechtigkeit, Ökologische Lebensfähigkeit und Partizipation." In **Beton ist Beton**, pp. 205-15. See Wensierski and Büscher, 1981.

_______. "Kritische Partizipation im realen Sozialismus." In **Beton ist Beton**, pp. 197-203. See Wensierski and Büscher, 1981.

Fedenko, N. "Ekonomika vodnogo khoziaistva i tsena na vodu." **VEko**, no. 2 (1966), pp. 130-32.

Fedorenko, N. P. "Zadachi ekonomicheskoi nauki v svete reshenii XXVI S"ezda KPSS." **VEko**, no. 4 (1981), pp. 17-25.

_______, ed. **Ekonomicheskie problemy optimizatsii prirodopol'zovaniia**. Moscow: Nauka, 1973.

Fedorenko, N. P., and Gofman, K. G. "Problems of Optimization in the Planning and Control of the Environment." **VEko**, no. 10 (1972). Translated in **Soviet Review** 14 (Summer 1973): 24-38.

_______. "Upravlenie sotsialisticheskim prirodopol'zovaniem v sisteme obshchestvennogo vosproizvodstva." In **Sotsialisticheskoe prirodopol'zovanie: ekonomicheskie i sotsial'nye aspekty**, pp. 10-30. Edited by N. N. Nekrasov and E. Mateev. Moscow: Ekonomika; Sofiia: Partizdat, 1980.

_______. "Vvedenie." In **Ekonomicheskie problemy optimizatsii prirodopol'zovaniia**, pp. 1-7. Edited by N. P. Fedorenko. Moscow: Nauka, 1973.

Fedorenko, N. P.; Lemeshev, M. Ia.; and Reimers, N. F. "Sotsial'no-ekonomicheskaia effektivnost' okhrany prirody." **Priroda**, no. 10 (1980), pp. 2-13.

Fedorenko, N. P., and Reimers, N. F. "Priroda, ekonomika, nauka." **Priroda**, no. 3 (1974), pp. 2-13.

______. "Sblizhenie ekonomicheskikh i ekologicheskikh tselei v okhrane prirody." **Priroda**, no. 9 (1981), pp. 3-12.

Fedorov, E. K. "Man and Nature." **World Marxist Review**, no. 5 (1968), pp. 25-31.

______. "Metodologicheskie problemy nauk o Zemle." **VFil**, no. 7 (1966), pp. 96-107.

______. "Ot opisaniia k proektirovaniiu prirody." **VFil**, no. 1 (1978), pp. 34-49.

______. **Vzaimodeistvie obshchestva i prirody**. Leningrad: Gidrometeoizdat, 1972.

Fedorov, E. K., and Novik, I. B. "Ecological Aspects of Social Progress." In **Society and the Environment: A Soviet View**, pp. 37-55. Moscow: Progress Publishers, 1977.

______. "Problemy vzaimodeistviia cheloveka s prirodnoi sredoi." **VFil**, no. 12 (1972), pp. 46-58.

Feitel'man, N. "Ekonomicheskaia otsenka prirodnykh resursov." **VEko**, no. 10 (1980), pp. 63-73.

______. "Ob ekonomicheskoi otsenke mineral'nykh resursov." **VEko**, no. 11 (1968), pp. 110-16.

______. "Socioeconomic Problems of Ecological Equilibrium in Western Siberia." **VEko**, no. 10 (1978). Translated in **Problems of Economics** 21 (April 1979): 10-22.

______. "Upravlenie prirodopol'zovaniem." **VEko**, no. 3 (1984), pp. 56-66.

Feitel'man, N., and Smagarinskii, I. "Regulirovanie ratsional'nogo prirodopol'zovaniia." **VEko**, no. 11 (1981), pp. 36-47.

Fel'zenbaum, V. "Cost Accounting and the Socio-economic Effect of New Technology." **VEko**, no. 5 (1981). Translated in **Problems of Economics** 25 (June 1982): 21-38.

Fleron, Frederic J., Jr. "Introduction," and "Afterword." In **Technology and Communist Culture: The Socio-Cultural Impact of Technology under Socialism**, pp. 1-67, 457-87. Edited by Frederic J. Fleron, Jr. New York: Praeger, 1977.

______. "The Western Connection: Technical Rationality and Soviet Politics." **Soviet Union/Union Soviétique** 4, part 1 (1977): 58-84.

Forrester, Jay Wright. **World Dynamics**. Cambridge, MA: Wright-Allen Press, 1971.

Forschungsheim, das Kirchliche (Wittenberg). "Briefe zur Orientierung im Konflikt Mensch-Natur," nos. 1-7. Wittenberg, January 1980 to February 1983. Hectograph.

______. "Die Erde ist zu retten. Umweltkrise, christliche Glaube, Handlungsmöglichkeiten." Wittenberg, 1982. Hectograph.

______. "Kein Sonnenschein ohn' Unterlass." In **Beton ist Beton**, pp. 51-84. See Wensierski and Büscher, 1981.

Fox, Irving K., ed. **Water Law in the Soviet Union**. Madison, WI:

The University of Wisconsin Press, 1971.

Franzy, Hans. **Landeskultur und Umweltschutz in der DDR.** Hamburg: Gesamtdeutsches Institut, 1971.

Fridman, L. "Upravlenie ispol'zovaniem i vosproizvodstvom prirodnykh resursov." **VEko,** no. 6 (1983), pp. 153-54.

Friedrich, Carl J., and Brzezinski, Zbigniew. **Totalitarian Dictatorship and Autocracy.** Cambridge, MA: Harvard University Press, 1956.

Friedrich, Gerd, and Schutz, Gerhard. "Effektivität, wissenschaftlich-technische Revolution und sozialistische Wirtschaftsführung." **Einheit** 28 (1973): 187-97.

Frolov, I. T. "Filosofiia global'nykh problem." **VFil,** no. 2 (1980), pp. 29-44.

_______. "Nauka i gumanisticheskie idealy v reshenii global'nykh problem." **VFil,** no. 6 (1979), pp. 116-18.

Frolov, I. T., and Zagladin, V. V. (Transliterated in the German as Frolow, I. T., and Sagladin, W. W.) "Globale Probleme und die Zukunft der Menschheit." **Einheit** 34 (1979): 713-21.

Füllenbach, Josef. **Umweltschutz zwischen Ost und West.** Bonn: Europa Union Verlag, 1977.

Gabor, Dennis, and Colombo, Umberto. **Beyond the Age of Waste.** New York: Pergamon Press, 1978.

Garkovenko, R. V.; Novik, I. B.; and Shatalov, A. T. **Obshchestvo i priroda.** Moscow: Znanie, 1974.

Gensichen, Peter. "Jahrzehnte zu spät . . . ?" In **Beton ist Beton,** pp. 99-101. See Wensierski and Büscher, 1981.

Gerasimov, I. P. "Constructive Geography: Aims, Methods, and Results." **Izvestiia Vsesoiuznogo Geograficheskogo Obshchestva,** no. 5 (1966). Translated in **SGRT** 9 (November 1968): 739-55.

_______. "Fifty Years of Development of Soviet Geographic Thought." **Izvestiia Akademii Nauk SSSR, seriia geograficheskaia,** no. 5 (1967). Translated in **SGRT** 9 (April 1968): 238-52.

_______. "Metodologicheskie problemy ekologizatsii sovremennoi nauki." **VFil,** no. 11 (1978), pp. 61-72.

_______. "Nauchno-tekhnicheskaia revoliutsiia i tendentsii razvitiia geograficheskoi nauki." **VFil,** no. 5 (1977), pp. 121-32.

_______. "Nuzhen general'nyi plan preobrazovaniia prirody nashei strane." **Kommunist,** no. 2 (1969), pp. 68-79.

_______, ed. **Man, Society, and the Environment.** USSR Academy of Sciences, Institute of Geography. Moscow: Progress Publishers, 1975.

Gerasimov, I. P.; Armand, D. L.; and Efron, K. M. **Natural Resources of the Soviet Union: Their Use and Renewal.** San Francisco: W. H. Freeman & Co., 1971.

Gerasimov, I. P., and Komar, I. V. "Future Environments of

America." **Izvestiia Akademii Nauk SSSR, seriia geograficheskaia**, no. 3 (1968). Translated in **SGRT** 9 (November 1968): 797-805.

Gerasimov, I. P., and Preobrazhenskii, V. S. "Osnovnye zadachi sovetskoi geograficheskoi nauki v svete reshenii XXV S"ezda KPSS." **Izvestiia Akademii Nauk SSSR, seriia geograficheskaia**, no. 4 (1976), pp. 5-17.

Gerlach, Günter, and Schubert, Kurt. "Die weitere Verwirklichung der territorialen Rationalisierung." **Staat und Recht** 27 (1978): 1083-94.

Gesamtdeutsches Institut, Bundesanstalt für gesamtdeutsche Aufgaben. **Die Rohstoffwirtschaft der DDR.** Bonn, 1980.

Gilbert, Alan. "Social Theory and Revolutionary Activity in Marx." **American Political Science Review** 73 (June 1979): 521-53.

Girusov, E. V. "Metodologicheskie voprosy teorii vzaimodeistviia prirody i obshchestva." In **Nauchno-tekhnicheskaia revoliutsiia, chelovek, ego prirodnaia i sotsial'naia sreda**, pp. 89-94. Edited by V. G. Marakhov. Leningrad: Iz. Leningradskogo Universiteta, 1977.

Gladkov, N. A. **Problema okhrany prirody i ee narodnokhoziaistvennoe znachenie.** Moscow: Iz. Moskovskogo Universiteta, 1968.

Gofman, K. G. **Ekonomicheskaia otsenka prirodnykh resursov v usloviiakh sotsialisticheskoi ekonomiki.** Moscow: Nauka, 1977.

Gofman, K. G., and Gusev, A. A. "Ekologicheskie izderzhki i kontseptsiia ekonomicheskogo optimuma kachestva okruzhaiushchei prirodnoi sredy." **EiMM** 17 (1981): 515-27.

Gokhberg, M. "Regional'noe prirodopol'zovanie." **VEko**, no. 6 (1980), pp. 75-86.

Gokhman, V. M.; Lappo, G. M.; and Pivovarov, Iu. L. "Urbanisation and Its Impact on the Environment." In **Man, Society, and the Environment**, pp. 280-304. Edited by I. P. Gerasimov. USSR Academy of Sciences, Institute of Geography. Moscow: Progress Publishers, 1975.

Goldman, Marshall. **The Spoils of Progress: Environmental Pollution in the Soviet Union.** Cambridge, MA: MIT Press, 1972.

Gordina, E. V. "Sotsial'no-politicheskaia ogranichennost' zapadnykh modelei ekologicheskogo krizisa." **EiMM** 13 (1977): 31-41.

Gorizontov, B. "Sotrudnichestvo stran SEV v okhrane okruzhaiushchei sredy i ispol'zovanii prirodnykh resursov." **VEko**, no. 9 (1974), pp. 68-78.

Gorizontov, B., and Prokudin, V. "Environmental Protection in COMECON Member Nations." **VEko**, no. 4 (1978). Translated in **Problems of Economics** 21 (December 1978): 24-40.

Gorokhov, V. G., and Kaganchuk, G. V. "Sotsial'no-filosofskie problemy ekologizatsii sovremennogo gornogo proizvodstva." **VFil**, no. 12 (1979), pp. 144-49.

Gorz, André. **Ecology as Politics**. Montreal: Black Rose Press, 1980.

Gosplan SSSR. **Tipovaia metodika razrabotki piatiletnego plana proizvodstvennogo ob'edineniia (kombinata), predpriiatiia.** Moscow: Ekonomika, 1975.

Gouldner, Alvin W. **The Coming Crisis in Western Sociology**. New York: Basic Books, 1970.

Graf, Dieter. "Naturpotentiale und Naturressourcen – Bemerkungen aus ökonomischer Sicht." **Petermanns Geographische Mitteilungen** 124 (1980): 53-57.

_______. "Naturressourcen und Nationalreichtum." **WW** 29 (June 1981): 670-85.

_______. "Von 'Grenzen des Wachstums' zu den Grenzen der Verschwendung." **WW** 25 (October 1977): 1551-55.

_______. "Zu einigen Grundfragen der ökonomischen Bewertung von Naturressourcen im entwickelten Sozialismus." **WW** 26 (July 1978): 811-23.

_______. "Zur ökonomischen Bewertung von Naturressourcen und Naturpotentialen." In **Beiträge zur planmässigen Gestaltung der Landschaft**, pp. 33-39. Edited by Hans Richter. Gotha: Leipzig: VEB Hermann Haack, 1978.

Graham, Loren R. "The Place of the Academy of Sciences in the Overall Organization of Soviet Sciences." In **Soviet Science and Technology: Domestic and Foreign Perspectives**, pp. 44-62. Edited by John R. Thomas and Ursula M. Kruse-Vaucienne. Washington, DC: National Science Foundation, 1977.

_______. **Science and Philosophy in the Soviet Union**. New York: Alfred A. Knopf, 1972.

Griffiths, Franklyn. "A Tendency Analysis of Soviet Policy-Making." In **Interest Groups in Soviet Politics**, pp. 335-77. Edited by H. Gordon Skilling and Franklyn Griffiths. Princeton: Princeton University Press, 1971.

Gringmuth, Werner. "Die Mensch-Umwelt Relation bei der Schaffung der entwickelten sozialistischen Gesellschaft im Licht der Marxschen Theorie." **WW** 21 (August 1973): 1230-39.

_______. "Zur ökonomischen Bedeutung von Naturpotentialen und Naturressourcen und zu Bedingungen ihrer Nutzung und Reproduktion." In **Beiträge zur planmässigen Gestaltung der Landschaft**, pp. 41-47. Edited by Hans Richter. Gotha/Leipzig: VEB Hermann Haack, 1978.

Gringmuth, Werner et al. **Reproduktion der natürlichen Umwelt**. Akademie der Wissenschaften der DDR, Zentralinstitut für

Wirtschaftswissenschaften, Forschungsberichte no. 8. Berlin: Akademie Verlag, 1974.

Gromov, B. V.; Zaitsev, V. A.; Laskorin, B. N.; Petrash, A. I.; Tsygankov, A. P.; and Iagodin, G. A. **Bezotkhodnoe promyshlennoe proizvodstvo – osnovnye printsipy bezotkhodnykh proizvodstv**. Itogi nauki i tekhniki: "Okhrana prirody i vosproizvodstvo prirodnykh resursov," vol. 9. Moscow: Vsesoiuznyi institut nauchnoi i tekhnicheskoi informatsii, 1981.

Grosse, Hermann, and Puschman, Manfred. **Qualität des Lebens: Ausweg oder Irreführung**. Berlin: Dietz Verlag, 1974.

Gruhn, Werner. "Aktuelle Aspekte der DDR-Umweltpolitik." **Deutsche Studien** 19, no. 76 (1981): 424-40.

______. "Kernkraftwerksicherheit aus DDR-Sicht." **Deutschland Archiv** 10 (1977): 806-10.

______. **Umweltpolitische Aspekte der DDR-Energiepolitik**. Erlangen, FRG: Institut für Gesellschaft und Wissenschaft, 1982.

______. "Umweltschutz als ökonomischer Faktor." Presented at the XVII. DDR-Forscher Tagung at the Europäische Akademie Lerbach. Bergisch-Gladbach, FRG, June 12-15, 1984.

______. "Umweltschutz in der DDR." **Deutschland Archiv** 5 (1972): 1038-50.

Grundmann, Siegfried. "Mensch und Umwelt." **DZfPh** 21 (1973): 190-206.

______. "Zur Soziologie des Territoriums." **WW** 23 (January 1975): 39-56.

Grundmann, Siegfried, and Stabmow, Erwin. "Beziehungen von Mensch und Umwelt." **WW** 19 (December 1971): 1774-84.

Gudozhnik, G. S. **Nauchno-tekhnicheskaia revoliutsiia i ekologicheskii krizis**. Moscow: Mezhdunarodnye otnosheniia, 1975.

Guliarov, M. S.; Vinberg, G. G.; and Chernov, Iu. I. "Ekologiia – zadachi i perspektivy." **Priroda**, no. 5 (1977), pp. 3-11.

Gusev, A. A. "Sotsial'no-ekonomicheskaia effektivnost' prirodookhrannykh meropriiatii." In **Sotsialisticheskoe prirodopol'zovanie: Ekonomicheskie i sotsial'nye aspekty**, pp. 134-86. Edited by N. N. Nekrasov and E. Mateev. Moscow: Ekonomika; Sofiia: Partizdat, 1980.

Gustafson, Thane. **Reform in Soviet Politics: Lessons of Recent Policies on Land and Water**. Cambridge, ENG: Cambridge University Press, 1981.

Gvishiani, D. M. "Methodological Problems in Modeling Global Development." **VFil**, no. 2 (1978). Translated in **Soviet Studies in Philosophy** 17 (Fall 1978): 3-29.

______. "Nauka i global'nye problemy sovremennosti." **VFil**, no. 3 (1981), pp. 97-108.

Gvishiani, D. M., and Sokoleva, V. E., eds. **Programma 'chelovek i**

**biosfera' v stranakh sotsializma.** Moscow: VNIISI, 1979.

Gzovskii, V. "Ispol'zovanie vozobnovliaemykh prirodnykh resursov v stranakh SEV." **VEko,** no. 5 (1983), pp. 108-18.

Haase, Jutta, and Haase, Günter. "Die Mensch-Umwelt Problematik: Gedanken zum Ausgangspunkt und zum Beitrag der geographischen Forschung." **Geographische Berichte** 16, no. 61 (1971): 243-70.

Hager, Kurt. **25 Jahre – Aufsteig des Sozialismus im Bruderband mit der Sowjetunion.** Berlin: Dietz Verlag, 1974.

_______. **Wissenschaft und Technologie im Sozialismus.** Berlin: Dietz Verlag, 1974.

Hammer, Darrell P. **USSR: The Politics of Oligarchy.** Hinsdale, IL: Dryden Press, 1974.

Harich, Wolfgang. "Ich hatte alle gegen mich." Interview. **Der Spiegel,** June 11, 1979, pp. 70-76.

_______. **Kommunismus ohne Wachstum? Babeuf und der 'Club of Rome.'** Reinbek bei Hamburg: Rowohlt, 1975.

Harris, Chauncy D. "Geographers and Geographic Institutions in the Soviet Union Visited by a Delegation of American Geographers August 17-September 13, 1961."

Havemann, Robert. **Morgen: Die Industriegesellschaft am Scheideweg.** Munich: R. Piper, 1980.

Hegewald, Helmar. "Moralische Triebkräfte bei der rationellen Nutzung der Naturressourcen im Sozialismus." **DZfPh** 30 (1982): 1457-67.

Heidenheimer, Arnold. **The Governments of Germany.** New York: Thomas Y. Crowell Co., 1971.

Heinrichs, Wolfgang. "Probleme der Vervollkommung der Materialökonomie im Prozess der verstärkten Intensivierung: Richtungen und Wege bei der schrittweisen Lösung des Energie- und Rofstoffbedarfs." **WW** 24 (July 1976): 1057-74.

Herrnberger, Ilona. "Wirtschaftsrechtliche Fragen des Umweltschutzes." **Staat und Recht** 32 (1983): 653-55.

Hintzky, B. W. "Probleme des Umweltschutzes." **Dokumentation der Zeit: Gesamtdeutsches Information Archiv,** no. 19 (1971), pp. 6-10.

Hoffman, Erik P., and Laird, Robbin F. **"The Scientific-Technological Revolution" and Soviet Foreign Policy.** Elmsford, NY: Pergamon Press, 1982.

Hoffmann, Manfred. **Instrumente zur Lenkung der landwirtschaftlichen Bodennutzung in der DDR.** Berlin (West): in Kommission bei Duncker und Humblot, 1977.

Hoffmann, Manfred, and Schinke, Eberhard. **Bodenrecht und Bodennutzung in der Landwirtschaft der DDR.** Bericht des Bundesinstituts für ostwissenschaftliche und internationale Studien, no. 57. Cologne, 1973.

Höhmann, Hans-Hermann; Seidenstecher, Gertraud; and Vajna,

Thomas. **Umweltschutz und ökonomisches System in Osteuropa: Drei Beispiele: Sowjetunion, DDR, Ungarn.** Stuttgart: Verlag W. Kohlhammer, 1973.

Holliday, George. "Summary of Discussion." In **Soviet Science and Technology: Domestic and Foreign Perspectives**, pp. 142-46. Edited by John R. Thomas and Ursula M. Kruse-Vaucienne. Washington, DC: National Science Foundation, 1977.

Hoosen, David J. M. "Introduction." In **Theoretical Problems of Geography** by V. A. Anuchin. Edited by Ronald J. Fuchs and George J. Demko. Columbus: Ohio State University, 1977.

______. "Methodological Clashes in Moscow." **Annals of the Association of American Geographers** 52 (1962): 469-75.

Horowitz, Bertrand, and Whitehouse, F. Douglas. "Soviet Land Value and Enterprise Location." **Land Economics** 49 (May 1973): 233-37.

Horsch, Gerd, and Leonhardt, Alfred. "Umweltkrise – Bewegungsform staatsmonopolistischer Kapitalverwendung." **WW** 22 (July 1974): 1030-42.

Hörz, Herbert. "Globale Probleme der Menschheitsentwicklung." **DZfPh** 30 (1982): 1301-22.

______. **Mensch contra Materie?** Berlin: VEB Deutscher Verlag der Wissenschaften, 1976.

______. "Natur, Naturwissenschaften, Kultur." **DZfPh** 31 (1983): 785-99.

Hösel, Dieter, and Schulze, Gerhard. "Zur Verantwortung der örtlichen Räte." **Staat und Recht** 29 (1980): 866-76.

Hough, Jerry. **The Soviet Union and Social Science Theory.** Cambridge, MA: Harvard University Press, 1977.

Hough, Jerry, and Fainsod, Merle. **How the Soviet Union is Governed.** Cambridge, MA: Harvard University Press, 1979.

Hugler, Oskar. "Rechtsfragen der internationalen Zusammenarbeit zum Schutz der Umwelt." **Neue Justiz** 37 (1983): 234-37, 244.

Huntington, Samuel. "Social and Institutional Dynamics of One-Party Systems." In **Authoritarian Politics in Modern Society**, pp. 3-47. Edited by Samuel Huntington and Clement Moore. New York: Basic Books, 1970.

Iakovets, Iu. V. "Intensification of Production and the Evaluation of Resources." **Ekonomika i organizatsiia promyshlennogo proizvodstva**, no. 9 (1979). Translated in **Problems of Economics** 22 (March 1980): 57-73.

Iakovlev, P. "Razvitie proizvodstvennoi infrastruktury." **VEko**, no. 9 (1981), pp. 150-53.

Il'ev, L. "Nature Conservation and Cost Accounting." **Ekonomicheskie nauki**, no. 11 (1977), pp. 63-66. Translated in

**Problems of Economics** 21 (April 1979): 3-9.

Il'ichev, L. F. "Metodologicheskie problemy estestvovaniia i obshchestvennykh nauk." **Vestnik Akademii Nauk SSSR**, no. 11 (1963), pp. 4-46.

Isachenko, A. G. "Geography and Historical Geography." **Izvestiia Vsesoiuznogo Geograficheskogo Obshchestva**, no. 6 (1976). Translated in **SGRT** 19 (March 1978): 180-85.

______. "On the Unity of Geography." **Izvestiia Vsesoiuznogo Geograficheskogo Obshchestva**, no. 4 (1971). Translated in **SGRT** 13 (April 1972): 195-219.

Iurchenko, A. "Kompleksnoe ispol'zovanie syr'ia, pererabotka vtorichnykh resursov – vazhnyi reserv ekonomiki." **PKh**, no. 11 (1981), pp. 46-54.

Izrael', Iu. A. "Fundamental'nye issledovaniia po okhrane i uluchsheniiu prirodnoi sredy i ratsional'nomy prirodopol'zovaniiu." **Vestnik Akademii Nauk SSSR**, no. 7 (1980), pp. 105-10.

______. "K strategii okhrany prirodnoi sredy i ratsional'nogo prirodopol'zovaniia v SSSR." **VFil**, no. 6 (1979), pp. 119-27.

Jackson, W. A. Douglas, ed. **Soviet Resource Management and the Environment.** Columbus, OH: American Association for the Advancement of Slavic Studies, 1978.

Jacob, Günter. "25 Jahre Geographische Gesellschaft der Deutschen Demokratischen Republik." **Petermanns Geographische Mitteilungen** 123 (1979): 145-49.

Jänicke, Martin. "Umweltpolitik in Osteuropa." In **Sozialismus in Theorie und Praxis.** Festschrift für Richard Lowenthal. Edited by Hannelore Horn, Alexander Schwan, and Thomas Weingartner. Berlin [West]: Walter de Gruyter, 1978.

______. **Wie das Industriesystem von seinen Missständen profitiert.** Opladen: Westdeutscher Verlag, 1979.

Johnson, Chalmers. "Comparing Communist Nations." In **Change in Communist Systems**, pp. 1-32. Edited by Chalmers Johnson. Stanford, CA: Stanford University Press, 1970.

Joravsky, David. **The Lysenko Affair.** Cambridge, MA: Harvard University Press, 1970.

Judy, Richard W. "The Economists." In **Interest Groups in Soviet Politics**, pp. 213-45. Edited by H. Gordon Skilling and Franklyn Griffiths. Princeton: Princeton University Press, 1971.

Jurk, Heinz et al. **Sekundärrohstoffe rationell erfassen und verwerten.** Berlin: Staatsverlag der DDR, 1979.

Kadetov, S. "Kompleksnoe ispol'zovanie mestorozhdenii poleznykh iskopaemykh." **VEko**, no. 3 (1980), pp. 33-38.

Kalesnik, S. V., and Davitaia, F. F. "The Task of Soviet Geography in Supporting Economic Development During the 10th Five-Year Plan (1976-80)." **Materialy VI s"ezda Geograficheskogo Obshchestva v SSSR.** Translated in **SGRT** 17 (April

1976): 217-34.
Kamshilov, M. M. "Chelovek i zhivaia priroda." **Priroda**, no. 3 (1969), pp. 28-37.
______. "Nauchno-tekhnicheskii progress i evoliutsiia biosfery." **VFil**, no. 4 (1972), pp. 69-77.
______. "Printsipy organizatsii zhivoi prirody." **Priroda**, no. 2 (1969), pp. 28-38.
Kantor, K. M. "Ekologiia i progress." **VFil**, no. 8 (1977), pp. 106-16.
Kantorovich, L. V. **The Best Use of Economic Resources.** Cambridge, MA: Harvard University Press, 1965.
Kapitsa, P. L. "According to the Laws of Physics." **Pravda**, Nov. 28, 1981. Translated (condensed) in **CDSP** 33, no. 48 (1981), p. 10.
______. "Energiia i fiziki." **Vestnik Akademii Nauk SSSR**, no. 1 (January 1976), pp. 34-43.
______. "Nauchnyi i sotsial'nyi podkhod k resheniiu global'nykh problem." **VFil**, no. 1 (1977), pp. 46-60.
Karliczek, Hans-Joachim, and Melzer, Helmut. "Komplexe Entwicklung der Territorien und die Funktionen der örtlichen Organe der Staatsmacht." **Staat und Recht** 27 (1978): 770-79.
Karnaukhova, E. "Ekonomicheskaia otsenka zemel' v sel'skom khoziaistve." **VEko**, no. 8 (1968), pp. 88-104.
Kästner, Rolf, and Pöthig, Bernd. "Zur ökonomischen Bewertung von Naturressourcen im Sozialismus." **WW** 29 (November 1981): 1374-77.
Katsura, A. V., and Krasnopol'skaia, L. N. "Filosofskie problemy global'noi ekologii." **Filosofskie nauki**, no. 2 (1981), pp. 155-58.
Kazantsev, N. D. "Iz istorii nauki sovetskogo zemel'nogo i prirodookhranitel'nogo prava." **Pravovedenie**, no. 1 (1968), pp. 75-83.
Kaznacheev, V. P. "Voprosy dialektiki prirody v rabotakh V.I. Vernadskogo." **VFil**, no. 8 (1981), pp. 100-109.
Kelle, V. Zh. "The Leninist Conception of Scientific Ideology and Its Critics." **VFil**, no. 4 (1970). Translated in **Soviet Studies in Philosophy** 9 (Fall 1970): 99-120.
Kelley, Donald R. "Economic Growth and Environmental Quality in the U.S.S.R.: Soviet Reaction to 'The Limits to Growth.'" **Canadian Slavonic Papers** 18 (1976): 266-83.
______. "Environmental Policy-Making in the USSR: The Role of Industrial and Environmental Interest Groups." **Soviet Studies** 28 (October 1976): 570-89.
Kelley, Donald R.; Stunkel, Kenneth R.; and Wescott, Richard R. **The Economic Superpowers and the Environment: The United States, the Soviet Union and Japan.** San Francisco:

W. H. Freeman & Co., 1976.

Khachaturov, T. "Economic Problems of Ecology." **VEko,** no. 6 (1978). Translated in **Problems of Economics** 22 (May 1979): 3-22.

_______. **Ekonomika prirodopol'zovaniia.** Moscow: Ekonomika, 1982.

_______. "Natural Resources and National Economic Planning." **VEko,** no. 8 (1973). Translated in **Problems of Economics** 16 (March 1974): 3-28.

_______. "Ob ekonomicheskoi otsenke prirodnykh resursov." **VEko,** no. 1 (1969), pp. 66-74.

Khil'mi, G. F. **Foundations of the Physics of the Biosphere.** Leningrad: Hydrometeorological Publishing House, 1967.

Khorev, B. S. "Ekologicheskaia politika i rol' geograficheskoi nauki." **Geografiia i prirodnye resursy,** no. 1 (1981), pp. 174-79.

Klare, Hermann. "Briefe." **Spektrum,** June 1973, pp. 2-3.

Klein, Dieter. "Kritisches zur 'Qualität des Lebens' im Kapitalismus." **Einheit** 29 (February 1974): 222-29.

Klinger, Harry, and Leunert, Wolfram. "Möglichkeiten und Grenzen der ökonomisch-ökologischen Beurteilung betrieblicher Umweltschutzmassnahmen unter besonderer Berücksichtigung der Sekundärrohstoffnutzung." **Wissenschaftliche Zeitschrift der Technischen Universität Dresden** 32 (1983): 181-84.

Kliushkin, E. A. "Chto takoe okhrana prirody?" In **Okhrana prirody Turkmenii i nashi zadachi,** pp. 14-17. Akademiia Nauk Turkmenskoi SSR. Ashkhabad: Ylym, 1972.

Knabe, Hubertus. "Gesellschaftlicher Dissens im Wandel. Ökologische Diskussionen und Umweltengagement in der DDR." Presented at the XVII. DDR-Forscher-Tagung at the Europäische Akademie Lerbach. Bergisch-Gladbach, FRG, June 12-15, 1984.

_______. "'Der Mensch mordet sich selbst': Ökologiekritik in der erzählenden DDR-Literatur." **Deutschland Archiv** 16 (1983): 954-73.

Knobloch, Gerd, and Steinitz, Klaus. "Akkumulation und Gestaltung einer effektiven Struktur der Volkswirtschaft." **WW** 19 (July 1971): 1017-27.

Köhler, Johann. "Der Charakter der Kosten für den Umweltschutz." In **Ausgewählte Probleme der Leitung und Planung des Umweltschutzes: Vorträge des Berg- und Hüttenmännischen Tages 1975 in Freiberg,** pp. 47-66, Freiberger Forschungshefte D102. Leipzig: VEB Deutscher Verlag für Grundstoffindustrie, 1976.

_______. "Zu der Problematik der produktiven und der unproduktiven Arbeit sowie der Dienstleistungen." **WW** 22 (June

1974): 852-86.

Kolbasov, O. S. "Iskliuchitel'naia sobstvennost' gosudarstva na ob"ekty prirody." In **Gosudarstvo, pravo, ekonomika**, pp. 218-47. Edited by Viktor M. Chkhikvadze. Akademiia Nauk SSSR, Institut gosudarstva i prava. Moscow: Iuridicheskaia literatura, 1970.

______. "Legislation on Water Use in the USSR." In **Water Law in the Soviet Union**, pp. 93-218. Edited by Irving K. Fox. Madison, WI: The University of Wisconsin Press, 1971.

______. "Okhrana prirody." **SGiP**, no. 2 (1972), pp. 12-20.

______. "Osnovnye napravleniia pravotvorchestva v oblasti okhrany okruzhaiushchei sredy." **SGiP**, no. 3 (1980), pp. 69-77.

______. "Pravovaia okhrana prirody v SSSR." **SGiP**, no. 9 (1967), pp. 41-49.

______. "Protect Nature." **Izvestiia**, Jan. 16, 1980. Translated (excerpts) in **CDSP** 32, no. 2 (1980), pp. 22-23.

Komar, I. V. "Resource Cycles, Their Improvement to Optimum Levels and Forecasting." In **Man, Society, and the Environment**, pp. 234-59. Edited by I. P. Gerasimov. USSR Academy of Sciences, Institute of Geography. Moscow: Progress Publishers, 1975.

Komarov, Boris (pseud.). **Destruction of Nature in the Soviet Union**. White Plains, NY: M. E. Sharpe, 1980. Originally published as **Unichtozhenie prirody: Obostrenie ekologicheskogo krizisa v SSSR**. Frankfurt am Main: Possev Verlag, 1978.

Komarov, D. V. "Tekhnosfera kak produkt i uslovie zhiznedeiatel'nosti obshchestva." In **Nauchno-tekhnicheskaia revoliutsiia, chelovek, ego prirodnaia i sotsial'naia sreda**, pp. 129-31. Edited by V. G. Marakhov. Leningrad: Iz. Leningradskogo universiteta, 1977.

Konstantinidi, S. S. "On Public and Governmental-Legal Measures for Improving Conservation in the Kazakh Soviet Socialist Republic." **Izvestiia Akademii Nauk KazSSR, seriia obshchestvennaia**, no. 5 (1970). Translated in **Soviet Review** 12 (Fall 1971): 261-73.

Korbonski, A. "The Prospects for Change in Eastern Europe." **Slavic Review** 33 (June 1974): 219-40.

Kosing, Alfred. "Mensch-Gesellschaft-Umwelt." **Einheit** 31 (1976): 790-801.

Kovalev, E. E.; Ivanov, V. I.; Pakhomov, B. Ia.; and Ivanova, A. A. "Novaia tekhnika i problema bezopasnosti cheloveka." **VFil**, no. 5 (1981), pp. 49-59.

Kovskii, Vadim. "'Romantika Kochev'ia' i real'nost' narodnoi zhizni." **Literaturnoe obozrenie**, no. 2 (March 1974), pp. 37-40.

Kozhinov, Vadim. "Avtoritet istorii." **Literaturnoe obozrenie**, no. 3 (March 1977), pp. 63-69.

Kozyr', M. I. "Gosudarstvennoe upravlenie zemel'nym fondom v SSSR." **SGiP**, no. 8 (1968), pp. 49-58.

Kozyreva, E. K., and Narkevich, O. E. **Problemy okhrany vneshnei sredy v nekotorykh kapitalisticheskikh stranakh.** Moscow: Ministerstvo zdravokhraneniia SSSR, 1974.

Krack, Erhard. "Bürger, Gemeinde, Umwelt." **Stadt und Gemeinde** 18 (August-September 1974): 12-20.

Kramer, John M. "The Politics of Conservation and Pollution in the USSR." Ph.D. dissertation, University of Virginia, 1973.

______. "Prices and Conservation of Natural Resources in the Soviet Union." **Soviet Studies** 24 (January 1973): 364-73.

Krasnov, N. I. "Pravovoe obespechenie ratsional'nogo ispol'zovaniia zemli." **SGiP**, no. 3 (1967), pp. 46-53.

Krianev, Iu. B. "Ekumenicheskie kontseptsii global'nykh problem." **VFil**, no. 5 (1981), pp. 141-48.

Krotkov, F. G. "Zagriaznenie okruzhaiushchei sredy i problemy gigieny." **Priroda**, no. 5 (1975), pp. 63-69.

Krylova, N. S., ed. **Priroda i ekonomika: sbornik statei.** Moscow: Znanie, 1975.

Ksintaris, V. "Ispol'zovanie vtorichnykh material'nykh resursov v narodnom khoziaistve." **PKh**, no. 6 (1979), pp. 38-44.

______. "Ispol'zovanie vtorichnykh resursov." **PKh**, no. 6 (1981), pp. 80-83.

Kuczynski, Jürgen. "Briefe." **Spektrum**, June 1973, p. 2.

______. **Das Gleichgewicht der Null: Zu den Theorien des Nullwachstums.** Frankfurt am Main: Verlag Marxistische Blätter, 1973.

______. "Der unbefangene Wissenschaftler." **Das Magazin**, no. 2 (1974), p. 54.

Kudrova, E. S. **Statistika natsional'nogo dokhoda evropeiskikh sotsialisticheskikh stran.** Moscow: Statistika, 1969.

Kulow, Hans. "Das Verhältnis von Natur und Gesellschaft und der gesellschaftliche Fortschritt in unserer Epoche." **DZfPh** 27 (1979): 603-9.

Kunert, Günter, and Girnus, Wilhelm. "Anlässlich Ritsos – Ein Briefwechsel." **Sinn und Form** 31 (1979): 850-64.

Kushnirsky, Fyodor I. "The Limits of Soviet Economic Reform." **Problems of Communism** 33 (July-August 1984): 33-43.

Kutzschbauch, Kurt. "Ökonomische Aspekte stoff- und energiewirtschaftlich effektiverer Technologien und einer verstärkten wirtschaftlichen Nutzung von Abprodukten als Sekundärrohstoffe." **WW** 29 (June 1981), pp. 656-69.

Kuznetsov, I. V. "Estestvovanie, filosofiia i stanovlenie noosfery." **VFil**, no. 12 (1974), pp. 128-38.

Laptev, I. D. "Soznatel'noe i stikhiinoe vo vzaimodeistvii obshchestva i prirody." **VFil**, no. 7 (1976), pp. 82-92.

Laszlo, Ervin et al. **Goals for Mankind: A Report to the Club of Rome on the New Horizons of the Global Community.** New York: New American Library, 1977.

Lavrov, S. B., and Agafonov, N. T. "Conceptual Issues in Economic Geography at the 23d International Geographical Congress." **Izvestiia Vsesoiuznogo Geograficheskogo Obshchestva**, no. 1 (1977). Translated in **SGRT** 19 (May 1978): 324-33.

Lavrov, S. B., and Khorev, B. S. Review of **Osnovy prirodopol'zovaniia. Teoreticheskii aspekt**, by V. A. Anuchin. **Izvestiia Vsesoiuznogo Geograficheskogo Obshchestva**, no. 6 (1979). Translated in **SGRT** 21 (April 1980): 235-41.

Lehmann, Edgar. "Quantitative nicht erfassbare Werte in der Umweltforschung." **Geographische Berichte** 18, no. 68 (1973): 168-73.

Leksin, V. N. "National-Economic Effectiveness and Branch Interests." **Ekonomika i organizatsiia promyshlennogo proizvodstva**, no. 4 (July-August 1977). Translated (abstract) in **CDSP** 29, no. 36 (1977), pp. 1-2.

Lemeshev, M. Ia. "Problemy optimizatsii sotsialisticheskogo prirodopol'zovaniia: regional'nye aspekty." **Izvestiia Akademii Nauk SSSR, seriia geograficheskaia**, no. 2 (1979), pp. 17-28.

Lemeshev, M. Ia., and Kozerskaia, N. S. Review of **Aktual'nye problemy okhrany okruzhaiushchei sredy: ekonomicheskie aspekty**, by N. G. Chumachenko et al. **EiMM** 17 (1981): 1227-28.

Lemeshev, M. Ia., and Ushakov, E. P. "O prirodookhrannykh zadachakh ekonomicheskogo razvitiia." **EiMM** 15 (1979): 21-30.

Leonhardt, Alfred. "Marx' Antwort auf Umweltfragen." **IPW Bericht**, no. 7 (1974), pp. 39-45.

Leonhardt, Alfred, and Speer, Gerhard. **Umweltreproduktion im staatsmonopolistischen Kapitalismus.** Berlin: Akademie Verlag, 1977.

Leonov, Leonid. "Concerning Large Chips of Wood." **Literaturnaia gazeta**, March 30, 1965. Translated in **CDSP** 17, no. 20 (1965), pp. 13-15.

Leonov, Leonid, and Riabinin, Boris. "'Preserve' Means Inviolable." **Sovetskaia Rossiia**, May 24, 1970. Translated in **CDSP** 22, no. 23 (1970), pp. 5, 7.

**Lexikon der Wirtschaft: Industrie.** Berlin: Verlag Die Wirtschaft, 1970.

**Lexikon der Wirtschaft: Rechnungsführung und Statistik.** Berlin: Verlag Die Wirtschaft, 1974.

Liebman, Marcel. **Leninism under Lenin.** London: Merlin Press,

1975.

Linden, Carl A. **Khrushchev and the Soviet Leadership.** Baltimore: Johns Hopkins Press, 1966.

Lipets, Iu., and Chizhov, N. N. "Kriticheskaia sotsial'no-ekonomicheskaia otsenka nekotorykh prognozov v global'noi ekologii." In **Priroda i ekonomika: sbornik statei**, pp. 47-57. Edited by N. A. Krylova. Moscow: Znanie, 1975.

Lohs, Karlheinz, and Döring, Sonnhild, eds. **Im Mittelpunkt der Mensch: Umweltgestaltung-Umweltschutz.** Berlin: Akademie Verlag, 1975.

Loiter, M. N. "Effektivnost' kapital'nykh vlozhenii v okhranu prirodnoi sredy." **VEko**, no. 1 (1976), pp. 28-38.

_______. "Ekonomicheskie mery po ratsional'nomu ispol'zovaniiu vodnykh resursov." **VEko**, no. 12 (1967), pp. 75-86.

_______. **Prirodnye resursy i effektivnost' kapital'nykh vlozhenii.** Moscow: Nauka, 1974.

Los', V. A. "Sotsial'no-filosofskie problemy biosfery." **VFil**, no. 9 (1980), pp. 168-69.

Los', V. A., and Faddeev, E. T. "Chelovek, obshchestvo i priroda v vek NTR." **VFil**, no. 12 (1981), pp. 47-57.

Löther, Rolf. "Die Dialektik von Biologischem und Gesellschaftlichem im wissenschaftlichen Menschenbild." **Einheit** 29 (1974): 789-98.

_______. "Philosophische Probleme der Biologie." **DZfPh** 14 (1966): 315-27.

Lovins, Amory B. **Soft Energy Paths: Toward a Durable Peace.** Cambridge, MA: Ballinger Publishing Co., 1977.

Luchitskii, I. V.; Vorob'ev, V. V.; and Ermikov, V. D. "Kompleksnaia programma 'Sibir'' i okhrana okruzhaiushchei sredy." **Geografiia i prirodnye resursy**, no. 2 (1980), pp. 3-13.

Lüdemann, Heinz. "Geographische Forschung in der DDR — Entwicklung und Perspektiven." **Petermanns Geographische Mitteilungen** 124 (1980): 97-103.

_______. "Perspektiven der geographischen Forschung." **Spektrum**, no. 9 (1981), pp. 11-13.

Ludz, Peter Christian. **Deutschlands doppelte Zukunft: Bundesrepublik und DDR in der Welt von morgen.** Munich: Carl Hanser Verlag, 1974.

McAuley, Mary. **Politics and the Soviet Union.** Harmondsworth, Middlesex, ENG: Penguin Books Ltd., 1977.

McCauley, Martin. **Marxism-Leninism in the German Democratic Republic.** London: Macmillan Press, 1979.

McIntyre, Robert J., and Thornton, James R. "On the Environmental Efficiency of Economic Systems." **Soviet Studies** 30 (April 1978): 173-92.

Maier, Harry. "Gibt es Grenzen des ökonomischen Wachstums?" **Spektrum**, no. 6 (1976), pp. 2-4.

______. "Ökonomisches Wachstum und gesellschaftlicher Fortschritt." **WW** 26 (January 1978): 28-47.

Maier, Harry, and Roos, Hans. "Die Mensch-Umwelt Beziehung als politökonomisches Problem." In **Reproduktion der natürlichen Umweltbedingungen.** By Werner Gringmuth et al. Akademie der Wissenschaften der DDR, Zentralinstitut für Wirtschaftswissenschaften, Forschungsberichte no. 8. Berlin: Akademie Verlag, 1974.

Mamedov, N. M. "Ekologicheskaia problema i tekhnicheskie nauki." **VFil**, no. 5 (1980), pp. 111-20.

______. "Global'nye aspekty ekologicheskogo vzaimodeistviia." **VFil**, no. 11 (1981), pp. 74-83.

Mandel, William. "The Soviet Ecology Movement." **Science and Society** 36 (Winter 1972): 385-416.

Mannheim, Karl. **Ideology and Utopia: An Introduction to the Sociology of Knowledge.** New York: Harcourt, Brace, & World, 1968.

Marakhov, V. G. "Nauchno-tekhnicheskaia revoliutsiia i prirodnaia sreda." **VFil**, no. 8 (1974), pp. 91-101.

______, ed. **Nauchno-tekhnicheskaia revoliutsiia, chelovek, ego prirodnaia i sotsial'naia sreda.** Leningrad: Iz. Leningradskogo universiteta, 1977.

Marakhov, V. G., and Meleshchenko, Iu. S. "Sovremennaia nauchno-tekhnicheskaia revoliutsiia i ee sotsial'nye posledstviia v usloviiakh sotsializma." **VFil**, no. 3 (1966), pp. 129-40.

Marcuse, Herbert. **Soviet Marxism: A Critical Analysis.** New York: Columbia University Press, 1958.

Markarian, E. S. "Lokal'nye istoricheskie osobennosti kul'tury i protsessy ekologicheskoi adaptatsii." **Vestnik Akademii Nauk SSSR**, no. 1 (1981), pp. 111-17.

"Marksistsko-leninskaia teoriia nauchno-tekhnicheskoi revoliutsii." **VFil**, no. 2 (1974), pp. 3-14.

Maron, Monika. **Flugasche.** Frankfurt am Main: S. Fischer Verlag, 1981.

Martynov, F. "Metodologicheskie problemy regional'nogo prirodopol'zovaniia (na primere Kuzbasskogo TPK)." **VEko**, no. 11 (1979), pp. 53-60.

Marx, Karl. **Capital**, 3 vols. New York: International Publishers, 1967.

Marx, Karl, and Engels, Frederick. **The Communist Manifesto.** New York: Monthly Review Press, 1964.

______. **The German Ideology.** New York: International Publishers, 1970.

Maslennikov, S., and Shkatov, V. "Plata za prirodnye resursy v evropeiskikh sotsialisticheskikh stranakh." **VEko**, no. 8 (1974), pp. 70-80.

Matley, Ian M. "The Marxist Approach to the Geographical Environment." **Annals of the Association of American Geographers** 56, no. 1 (March 1966): 97-111.

_______. Review of **Theoretical Problems of Geography**, by V. A. Anuchin. In **SGRT** 19 (October 1978): 576-79.

Mayer, Christian [Carl Amery]. **Das Ende der Vorsehung: Die gnadenlosen Folgen des Christentums.** Reinbek bei Hamburg: Rowohlt, 1972.

Meadows, Donella H.; Meadows, Dennis L.; Randers, Jørgen; and Behrens, William W. III. **The Limits to Growth: A Report for the Club of Rome's Project on the Predicament of Man.** New York: Universe Books, 1972.

Medunin, A. E. "Vliianie nauchno-tekhnicheskoi revoliutsii na prirodu Zemli." **VFil**, no. 3 (1969), pp. 23-33.

Medvedev, Roy, and Medvedev, Zhores. "The USSR and the Arms Race." **New Left Review**, no. 130 (November-December 1981), pp. 5-22.

Medvedev, Zhores A. **The Medvedev Papers.** London: Macmillan, St. Martin's Press, 1971.

_______. **Nuclear Disaster in the Urals.** New York: W. W. Norton & Co., 1979.

_______. **The Rise and Fall of T. D. Lysenko.** New York: Columbia University Press, 1969.

Meissner, Herbert. "Ein neues 'Weltmodell' für eine alte Ideologie." **Einheit** 28 (1973): 1232-41.

_______. "Das 'Wachstumsdilemma' der bürgerlichen politischen Ökonomie." **WW** 23 (March 1975): 406-23.

Meleshkin, M. T.; Zaitsev, A. P.; and Marinov, Kh. **Ekonomika i okruzhaiushchaia sreda: vzaimodeistvie i upravlenie.** Moscow: Ekonomika, 1979.

Mel'nik, L. "Accounting for Ecological Indicators in Solving Economic Problems." **Ekonomicheskie nauki**, no. 2 (1983), pp. 58-63. Translated in **Problems of Economics** 26 (March 1984): 15-26.

Melzer, Helmut. "Komplexe Entwicklung der Territorien und die Funktionen der örtlichen Organe der Staatsmacht." **Staat und Recht** 27 (1978): 770-79.

Meyer, Alfred G. "The Functions of Ideology in the Soviet Political System." **Soviet Studies** 17 (January 1966): 273-85.

_______. **Leninism.** Cambridge, MA: Harvard University Press, 1957.

_______. "USSR Incorporated." **Slavic Review** 20 (October 1961): 369-76.

Mikhailov, Iu. P. "Geograficheskie grani protsessa prirodopol'zovaniia." **Geografiia i prirodnye resursy**, no. 3 (1980), pp. 159-64.

Mikhasiuk, I. "The Economic Evaluation of Land and the Improvement of Rental Relationships in the USSR." **Ekonomicheskie nauki**, no. 7 (1978), pp. 40-47. Translated in **Problems of Economics** 21 (February 1979): 65-78.

Mints, A. A. **Ekonomicheskaia otsenka estestvennykh resursov.** Moscow: Mysl', 1972.

Mints, A. A., and Preobrazhenskii, V. S. "Timely and Controversial Problems in the Systems Orientation of Geography." **Izvestiia Akademii Nauk SSSR, seriia geograficheskaia**, no. 6 (1973), pp. 107-18. Translated in **SGRT** 16 (February 1975): 61-74.

Mittag, Günter. "Ökonomische Strategie der Partei – klares Konzept für weiteres Wachstum." **Einheit** 38 (1983): 1003-21.

Mohrig, Werner. "Populationskinetik und menschliche Bevölkerungsentwicklung." **DZfPh** 22 (1974): 881-89.

Mohs, Hans; Richter, Hans; Scholz, Dieter; Harke, Hellmut; and Hermann, Siegfried. "Forschung und Wissenschaftsentwicklungen an der Sektion Geographie der Martin-Luther Universität Halle-Wittenberg." **Geographische Berichte** 19, no. 72 (1974): 186-98.

Moiseev, N. N. "Aktual'nye voprosy ekologicheskoi evoliutsii i 'mirovaia dinamika' Dzheia Forrestera." **VFil**, no. 7 (1978), pp. 171-82.

_______. "Koevoliutsiia cheloveka i biosfery v vek komp'iuterov." **Vestnik Akademii Nauk SSSR**, no. 3 (1982), pp. 90-97.

Moore, Barrington. **Soviet Politics: The Dilemma of Power: The Role of Ideas in Social Change.** Cambridge, MA: Harvard University Press, 1950.

Mottek, Hans. "Umweltschutz – ökonomisch betrachtet." In **Im Mittelpunkt der Mensch: Umweltgestaltung und Umweltschutz**, pp. 1-15. Edited by Karlheinz Lohs and Sonnhild Döring. Berlin: Akademie Verlag, 1975.

_______. "Zu einigen Grundfragen der Mensch-Umwelt Beziehung." **WW** 20 (January 1972): 36-43.

Motylev, V. "Forecasts of the 'Club of Rome': Reality and Prophecy." **VEko**, no. 4 (1977). Translated in **Problems of Economics** 20 (October 1977): 75-94.

Mullagaliamov, M. "The City's Air." **Trud**, Dec. 2, 1977. Translated (condensed) in **CDSP** 29, no. 49 (1977), pp. 7-8.

Müller, Eva. "Nichtmaterielle Güter und Naturgüter als zwei wichtige Bestandteile des Nationalreichtums." **WW** 29 (August 1981): 949-64.

Müller, Klaus. "Subjektiver Faktor und Zusammenhang von Gesellschaft und Natur." **DZfPh** 22 (1974): 584-97.

**Nauchno-teoreticheskie i organizatsionnye voprosy okhrany okruzhaiushchei sredy.** Itogi nauki i tekhniki: "Okhrana prirody i vosproizvodstvo prirodnykh resursov," vol. 4.

Moscow: Vsesoiuznyi institut nauchnoi i tekhnicheskoi informatsii, 1978.

"Nauka i global'nye problemy sovremennosti." Round table. **VFil**, no. 8 (1974), pp. 102-24; no. 9 (1974), pp. 67-83; no. 10 (1974), p. 46-60; and no. 11 (1974), pp. 60-75.

Nekrasov, N. N. (Transliterated in the German as Nekrassow, N. N.). **Territorialökonomie: Theorie-Probleme-Aufgaben**. Berlin: Verlag Die Wirtschaft, 1979. Originally published in the Soviet Union as **Regional'naia ekonomika**, 2nd ed. Moscow: Ekonomika, 1978.

Nekrasov, N. N., and Mateev, E., eds. **Sotsialisticheskoe prirodopol'zovanie: Ekonomicheskie i sotsial'nye aspekty**. Moscow: Ekonomika; Sofiia: Partizdat, 1980.

Nesterov, L. I., and Bukhval'd, E. M. "Prirodnye resursy kak element sovokupnogo obshchestvennogo bogatstva." **Izvestiia Akademii Nauk SSSR, seriia ekonomicheskaia**, no. 3 (1980), pp. 20-32.

Nick, Harry. "Für rationellere Beziehungen zwischen Gesellschaft und Natur." **Urania**, no. 7 (1977), pp. 40-43.

______. "Mensch und Umwelt." **Einheit** 34 (1979): 702-12.

______. "Wirtschaftswachstum – seine Perspektiven im Sozialismus und seine bornierten Kritiker." **Einheit** 31 (1976): 586-96.

Nikolaev, A. "Ob odnoi aktual'noi probleme." **VEko**, no. 11 (1971), pp. 133-36.

Nordhaus, William D., and Tobin, James. "Measures of Economic Welfare." In **Economics of the Environment: Selected Readings**, 2nd edition, pp. 479-88. Edited by Robert and Nancy S. Dorfman. New York: W. W. Norton & Co., 1977.

Norhara-Schnabel, Ilka. "Zur Entwicklung der Umweltpolitik in der DDR." **Deutschland Archiv** 9 (1976): 809-29.

Nove, Alec. **The Soviet Economic System**. London: Allen & Unwin, 1977.

______. **The Soviet Economy: An Introduction**. New York: Praeger Publishers, 1969.

Novik, I. B. "Budushchee nauki i ekologicheskaia problema." **VFil**, no. 1 (1975), pp. 110-15.

______. "Metodologicheskie aspekty optimizatsii biosfery." **Priroda**, no. 9 (1972), pp. 5-9.

Novik, I. B., and Los', V. A. Review of **Chelovek, obshchestvo i okruzhaiushchaia sreda**, edited by I. P. Gerasimov. **VFil**, no. 4 (1974), pp. 172-73.

Novikov, I. "Okruzhaiushchei srede – sotsialisticheskoe otnoshenie," **Kommunist**, no. 14 (1982), pp. 31-42.

Novikov, V., and Lipsits, I. "Sovershenstvovanie metodologii i metodiki tsenoobrazovaniia na vtorichnoe syr'e." **PKh**, no. 12 (1983), pp. 48-53.

"Obshchestvo i priroda." Editorial. **VFil**, no. 4 (1973), pp. 42-56.

"Obsuzhdenie problem ekonomicheskoi otsenki prirodnykh resursov." **EiMM** 11 (1975): 601-4.

"O deiatel'nosti Otdeleniia ekonomiki Akademii Nauk SSSR." **VEko**, no. 9 (1983), pp. 153-57.

Oehler, Ellenor. "Aktuelle Fragen der Lehren von Karl Marx zur Gestaltung der Umweltbeziehungen." **Staat und Recht** 32 (1983): 305-14.

______. "Die genossenschaftliche Bodennutzung und ihre rechtliche Gestaltung im neuen ökonomischen System der Planung und Leitung der Volkswirtschaft in der Landwirtschaft." **Staat und Recht** 14 (1965): 263-69.

______. "Zu staatlichen und rechtlichen Problemen der Landschaftsplanung." In **Beiträge zur planmässigen Gestaltung der Landschaft**, pp. 17-26. Edited by Hans Richter. Gotha/Leipzig: VEB Hermann Haack, 1978.

**Ökonomisches Lexicon**. 2 vols. Berlin: Verlag Die Wirtschaft, 1969.

Oldak, P. G. "Environmental Protection is a Comprehensive Task." **Izvestiia**, Feb. 4, 1976. Translated (condensed) in **CDSP** 28, no. 5 (1976), p. 5.

______. "Novoe mesto prirody v ekonomicheskikh issledovaniiakh." **Priroda**, no. 2 (1973), pp. 33-41.

______. "Scientific-Technical Progress and the Limits of Economic Analysis." **Mirovaia ekonomika i mezhdunarodnye otnosheniia**, no. 8 (1971). Translated in **Problems of Economics** 14 (April 1972): 3-23.

______. **Sovremennoe proizvodstvo i okruzhaiushchaia sreda**. Novosibirsk: Nauka, sibirskoe otdelenie, 1979.

______. "Voprosy ekonomicheskoi otsenki resursov biosfery." In **Ekonomicheskie problemy optimizatsii prirodopol'zovaniia**, pp. 53-60. Edited by N. P. Fedorenko. Moscow: Nauka, 1973.

"On Additional Measures to Intensify Nature Conservation and Improve the Utilization of Natural Resources." **Pravda**, Jan. 6, 1979. Condensed in **CDSP** 31, no. 1 (1979), pp. 1-3.

Ophuls, William. **Ecology and the Politics of Scarcity**. San Francisco: W. H. Freeman & Co., 1977.

Osadchenko, S. "Opyt primeneniia ekonomicheskikh rychagov i stimulov." **VEko**, no. 5 (1980), pp. 42-50.

Osipovich, L. Ia. "Khozraschet i ispol'zovanie resursov." **PKh**, no. 9 (1979), pp. 79-86.

"Osnovnye napravleniia raboty Otdeleniia ekonomiki AN SSSR po razvitiiu ekonomicheskoi nauki v svete reshenii XXVI S"ezda KPSS." **VEko**, no. 4 (1981), pp. 9-16.

"Ot redaktsionnoi kollegii." **Geografiia i prirodnye resursy**, no. 1 (1980), pp. 3-4.

Panfilov, E. G., ed. **Ekologiia i politika: Problemy zashchity prirodnoi sredy**. Prague: Mir i sotsializm, 1972.

Parsons, Howard L. "Mal'tuzianstvo i sotsializm." **VFil**, no. 1 (1977), pp. 60-70.

______. **Marx and Engels on Ecology**. Westport, CT: Greenwood Press, 1977.

______. Review of **Man, Society, and the Environment**, edited by I. P. Gerasimov. **Filosofskie nauki**, no. 1 (1978), pp. 172-75. Translated from the English by N. I. Deminoi.

Passmore, John. **Man's Responsibility for Nature: Ecological Problems and Western Traditions**. New York: Charles Scribner's Sons, 1974.

Paucke, Horst. "Globale Probleme und der Menschheitsfortschritt." **DZfPh** 30 (1982): 1497-1508.

______. "Zum Verhältnis von Wirtschaftswachstum und Umweltbelastung." **Zeitschrift für den Erdkundeunterricht** 34, no. 2/3 (1982): 49-55.

Paucke, Horst, and Bauer, Adolf. "Feststellungen und Fragen zur ökologischen Krisenpolitik." **DZfPh** 25 (1977): 473-87.

______. **Umweltprobleme – Herausforderung der Menschheit**. Frankfurt am Main: Verlag Marxistische Blätter, 1980. Originally published in Berlin (East): Dietz Verlag, 1979.

______. "Zu einigen Prinzipien der Naturnutzung im Sozialismus." **DZfPh** 26 (1978): 1027-35.

______. "Zum Verhältnis von Natur- und Produktionskreisläufen." **DZfPh** 28 (1980): 905-14.

Paucke, Horst, and Kroske, Heinz. "Weltmodelle – neue Konvergenzvarianten zu Umweltproblemen." **DZfPh** 23 (1975): 1501-12.

______. "Zum Kosten-Nutzen Problematik von Massnahmen in Umweltschutz und Umweltgestaltung." In **Gesellschaft und Umwelt: Hans Mottek zum 65. Geburtstag**, pp. 61-67. Akademie der Wissenschaften der DDR, Sitzungsbericht, no. 2/G. Berlin: Akademie Verlag, 1976.

Paucke, Horst, and Streibel, Günter. "Gratisdienste der Natur – wichtiger Effektivitätsfaktor der sozialistischen Volkswirtschaft." **WW** 31 (September 1983): 1317-32.

______. "Zur ökonomischen Bewertung von Naturressourcen." **WW** 29 (September 1981): 1063-80.

______. "Zur Verflectung von Naturprozessen und volkswirtschaftlichem Reproduktionsprozess." **WW** 28 (April 1980): 405-21.

______. "Zur Wechselbeziehung von Materialökonomie, Technologie, und Umweltschutz." **WW** 25 (October 1977): 1467-81.

______. "Zu Versuchen kapitalistischer Systemstabilisierung unter dem Aspekt der Umweltreproduktion." **WW** 26 (August 1978): 924-38.

Pavlenko, V. "Novyi etap v sovershenstvovanii territorial'nogo planirovaniia," **PKh**, no. 8 (1980), pp. 7-14.

Pestel, Eduard, and Mesarovic, Mihajlo. **Mankind at the Turning Point**. New York: Dulton, 1974.

Petrakov, N. Ia., and Khrabrov, I. M. "Nekotorye voprosy ekonomicheskoi otsenki prirodnykh resursov v svete marksistsko-leninskoi teorii zemel'noi renty." **EiMM** 6 (1970): 355-62.

Petropavlovskaia, N., and Fal'tsman, V. "Vtorichnye resursy metalla kak faktor ekonomicheskogo rosta." **PKh**, no. 9 (1983), pp. 44-51.

Petrov, V. V. "Ob"ekt i predmet pravovoi okhrany prirody v SSSR." **SGiP**, no. 4 (April 1976), pp. 58-64.

Petrova, E. "Obsuzhdenie problem ekonomicheskoi otsenki zemli." **VEko**, no. 6 (1967), pp. 145-49.

Pitkin, Hanna Fenichel. **Wittgenstein and Justice: On the Significance of Ludwig Wittgenstein for Social and Political Thought**. Berkeley: University of California Press, 1972.

Planer-Friedrich, Götz. "Schranken instrumenteller Vernunft." In **Beton ist Beton**, pp. 119-28. See Wensierski and Büscher, 1981.

Pletnikov, J. K. (Transliterated as Pletnikow, J. K. in the German.) "Die Wechselwirkung von Gesellschaft und Natur und Probleme der Noosphäre." **DZfPh** 24 (1976): 148-57.

Pokaraev, L., and Zatsev, A. "Voprosy sovershenstvovaniia normirovaniia material'nykh resursov." **PKh**, no. 8 (1979), pp. 72-81.

Polataev, P. "Ecology and Economic Development." **PKh**, no. 12 (1979). Translated in **Soviet Review** 22 (Spring 1981): 92-106.

______. "Perspektivy razvitiia okhrany prirody." **PKh**, no. 1 (1982), pp. 38-46.

Polianskaia, G. N. "Poniatie okhrany prirody i prirodookhranitel'noe zakonodatel'stvo." **SGiP**, no. 9 (1975), pp. 69-76.

**Politicheskaia ekonomiia: slovar'**. Moscow: Iz. Politicheskoi literatury, 1979.

**Politicheskaia ekonomika: Uchebnik**. 2 volumes. Moscow: Politizdat, 1979.

Pool, Ithiel de Sola. "Communication in Totalitarian Societies." In **Handbook of Communication**, pp. 462-511. Edited by Ithiel de Sola Pool et al. Chicago: Rand McNally College Publishing Company, 1973.

Popov, B. A. Review of **Chelovek, obshchestvo i okruzhaiushchaia sreda**, edited by I. P. Gerasimov. **Izvestiia Vsesoiuznogo Geograficheskogo Obshchestva**, no. 2 (1975). Translated in **SGRT** 17 (December 1976): 700-703.

Popov, G. "Improving Administration." **Pravda**, July 6, 1977. Translated (condensed) in **CDSP** 29, no. 27 (1977), pp. 12, 20.

Porjasow, D. et al. **Nationaleinkommen im Sozialismus.** Berlin: Verlag Die Wirtschaft, 1976.
Povileiko, R. P. "Ecology and Economics (A Survey of Ideas and the Literature)." **Ekonomika i organizatsiia promyshlennogo proizvodstva**, no. 9 (1982), pp. 202-4. Translated in **Problems of Economics** 26 (September 1983): 3-14.
Powell, David E. "The Social Costs of Modernization: Ecological Problems in the USSR." **World Politics** 23 (July 1971): 618-34.
"Pravovaia okhrana okruzhaiushchei sredy." Round table. **SGiP**, no. 9 (1979), pp. 98-119.
Prigozhin, A. I. "Rimskii klub pered sotsial'nymi problemami." **VFil**, no. 5 (1978), pp. 164-71.
Pritzel, Konstantin. "Die Umweltpolitik in den intereuropäischen und innerdeutschen Beziehungen." **Deutschland Archiv** 13 (1980): 834-43.
"Problemy planirovaniia i upravleniia prirodnymi resursami." **EiMM** 12 (1976): 194-99.
**Programme of the Socialist Unity Party of Germany.** Ninth Congress of the SED of Germany. Dresden: Verlag Zeit im Bild, May 1976.
**Programm und Statut der SED vom 22. Mai 1976.** Cologne: Verlag Wissenschaft und Politik, 1976.
Pryde, Philip R. **Conservation in the Soviet Union.** Cambridge, ENG: Cambridge University Press, 1972.
______. "Nuclear Energy in the Soviet Union." **Current History** 77 (October 1979): 115-18.
Przeworksi, Adam, and Teune, Henry. **The Logic of Comparative Social Inquiry.** New York: John Wiley & Sons, 1970.
Quarg, Martin, ed. **ABC Umweltschutz.** Leipzig: Deutscher Verlag für Grundstoffindustrie, 1976.
Raestrup, Reiner, and Weymar, Thomas. "'Schuld ist allein der Kapitalismus': Umweltprobleme und ihre Bewältigung in der DDR." **Deutschland Archiv** 15 (1982): 832-44.
Raiatskas, R. L., and Sutkaitis, V. P. "Modelirovanie ekonomicheskoi dinamiki s uchetom zagriazneniia okruzhaiushchei sredy." **EiMM** 15 (1979): 45-57.
______. "Sistema ekologo-ekonomicheskikh modelei dlia planirovaniia i upravleniia sostoianiem okruzhaiushchei sredy." **EiMM** 16 (1980): 1081-93.
Rakitskii, B. "Ekonomicheskie funktsii platy za resursy." **VEko**, no. 12 (1966), pp. 17-28.
"Ratsional'no ispol'zovat' resursy proizvodstva." **PKh**, no. 9 (1981), pp. 3-11.
Rauchfluss, Wolfgang. "Materialökonomie – dringendes Gebot." **Einheit** 30 (1975): 162-68.
Reichelt, Hans. "Die Aufgaben der sozialistischen Landeskultur

und des Umweltschutzes im Volkswirtschaftsplan 1973." **Die Wirtschaft**, Ausgabe A. No. 30 (July 26, 1972), p. 8.

_______. "Der Beitrag der Kammer der Technik bei der Gestaltung der sozialistischen Landeskultur und zum Schutz der Umwelt." **Die Wirtschaft**, Ausgabe A. Beilage 19/72, no. 42 (October 18, 1972), pp. 4-11.

Reinhold, Otto. "Der Charakter der gegenwärtige Krise der kapitalistischen Weltwirtschaft und die imperialistische Apologetik." **Einheit** 31 (1976): 79-86.

**Report of the Central Committee of the CPSU to the XXVI Congress of the Communist Party of the Soviet Union and the Immediate Tasks of the Party in Home and Foreign Policy.** Delivered by L. I. Brezhnev on Feb. 23, 1981. Moscow: Novosti Press Agency Publishing House, 1981.

Richter, Hans. **Geographische Aspekte der sozialistischen Landeskultur.** Studienbücherei Geographie für Lehrer, vol. 17. Gotha/Leipzig: VEB Hermann Haack, 1979.

Richter, Hans, ed. **Beiträge zur planmässigen Gestaltung der Landschaft.** Geographische Gesellschaft der DDR. Gotha/Leipzig: VEB Hermann Haack, 1978.

Richter, Peter. "Veto gegen Missbrauch der Natur." **Forum** 29, no. 3 (1975), pp. 10-11.

Riedel, H. G., and Enders, K. L. "Ökonomische und betriebswirtschaftliche Aspekte technischer Umweltschutzmassnahmen im industriellen Bereich Energie- und Chemiewirtschaft." In **Ausgewählte Probleme der Leitung und Planung des Umweltschutzes: Vorträge des Berg- und Hüttenmännischen Tages 1975 in Freiberg,** pp. 23-46. Freiberger Forschungshefte D102. Leipzig: VEB Deutscher Verlag für Grundstoffindustrie, 1976.

Ritschel, O. "Leitung und Planung des Umweltschutzes in VEB Bergbau- und Hüttenkombinat 'Albert Funk' Freiberg." In **Ausgewählte Probleme der Leitung und Planung des Umweltschutzes: Vorträge des Berg- und Hüttenmännischen Tages 1975 in Freiberg,** pp. 7-22. Freiberger Forschungshefte D102. Leipzig: VEB Deutscher Verlag für Grundstoffindustrie, 1976.

Rohde, Günther, ed. **Bodenrecht: Lehrbuch.** Berlin: Staatsverlag der DDR, 1976.

Roos, Hans; Streibel, Günter; Gringmuth, Werner; and Kutzschbauch, Kurt. **Umweltgestaltung und Naturressourcen.** Berlin: Verlag Die Wirtschaft, 1979.

Roth, Gerhard. "Betriebswirtschaftliche Probleme der Ermittlung des einmaligen Aufwandes für Umweltschutzmassnahmen." **Wissenschaftliche Zeitschrift der Karl-Marx-Universität Leipzig. Matematisch-Naturwissenschaftliche Reihe** 30 (1981): 100-105.

Rutkevich, M. N., and Shvarts, S. S. "Filosofskie problemy upravleniia biosferoi." **VFil**, no. 10 (1971), pp. 52-63.

Ryvkin, A. A., and Amirov, I. Sh. "Kriticheskii analiz modelei global'nogo proekta 'Chelovechestvo pered vyborom.'" **EiMM** 13 (1977): 440-56.

Sakharov, A. D. **Progress, Coexistence, and Intellectual Freedom.** New York: W. W. Norton & Co., 1968.

Sandig, Hans, and Wolf, Siegfried. "Einige Probleme der Wirkung verstärkter Produktion einheimischer mineralischen Rohstoffe auf die Effektivität der gesellschaftlichen Produktion in der DDR." **WW** 27 (January 1979): 61-70.

Saushkin, Iu. G. "Geograficheskoe issledovanie 'ochelovechennoi,' ili 'istoricheskoi' prirody." **Voprosy geografii** 106 ("Vliianie cheloveka na landshaft"), pp. 27-37. Moscow: Mysl', 1977.

_______. "Geographic Prediction of Anthropogenic Processes in Nature." **Geografiia v Shkole**, no. 3 (1968), pp. 6-10. Translated in **SGRT** 9 (November 1968): 776-84.

_______. "A History of Soviet Economic Geography." In **Ekonomicheskaia geografiia v SSSR**. Edited by N. N. Baranskii, N. P. Nikitin, V. V. Pokshishevskii, and Iu. G. Saushkin. Moscow: Prosveshchenie, 1965. Translated in **SGRT** 7 (October 1966): 3-104.

_______. "The Second Time Around." **Pravda**, Jan. 26, 1978. Translated (condensed) in **CDSP** 30, no. 41 (1978), pp. 14-15.

_______. "The Today and Tomorrow of Geography." **Literaturnaia gazeta**, June 17, 1965. Translated in **SGRT** 6 (September 1965): 50-56.

Schäfer, Klaus-Dieter. "Dominierende Standpunkte bürgerlicher Ideologen zu ökologischen Problemen der Gegenwart." **DZfPh** 27 (1979): 610-16.

Scherf, Konrad. "Landschaftsplanung aus der Sicht der Territorialplanung und Territorialforschung." In **Beiträge zur planmässigen Gestaltung der Landschaft**, pp. 11-16. Edited by Hans Richter. Gotha/Leipzig: VEB Hermann Haack, 1978.

_______. "Zum wissenschaftlichen Beitrag der Geographie bei der Erarbeitung eines Generalschemas für die Standortverteilung der Produktivkräfte in der DDR." **Wissenschaftliche Zeitschrift der Universität Halle. Matematisch-Naturwissenschaftliche Reihe** 31 (1982): 133-41.

Schliesser, Walfried. "Probleme der Warenproduktion und des Wertgesetzes im Sozialismus." **WW** 15 (June 1967): 901-21.

Schmidt, Helmut. "Über die volkswirtschaftliche Bedeutung von Sekundärrohstoffen und Abprodukte." **Einheit** 39 (1984): 370-73.

Schmidt-Renner, Gerhard. "Sozialistischer Umweltschutz unter volkswirtschaftlichen Gesichtspunkten." **Petermanns Geo-**

**graphische Mitteilungen** 118 (1974): 8-18.

________. "Über den Umweltbegriff als Kombination verschiedener Umweltaspekte." **Petermanns Geographische Mitteilungen** 122 (1978): 209-11.

________. "Über Regelungen zum Schutz von Naturressourcen und -potenzen der gesellschaftlichen Reproduktion und Rekreation vor Störfunktionen in der DDR." **Petermanns Geographische Mitteilungen** 116 (1972): 99-116.

________. "Zum räumlichen Aspekt der Umweltverschmutzung." **Geographische Berichte** 23, no. 88 (1978): 161-76.

Schnurrbusch, Gottfried. "Landwirtschaft und Landeskultur." In **Im Mittelpunkt der Mensch**, pp. 302-15. Edited by Karlheinz Lohs and Sonnhild Döring. Berlin: Akademie Verlag, 1975.

Schramm, Edgar. "Die Bodennutzungsgebühr – ein Stimulus zum Schutz des land- und forstwirtschaftlichen Bodenfonds." **WW** 17 (May 1969): 685-708.

Schubert, Manfred. **Zur Entwicklung abproduktarmer bzw. -freier Verfahren in Einheit von Umweltschutz, Material- und Energieökonomie.** Sitzungsberichte der Akademie der Wissenschaften der DDR. Mathematik – Naturwissenschaften – Technik, no. 21/N. Berlin: Akademie Verlag, 1982.

Schumacher, E. F. **Small Is Beautiful.** New York: Harper & Row, 1973.

Schuster, Erhard. "Charakter und Stellung der Umweltprobleme und einige sich daraus ergebende Konsequenzen." **Wissenschaftliche Zeitschrift der Technischen Universität Dresden** 31 (1982): 211-14.

Schwartau, Conrad. "Zunehmende Luftverschmutzung in der DDR durch Renaissance der Braunkohle." **DIW-Wochenbericht** 50, no. 4 (January 1983): 43-49.

Schwartz, Joel J., and Keech, William R. "Group Influence and the Policy Process in the Soviet Union." In **Communist Studies and the Social Sciences**, pp. 298-317. Edited by Frederic J. Fleron, Jr. Chicago: Rand McNally & Co., 1969.

Seleznev, G. "Godichnoe sobranie Leningradskikh filialov sovetov AN SSSR po ekonomicheskim problemam." **VEko**, no. 8 (1981), pp. 158-59.

Selivanovskii, V. "O planirovanii ispol'zovaniia vtorichnykh resursov." **PKh**, no. 6 (1983), pp. 83-89.

Sergeev, E. M. "Ratsional'noe ispol'zovanie geologicheskoi sredy." **Priroda**, no. 1 (1977), pp. 85-92.

Shakhnovich, M. I. "Kritika teologicheskoi interpretatsii problem ekologii." In **Nauchno-tekhnicheskaia revoliutsiia, chelovek, ego prirodnaia i sotsial'naia sreda**, pp. 150-54. Edited by V. G. Marakhov. Leningrad: Iz. Leningradskogo universiteta, 1977.

Shapiro, A. "Global'nye problemy i perspektivy mirovoi ekonomiki v burzhuaznykh kontseptsiiakh." **Mirovaia ekonomika i mezhdunarodnye otnosheniia**, no. 5 (1981), pp. 100-114.

Sharlet, Robert. **The New Soviet Constitution of 1977: Analysis and Text**. Brunswick, OH: King's Court Communications, Inc., 1978.

Sheremet, K. F. (Transliterated as Scheremet, K. F. in the German.) "Die neue Verfassung der UdSSR und die Entwicklung des Territorialprinzips der staatlichen Leitung." **Staat und Recht** 29 (1980): 836-44.

Shilin, K. I. "'Ekologicheskaia revoliutsiia' v sovremennoi zapadnoi nauke." **VFil**, no. 11 (1972), pp. 122-28.

Shkatov, V. "Tseny na prirodnye bogatstva i sovershenstvovanie planovogo tsenoobrazovaniia." **VEko**, no. 9 (1968), pp. 67-77.

Shvarts, S. S. "Problemy ekologii cheloveka." **VFil**, no. 9 (1974), pp. 102-10.

Shvedov, S. S. "American Social Philosophy and the Sources of the Ecological Crisis." **VFil**, no. 8 (1978). Translated in **Soviet Studies in Philosophy** 17 (Spring 1979): 69-87.

Sidorova, V. "Ratsional'no ispol'zovat' prirodnye resursy." **PKh**, no. 1 (1983), pp. 119-22.

"Simpozium po ekonomicheskim problemam prirodopol'zovaniia." **EiMM** 10 (1974): 381-85.

Skilling, H. Gordon. "Group Conflict in Soviet Politics: Some Conclusions." In **Interest Groups in Soviet Politics**, pp. 379-416. Edited by H. Gordon Skilling and Franklyn Griffiths. Princeton: Princeton University Press, 1971.

_______. "Groups in Soviet Politics." In **Interest Groups in Soviet Politics**, pp. 19-46. Edited by H. Gordon Skilling and Franklyn Griffiths. Princeton: Princeton University Press, 1971.

Skilling, H. Gordon, and Griffiths, Franklyn, eds. **Interest Groups in Soviet Politics**. Edited by H. Gordon Skilling and Franklyn Griffiths. Princeton: Princeton University Press, 1971.

Singleton, Fred, ed. **Environmental Misuse in the Soviet Union**. New York: Praeger, 1976.

Skzolny, Michael (pseud.). "Revolution in Poland." **Monthly Review** 33, no. 2 (June 1981): 1-21.

**Society and the Environment: A Soviet View**. Moscow: Progress Publishers, 1977.

Solodkov, M. V. **Neproizvodstvennaia sfera**. Moscow: Mysl', 1981.

_______. **Neproizvodstvennaia sfera pri sotsializme**. Moscow: Mysl', 1978.

_______, ed. **Ekonomika neproizvodstvennoi sfery**. Put out by the Ministry of Higher and Middle Special Education of the

USSR as a textbook for students of economics in universities. Moscow: Iz. Moskovskogo universiteta, 1980.
Sontheimer, Kurt, and Bleek, Wilhelm. **The Government and Politics of East Germany**. London: Hutchinson University Library, 1975.
"Sotsial'no-filosofskie problemy nauchno-tekhnicheskoi revoliutsii (issledovaniia 1971-1975 gg.)." **VFil**, no. 2 (1976), pp. 37-53.
**Sozialistische Landeskultur: Umweltgestaltung-Umweltschutz.** Brockhaus Handbuch. Leipzig: VEB F. A. Brockhaus Verlag, 1977.
Spindler, Bernd. **Zum Stand des Umweltschutzes in der DDR.** Bonn: Gesamtdeutsches Institut, 1979.
Staatliche Zentralverwaltung für Statistik. **Statistisches Jahrbuch der DDR 1983.** Berlin: Staatsverlag der DDR, 1983.
"Staatmonopolistischer Reproduktionsprozess und Umwelt." **WW** 22 (October 1974): 1553-58.
Steinitz, Klaus. "Theoretische und praktische Probleme des ökonomischen Wachstums im Sozialismus." **Einheit** 22 (1967): 858-70.
Stoof, Siegfried. "Fragen des Verhältnisses von sozialistischer Gesellschaft und Natur." **DZfPh** 27 (1979): 587-92.
Stoph, Willi. **Report on the Directives of the Eighth Congress of the SED and the Five-Year Plan for the Development of the National Economy of the GDR in the Period 1971-1975.** Dresden: Verlag Zeit im Bild, 1971.
Strauss, Wolfgang, and Dittmar, Armin. "Zur Berücksichtigung ökologischer Aspekte bei der Vorbereitung von Investitionen." **Wissenschaftliche Zeitschrift der Technischen Universität Dresden** 32 (1983): 135-37.
Streibel, Günther. "Die ökonomische Stimulierung der rationellen Nutzung natürlicher Gebietsressourcen." **WW** 14 (December 1966): 1968-90.
________. "Umweltschutz und Umweltgestaltung als volkswirtschaftliche Aufgabe." **WW** 23 (August 1975): 1139-65.
Strumilin, S. "O tsene 'darovykh blag' prirody." **VEko**, no. 8 (1967), pp. 60-72.
Sukachev, V., and Dylis, N. **Fundamentals of Forest Biogeocoenology**. Edinburgh: Oliver & Boyd, 1964.
Sukhotin, Iu. "Ob otsenkakh prirodnykh resursov." **VEko**, no. 12 (1967), pp. 87-98.
Supranowitz, Stephen et al. **Landeskulturgesetz: Kommentar zum Gesetz über die planmässige Gestaltung der sozialistischen Landeskultur in der DDR vom 14. Mai 1970.** Berlin: Staatsverlag der DDR, 1973.
Syrodoev, N. A. "Leninskie idei ob okhrane prirody i prirodookhranitel'noe zakonodatel'stvo." **SGiP**, no. 4 (1974), pp. 67-74.

Taga, Leonore Shever. "Externalities in a Command System." In **Environmental Misuse in the Soviet Union**, pp. 75-100. Edited by Fred Singleton. New York: Praeger, 1976.

Theologische Studienabteilung beim Bund der Evangelischen Kirchen in der DDR. "Beiträge, Referatdienst. Umwelt-Zukunft-Glaube." Berlin, June 1979. Hectograph.

Thiele, K. "Sonderheft: Waldsterben." **Information 1/83.** Bayerische Staatsforstverwaltung, 1983.

Thomas, John R. "A Current Assessment of Soviet Science." In **Soviet Science and Technology: Domestic and Foreign Perspectives**, pp. 44-62. Edited by John R. Thomas and Ursula M. Kruse-Vaucienne. Washington, DC: National Science Foundation, 1977.

Thomas, John R., and Kruse-Vaucienne, Ursula M. "Toward Understanding Soviet Science and Technology: An Overview." In **Soviet Science and Technology: Domestic and Foreign Perspectives**, pp. v-xxxiii. Edited by John R. Thomas and Ursula M. Kruse-Vaucienne. Washington, DC: National Science Foundation, 1977.

_______, eds. **Soviet Science and Technology: Domestic and Foreign Perspectives.** Washington, DC: National Science Foundation, 1977.

Thornton, Judith A. "Resources and Property Rights in the Soviet Union." In **Soviet Resource Management and the Environment**, pp. 1-12. Edited by W. A. Douglas Jackson. Columbus, OH: American Association for the Advancement of Slavic Studies, 1978.

_______. "Soviet Methodology for the Valuation of Natural Resources." **Journal of Comparative Economics** 2 (1978): 321-33.

Timofeev, Leonid. "Etalon chelovechnosti." **Literaturnoe obozrenie**, no. 6 (1977), pp. 66-69.

Timofeev, T. T. "O nekotorykh sotsial'nykh aspektakh vzaimosviazi cheloveka i okruzhaiushchei sredy." **VFil**, no. 12 (1976), pp. 40-51.

Tsagolov, N. A., ed. **Kurs politischeskoi ekonomii.** 2 vols. Moscow: Iz. Ekonomicheskoi Literatury, 1963.

Tsaregorodstev, G. I. "Ekologiia i meditsina." **VFil**, no. 7 (1979), pp. 166-69.

_______. "Obshchestvo, okruzhaiushchaia sreda i meditsina." **VFil**, no. 2 (1975), pp. 63-69.

_______. "'Tekhnizatsiia' sredy i zdorov'e cheloveka." **VFil**, no. 10 (1972), pp. 59-70.

Tsaregorodtsev, G. I., and Epokhin, V. G. "Ekologiia cheloveka v sisteme meditsinskogo znaniia." **VFil**, no. 9 (1980), pp. 73-86.

Tucker, Robert C. "The Deradicalization of Marxist Movements."

**American Political Science Review** 61 (June 1967): 343-58.

_______. "Stalinism as Revolution from Above." In **Stalinism: Essays in Historical Interpretation**, pp. 77-108. Edited by Robert Tucker. New York: W. W. Norton & Co., 1977.

Tupytsia, Iu. "Nature Conservation as a Problem of Economic Science." **Ekonomicheskie nauki**, no. 11 (1977). Translated in **Problems of Economics** 21 (December 1978): 17-23.

Unger, Aryeh L. **The Totalitarian Party: Party and People in Nazi Germany and Soviet Russia.** London: Cambridge University Press, 1974.

United Nations, Economic Commission for Europe. "A Note on Some Aspects of National Accounting Methodology in Eastern Europe and the Soviet Union." In **Readings in the Concept and Measurement of Income**, pp. 370-98. Edited by R. H. Parker and G. C. Harcourt. Cambridge, ENG: Cambridge University Press, 1969.

Urban, Detlef. "Kirchen treten an die Öffentlichkeit." In **Die evangelischen Kirchen in der DDR: Beiträge zu einer Bestandsaufnahme.** Edited by Reinhard Henkys. Munich: Chr. Kaiser Verlag, 1982.

Vagt, Erhard. "Harmonieren statt erobern." **Forum** 33, no. 1 (1979), p. 5.

"Vernost' printsipam Leninizma." **VFil**, no. 9 (1968), pp. 3-8.

Vilenskii, M. "Zemle – denezhnuiu otsenku." **VEko**, no. 3 (1968), pp. 104-11.

Vishnev, S. M. "O modelirovanii vzaimosviazei obshchestva i prirody." **EiMM** 10 (1974): 481-93.

Volgyes, Ivan, ed. **Environmental Deterioration in the Soviet Union and Eastern Europe.** New York: Praeger, 1974.

Voronov, A., and Zusman, L. "K voprosu ob ispol'zovanii i otsenke vtorichnykh resursov loma i metallootkhodov." **PKh**, no. 2 (1984), pp. 125-26.

Voronovitskii, M. M.; Gofman, K. G.; Gusev, A. A.; and Spivak, V. A. "Ekonomicheskie osnovy platy za zagriaznenie okruzhaiushchei sredy." **EiMM** 11 (May-June 1975): 482-90.

"V Tsentral'nom Komitete KPSS." **Ekonomicheskaia gazeta**, no. 24 (June 1980), p. 4.

"Vzaimodeistvie estestvennykh i obshchestvennykh nauk na sovremennom etape." Round table. **VFil**, no. 10 (1973), pp. 39-53.

Wensierski, Peter. "Die Gesellschaft für Natur und Umwelt." Presented at the XVII. Forscher-Tagung at the Europäische Akademie Lerbach. Bergisch-Gladbach, FRG, June 12-15, 1984.

Wensierski, Peter, and Büscher, Wolfgang, eds. **Beton ist Beton: Zivilisationskritik aus der DDR.** Hattingen, FRG: Scandica Verlag, 1981.

White, Lynn. "The Historical Roots of Our Ecological Crisis." **Science** 155 (March 10, 1967): 1203-7.

White, Stephen. **Political Culture in the USSR.** London: Macmillan Press, 1979.

Wilczynski, J. "Towards Rationality in Land Economics under Central Planning." **The Economic Journal** 79 (September 1969): 540-59.

**Wörterbuch der Ökonomie Sozialismus.** Berlin: Dietz Verlag, 1979.

Wright, Arthur. "Environmental Disruption: An Attempt at an Analytical Framework." **ASTE Bulletin** 13, no. 1 (Spring 1971): 1-16.

Zagladin, V. V. "Metodologicheskie problemy issledovaniia global'nykh protsessov mirovogo razvitiia." **VFil,** no. 9 (1981), pp. 17-31.

Zaikov, G. "Ob ekologicheskom aspekte khoziaistvennogo rascheta." **VEko,** no. 8 (1978), pp. 129-31.

Zenchenko, N. "Kompleksnost' ekonomicheskogo i sotsial'nogo razvitiia regionov." **PKh,** no. 4 (1982), pp. 26-36.

_______. "Rol' mestnykh organov v territorial'nom planirovanii." **PKh,** no. 2 (1979), pp. 58-67.

Zhakipov, N. "Nekotorye voprosy teorii sovetskogo vodnogo zakonodatel'stva." **SGiP,** no. 1 (1965), pp. 124-30.

Zhamin, V. "Ecology and Economics." **VEko,** no. 7 (1975), pp. 91-101. Translated in **Problems of Economics** 20 (July 1977): 74-93.

_______. "Infrastruktura pri sotsializme." **VEko,** no. 2 (1977), pp. 14-23.

Zharikov, E. S. "Filosofskii podkhod k postroeniiu sotsial'no-ekologicheskoi kontseptsii." In **Nauchno-tekhnicheskaia revoliutsiia, chelovek, ego prirodnaia i sotsial'naia sreda,** pp. 80-89. Edited by V. G. Marakhov. Leningrad: Iz. Leningradskogo universiteta, 1977.

Zhavoronkova, I. "Ob uluchshenii ispol'zovaniia mineral'no-syr'evykh resursov." **VEko,** no. 8 (1970), pp. 24-33.

Zhavoronkova, I., and Leksin, V. "Problemy ratsional'nogo ispol'-zovaniia syr'evykh resursov." **VEko,** no. 9 (1974), pp. 14-24.

Zhavoronkova, N. G. "Planirovanie gorodskimi sovetami okhrany okruzhaiushchei sredy." **Vestnik Moskovskogo Universiteta, seriia 11, pravo,** no. 5 (1981), pp. 68-73.

Ziegler, Charles E. "Policy Alternatives in Soviet Environmental Protection." The Carl Beck Papers in Russian and East European Studies, paper 102. Pittsburgh: University of Pittsburgh, n.d.

_______. "Soviet Environmental Policy and Soviet Central Planning: A Reply to McIntyre and Thornton." **Soviet Studies** 32 (January 1980): 124-34.

_______. "Soviet Environmental Policy Parameters: The Micro-Value Framework." **Studies in Soviet Thought** 23 (April 1982): 187-204.

Zile, Zigurds L. "Kolbasov's **Legislation on Water Use in the USSR** from the Perspective of Recent Trends in Soviet Law." In **Water Law in the Soviet Union**, pp. 75-90. Edited by Irving Fox. Madison, WI: University of Wisconsin Press, 1971.

_______. "Lenin's Contribution to Law: The Case of Protection and Preservation of the Natural Environment." In **Lenin and Leninism: State, Law, and Society**. Edited by Bernard W. Eissenstat. Lexington, MA: Lexington Press, 1971.

_______. "Soviet Struggle for Environmental Quality: The Limits of Environmental Law under Central Planning." In **Contemporary Soviet Law: Essays in Honor of John N. Hazard**, pp. 124-57. Edited by Donald D. Barry, William E. Butler, and George Ginsburgs. The Hague: Martinus Nijhoff, 1974.

Zubakov, V. A. "On the Content and the Tasks of Historical Geography (as the History of Nature Management)." **Izvestiia Vsesoiuznogo Geograficheskogo Obshchestva**, no. 6 (1976). Translated in **SGRT** 19 (March 1978): 170-80.

Zvonkova, T. V., and Saushkin, Iu. G. "Interaction Between Physical and Economic Geography." **Vestnik Moskovskogo Universiteta, geografiia**, no. 3 (1976), pp. 3-9. Translated in **SGRT** 18 (April 1977): 245-50.

_______. "Problems of Long-Term Geographic Prediction." **Vestnik Moskovskogo Universiteta, geografiia**, no. 4 (1968). Translated in **SGRT** 9 (November 1968): 755-65.

Zykov, M. B.; Krylova, I. A.; Podol'skii, I. Ia.; and Rozanov, S. I. "Problemy global'noi ekologii." **VFil**, no. 1 (1981), pp. 165-68.

## NEWSPAPERS AND POPULAR SCIENCE MAGAZINES CONSULTED

The following list includes the most important sources consulted. Articles from other magazines and newspapers have also been cited in the notes, but these sources were not systematically examined as were those listed here.

**Current Digest of the Soviet Press (CDSP)**, 1965-1983. This includes translations of articles from a wide range of Soviet publications including **Pravda, Izvestiia, Literaturnaia gazeta, Ekonomicheskaia gazeta, Ekonomika i organizatsiia promyshlennogo proizvodstva, Komsomol'skaia pravda, Sovetskaia Rossiia**, and others.

**Forum** (the organ of the Freie Deutsche Jugend in the GDR),

January 1975-April 1982.
**Leipziger Volkszeitung**, October-December 1973.
**Literaturnaia gazeta**, January 1981-April 1982.
**Neues Deutschland**, October-December 1969; April-June 1970; May-August 1974; February, March, May-December 1977; August 1980-April 1981; January 1983-March 1984.
**Pravda**, January-May 1982.
**Priroda**, 1965-March 1982.
**Urania**, 1965-1983.
**Die Wirtschaft**, Ausgabe A, 1972.

# Index

In general, subjects are not listed as sub-entries under the main entries "German Democratic Republic" and "Union of Soviet Socialist Republics." Rather, the reader should consult specific subject listings. Journals are listed under their Russian or German titles (as sub-headings under "Periodicals"), but names of institutes and organizations have been translated into English. Please consult the List of Abbreviations (pages 275 and 276) for full names of abbreviated organizations.